GUIDELINES FOR

Chemical Transportation Risk Analysis

Publications Available from the
CENTER FOR CHEMICAL PROCESS SAFETY of the AMERICAN INSTITUTE OF CHEMICAL ENGINEERS

Guidelines for Chemical Transportation Risk Analysis
Guidelines for Safe Storage and Handling of Reactive Materials
Guidelines for Technical Planning for On-Site Emergencies
Guidelines for Process Safety Documentation
Process Operations and Maintenance
Guidelines for Process Safety Fundamantals in General Plant Operations
Guidelines for Chemical Reactivity Evaluation and Application to Process Design
Tools for Making Acute Risk Decisions with Chemical Process Safety Applications
Guidelines for Preventing Human Error in Process Safety
Guidelines for Evaluating the Characteristics of Vapor Cloud Explosions, Flash Fires, and BLEVEs
Guidelines for Implementing Process Safety Management Systems
Guidelines for Safe Automation of Chemical Processes
Guidelines for Engineering Design for Process Safety
Guidelines for Auditing Process Safety Management Systems
Guidelines for Investigating Chemical Process Incidents
Guidelines for Hazard Evaluation Procedures, Second Edition with Worked Examples
Plant Guidelines for Technical Management of Chemical Process Safety, Rev. Ed.
Guidelines for Technical Management of Chemical Process Safety
Guidelines for Chemical Process Quantitative Risk Analysis
Guidelines for Process Equipment Reliability Data, with Data Tables
Guidelines for Vapor Release Mitigation
Guidelines for Safe Storage and Handling of High Toxic Hazard Materials
Guidelines for Use of Vapor Cloud Dispersion Models
Understanding Atmospheric Dispersion of Accidental Releases
Expert Systems in Process Safety
Concentration Fluctuations and Averaging Time in Vapor Clouds
Safety, Health, and Loss Prevention in Chemical Processes: Problems for Undergraduate Engineering Curricula
Safety, Health, and Loss Prevention in Chemical Processes: Problems for Undergraduate Engineering Curricula—Instructor's Guide
Workbook of Test Cases for Vapor Cloud Source Dispersion Models
Proceedings of the International Conference and Workshop on Modeling and Mitigating the Consequences of Accidental Releases of Hazardous Materials, 1995.
Proceedings of the International Symposium and Workshop on Safe Chemical Process Automation, 1994
Proceedings of the International Process Safety Management Conference and Workshop, 1993
Proceedings of the International Conference on Hazard Identification and Risk Analysis, Human Factors, and Human Reliability in Process Safety, 1992
Proceedings of the International Conference/Workshop on Modeling and Mitigating the Consequences of Accidental Releases of Hazardous Materials, 1991.
Proceedings of the International Symposium on Runaway Reactions, 1989
CCPS/AIChE Directory of Chemical Process Safety Services

GUIDELINES FOR

Chemical Transportation Risk Analysis

CENTER FOR CHEMICAL PROCESS SAFETY

of the

American Institute of Chemical Engineers

345 East 47th Street • New York, New York 10017

Library of Congress Cataloging-in Publication Data
Guidelines for chemical transportation risk analysis.
p. cm.
Includes bibliographical references and index.
ISBN 0–8169–0626–2 : $140.00
1. Chemicals—Transportation—Safety measures. I. American Institute of Chemical Engineers. Center for Chemical Process Safety.
TP201.G85 1994
604'.7—dc20 94–2481
CIP

This book is available at a special discount when ordered in bulk quantities. For information, contact the Center for Chemical Process Safety of the American Institute of Chemical Engineers at the address shown above.

Contents

Preface *xi*

Acknowledgments *xiii*

Acronyms *xv*

Introduction *1*

Management Overview 1
Organization of the Guidelines 2

1. Transportation Risk Analysis *5*

1.1. TRA Definitions 9
1.2. General Framework of a TRA 11
1.3. Typical Reasons for Conducting a TRA 15
1.4. Transportation versus Fixed Facility Risk Evaluations 16
1.4.1. General Similarities to Fixed Facility Evaluations *16*
1.4.2. Differences from Fixed Facility Evaluations *17*
1.5. Scope and Depth of TRA Studies 21
1.5.1. The Study Cube *21*
1.5.2. Range of Incidents *24*
1.5.3. Selection of Initiating Events and Incidents *24*
1.6. Screening Techniques 26
1.7. Limitations 28
1.8. Current Practice 29
1.9. The Use of Existing TRA Studies 31
1.10. Utilization of TRA Results 33
1.11. Project Management 35
1.11.1. Study Objectives *36*
1.11.2. Depth of Study *36*

1.11.3. Project Plan and Schedule *37*
1.11.4. Review and Approvals *38*
1.12. Maintenance of Study Results 40
1.13. References 41

2. TRA Frequency Analysis **43**

2.1. Pipelines 45
2.1.1 Failure Modes *46*
2.1.2. Parameters Influencing Accident Rates *50*
2.1.3. Parameters Influencing Release Probabilities *55*
2.1.4. Containment Systems *57*
2.1.5. Accident Trends with Time *58*
2.1.6. Confidence in Data *59*
2.1.7. Mode-Specific Issues *59*
2.1.8. Releases Associated with Supporting Equipment *60*
2.1.9. General Calculation Procedure *60*
2.2. Rail 60
2.2.1. Failure Modes *62*
2.2.2. Parameters Influencing Accident Rates *64*
2.2.3. Parameters Influencing Release Probabilities *69*
2.2.4. Container Types *70*
2.2.5. Accident Trends with Time *71*
2.2.6. Confidence in Data *72*
2.2.7. Mode-Specific Issues *73*
2.2.8. Non–Accident-Initiated Releases *74*
2.2.9. General Calculation Procedure *75*
2.3. Road Transport 76
2.3.1 Failure Modes *78*
2.3.2. Parameters Influencing Accident Rates *79*
2.3.3. Parameters Influencing Release Probabilities *81*
2.3.4. Container Types *82*
2.3.5. Accident Trends with Time *82*
2.3.6. Confidence in Data *82*
2.3.7. Mode-Specific Issues *84*
2.3.8. Non–Accident-Initiated Releases *85*
2.3.9. General Calculation Procedure *85*
2.4. Barge 86
2.4.1 Failure Modes *87*
2.4.2. Parameters Influencing Accident Rates *89*
2.4.3. Parameters Influencing Release Probabilities *90*
2.4.4. Container Types *90*
2.4.5 Accident Trends with Time *91*
2.4.6. Confidence in Data *91*
2.4.7. Mode-Specific Issues *92*

2.4.8. Non–Accident-Initiated Releases *92*
2.4.9 General Calculation Procedure *93*
2.5. Ocean-Going Vessel 94
2.5.1. Failure Modes *95*
2.5.2. Parameters Influencing Accident Rates *96*
2.5.3. Parameters Influencing Release Probabilities *97*
2.5.4. Container Types *98*
2.5.5. Accident Trends with Time *99*
2.5.6. Confidence in Data *99*
2.5.7. Mode-Specific Issues *99*
2.5.8. Non–Accident-Initiated Releases *100*
2.5.9. General Calculation Procedure *100*
2.6. Intermodal Containers 100
2.6.1. Failure Modes *101*
2.6.2. Parameters Influencing Accident Rates *103*
2.6.3. Parameters Affecting Release Probabilities *103*
2.6.4. Container Types *104*
2.6.5. Accident Trends with Time *104*
2.6.6. Confidence in Data *104*
2.6.7. Mode-Specific Issues *104*
2.6.8. Non–Accident-Initiated Releases *105*
2.6.9. General Calculation Procedure *105*
2.7. Conducting a Frequency Analysis 106
2.8. References 109

3. Special Topics ***111***

3.1. Selection of Release Scenarios 113
3.1.1. Truck Data *113*
3.1.2. Marine Data *115*
3.1.3. Rail Data *116*
3.1.4. Summary *117*
3.2. Population Densities 119
3.3. Route Segmentation 122
3.4. Ignition Algorithms 125
3.5. Treatment of Classification Yards 130
3.6. Transloading Operations 131
3.7. Unique Consequence Analysis Models 132
3.8. Failure Frequency Models 137
3.8.1. Predictive and Testing Methods *138*
3.8.2. Mechanical Predictive Methods *139*
3.8.3. Logical Procedures *140*
3.9. Environmental Risks 142
3.10. Available Software 143
3.11. References 145

4. *TRA Data* **149**

4.1. Taxonomy for Accident Data 150
4.2. Historical Data Sources 150
4.3. Adjustments and Enhancements to Data 159
4.4. Sources of Additional Data 161
4.4.1. Population Data *161*
4.4.2. Climatic Data *163*
4.4.3. Route Data *165*
4.4.4. Shipment Data *166*
4.4.5 International Data *170*
4.5. Completeness and Accuracy of Data Sources 171
4.6. Data Normalization 175
4.7. Options in the Absence of Complete Data 178
4.8. Ongoing Data Collection Efforts 179
4.9. References 180
Transportation Data Source Sheets 181

5. *Measurement, Calculation, and Presentation of Risk Estimates* **223**

5.1. Risk Measures/Presentation 224
5.2. Risk Calculations 230
5.2.1. Individual Risk *230*
5.2.2. Societal Risk *235*
5.2.3. Other Risk Measures *237*
5.3. Risk Uncertainty, Sensitivity and Importance 238
5.4. References 241

6. *TRA Application Examples* **243**

6.1. Qualitative/Consequence TRA Examples 246
6.1.1. Qualitative/Consequence TRA Characterization *246*
6.1.2. Comparison of Different Forms of One Material by Consequence Analysis *246*
6.1.3. Comparisons against Results of More Detailed Studies *248*
6.2. Semiquantitative/Frequency TRA Examples 248
6.2.1. Semiquantitative/Frequency TRA Characterization *248*
6.2.2. Change in Supplier *248*
6.2.3. Change in Material Container *249*
6.3. Quantitative/Risk TRA Examples 250
6.3.1. Quantitative/Risk TRA Characterization *250*
6.3.2. Determination of Available Risk Reduction Options and Their Ranking *251*
6.3.3. Evaluation of Alternative Modes for Getting from Point A to Point B *251*

7. TRA Case Studies **253**
7.1. Pipeline Risk Assessment 254
7.1.1. Introduction 254
7.1.2. Description 255
7.1.3. Identification, Enumeration, and Selection of Incidents 255
7.1.4. Incident Consequence Estimation 257
7.1.5. Incident Frequency Estimation 260
7.1.6. Risk Estimation 262
7.1.7. Discussion of Results 263
7.1.8. References 263
7.2. Tank Truck Risk Assessment 263
7.2.1. Introduction 263
7.2.2. Description 264
7.2.3. Identification, Enumeration, and Selection of Incidents 265
7.2.4. Incident Consequence Estimation 266
7.2.5. Incident Frequency Estimation 267
7.2.6. Risk Estimation 269
7.2.7. Discussion of Results 273
7.3. Rail Risk Assessment 273
7.3.1. Introduction 273
7.3.2. Description 274
7.3.3. Identification, Enumeration, and Selection of Incidents 274
7.3.4. Incident Consequence Estimation 277
7.3.5. Incident Frequency Estimation 278
7.3.6. Risk Estimation 278
7.3.7. Discussion of Results 284
7.3.8. References 284
7.4. Barge Risk Assessment 285
7.4.1. Introduction 285
7.4.2. Description 285
7.4.3. Identification, Enumeration, and Selection of Incidents 287
7.4.4. Incident Consequence Estimation 288
7.4.5. Incident Frequency Estimation 290
7.4.6. Risk Estimation 291
7.4.7. References 295
7.5. Ocean-Going Vessel Risk Assessment 295
7.5.1. Introduction 295
7.5.2. Description 295
7.5.3. Identification, Enumeration, and Selection of Incidents 296
7.5.4. Incident Consequence Estimation 297
7.5.5. Incident Frequency Estimation 298
7.5.6. Risk Estimation 301
7.5.7. Discussion of Results 301
7.5.8. References 302

8. Future Developments **303**

8.1. General Risk Assessment Advances/Trends 303
8.2. Release Rates and Volumes 305
8.3. Release Probabilities for Specific Containers 305
8.4. Accident Rates 306
8.5. Integration with Network Models 307
8.6. Comparisons with Actual Performance 308
8.7. Risk Reduction Strategies 309
8.8. Environmental Risk 310
8.9. Increased Cooperation 311
8.10. References 311

Appendix A. Standard Container Illustrations and Specifications **313**

I. Highway Cargo Tanks 313
II. Intermodal Tanks 321
III. Railroad Tank Cars 322
IV. Drums 327
V. Barges 328
VI. Vessels 328

Appendix B. Information Needs for a Study and Sources of Data **330**

Appendix C. Sample Outline for TRA Reports **332**

Appendix D. Conversion Factors **334**

Appendix E. Qualitative Evaluations **338**

Transportation Risk Analysis Topical Bibliography **341**

Glossary **363**

Index **375**

Preface

The Center for Chemical Process Safety (CCPS) was established in 1985 as a Directorate of the American Institute of Chemical Engineers (AIChE) for the purpose of assisting the chemical and hydrocarbon processing industries in avoiding or mitigating chemical accidents. To fulfill this objective, the CCPS has set the following specific goals, as stated in the 1993 CCPS annual report:

- Establish and publish the latest scientific and engineering practices (not standards) for prevention and mitigation of incidents involving toxic and/or reactive materials
- Encourage the use of such information by dissemination through publications, seminars, symposia and continuing education programs for engineers
- Advance the state of the art in engineering practices through research in prevention and mitigation of catastrophic events
- Develop and encourage the use of undergraduate education curricula which will improve the safety knowledge and consciousness of engineers

The series of process safety *Guidelines* books fulfills the first of these objectives. In 1985, CCPS published *Guidelines for Hazard Evaluation Procedures (HEP Guidelines),* which describes qualitative tools for identifying, assessing and reducing process hazards. After achieving great success, the *HEP Guidelines* were updated and a revised second edition incorporating a series of worked example case studies was published in 1992. Although the tools described in the *HEP Guidelines* may be applicable to a variety of operations, the focus of that book is on application of these tools to chemical processing facilities.

In 1989, CCPS published *Guidelines for Chemical Process Quantitative Risk Analysis (CPQRA Guidelines)*, which describes tools and methods for the quantitative analysis of risk. As with the *HEP Guidelines*, the focus of the *CPQRA Guidelines* is on the application of quantitative risk analysis tools to a fixed chemical process site.

This book, *Guidelines for Chemical Transportation Risk Analysis*, extends the concepts of the *HEP Guidelines* and the *CPQRA Guidelines* to include evaluation and assessment of the risks associated with hazardous material transportation. The book concentrates on the special concerns and characteristics of transportation risk, and describes the special tools and techniques used to address these concerns and characteristics. The methods described build upon and extend those described in the *CPQRA Guidelines*. CPQRA techniques which are directly applicable to Transportation Risk Analysis (TRA) are not repeated in this book, but are referenced. Thus, this book is intended to be used as a companion volume to the *CPQRA Guidelines* when dealing with a quantitative transportation risk analysis methodology.

The methods described in this book are intended to develop a qualitative, semiquantitative, or quantitative measure of the risk of a hazardous material transportation operation. Some of the methods will also provide insight into the sources of that risk and their relative importance. This information may be useful to the engineer or manager in identifying cost effective ways to manage and reduce that risk. However, this book does not discuss how to evaluate the qualitative or quantitative risk results and how to determine if the risks are tolerable or if improvements need to be made. The CCPS book, *Tools for Making Acute Risk Decisions, with Chemical Process Applications* (1994), describes a variety of decision making tools which can be used to assist in evaluating the results of a TRA and deciding what specific actions should be taken to best manage the risk of the transportation operations. The *TRA Guidelines* discuss methods for obtaining qualitative or quantitative estimates of transportation risk, one of the important pieces of data required to make decisions on management of the risks of hazardous material transportation.

Acknowledgments

This book was written by Arthur D. Little, Inc. (ADL) in support of a project funded by the Center for Chemical Process Safety (CCPS) of the American Institute of Chemical Engineers. The CCPS Risk Assessment Subcommittee managed the project and provided direction and assistance to ADL.

The Arthur D. Little, Inc. team was led by Lisa M. Bendixen, and included Alan Bing, Roger Hill, John Moskal, and Lorelei King.

The CCPS Risk Assessment Subcommittee was chaired by Roger W. Ormsby (Air Products and Chemicals) and Dennis C. Hendershot (Rohm and Haas Company) during the final stages of the project. The committee included (in alphabetical order): Brian R. Dunbobbin (Air Products and Chemicals), Raymond A. Freeman (Monsanto), Raymond W. French (Exxon Chemical), S. Barry Gibson (du Pont, retired). Richard M. Gustafson (Texaco), Thomas Janicik (Solvay Polymers), William K. Lutz (Union Carbide), Carmine A. Master (Fluor Daniel), Arthur G. Mundt (Dow Chemical), Scott Ostrowski (Exxon Chemical), Joseph C. Sweeney (ARCO Chemical), and Donald L. Winter (Mobil Oil). Felix Freiheiter provided staff support for CCPS. Bob Perry, Les Wittenberg, Jack Weaver, and Tom Carmody (formerly) of CCPS helped guide the project.

The substantial contributions, both in time and resources, of the employer organizations of both the subcommittee members and ADL are gratefuly acknowledged.

We also thank the following people (in alphabetical order) who generously contributed their time and expertise in providing a peer review of the book: Donald J. Connolley (Akzo Nobel Chemicals, Inc.), Thomas O. Gibson (Dow Chemical), Joseph W. Kleffner (Monsanto), Michael Latino (Dow Chemical), John Marshall (Dow Chemical), William R. Rhyne (H&R Technical Associates), Gary Skipp (Texaco), John Shortreed (University of Waterloo), John D. Snell (Occidental Chemical Corporation), Stephen Van Rensselaer (Rohm and Haas Company), and David Webb (Arthur D. Little, Inc.).

Acronyms

AADT	Annual Average Daily Traffic
AAR	Association of American Railroads
ACMH	Advisory Commission on Major Hazards (U.K.)
AGA	American Gas Association
AIChE/CCPS	American Institute of Chemical Engineers—Center for Chemical Process Safety
AIChE/DIPPR	American Institute of Chemical Engineers—Design Institute for Physical Property Data
ALR	Adiabatic Lapse Rate
ANSI	American National Standards Institute
API	American Petroleum Institute
ATA	American Trucking Association
ASME	American Society of Mechanical Engineers
BLEVE	Boiling Liquid Expanding Vapor Explosion
BMCS	Bureau of Motor Carrier Safety
BS	British Standard
CAER	Community Awareness and Emergency Responses
Caltrans	California Department of Transportation
CASMAIN	Marine Casualty Information Reporting System
CCPS	Center for Chemical Process Safety
CEAM	Center for Exposure Assessment Modeling
CFR	Code of Federal Regulations
CGA	Compressed Gas Association
CMA	Chemical Manufacturers Association
CONCAWE	The oil companies' international study group for CONservation of Clean Air and Water—Europe
CPU	Computer Processing Unit
CPQRA	Chemical Process Quantitative Risk Analysis

CY	Calendar Year
DOC	U.S. Department of Commerce
DOD	U.S. Department of Defense
DOE	U.S. Department of Energy
DOT	U.S. Department of Transportation
DRI	Distribution Ranking Index
EF	Error Factor
ENF	Expected Number of Failures
EPA	U.S. Environmental Protection Agency
ERW	Electric Resistance/Induction Welded
ETA	Event Tree Analysis
FAR	Fatal Accident Rate
FEMA	Federal Emergency Management Agency
FMEA	Failure Modes and Effects Analysis
F-N	Frequency-Number
FRA	Federal Railroad Administration
FTA	Fault Tree Analysis
GIS	Geographic Information System
HAZMAT	Hazardous Material
HEP	Hazard Evaluation Procedures
HMIR	Hazardous Materials Incident Reports
HMIS	Hazardous Materials Information System
HMSO	Her Majesty's Stationary Office (U.K.)
HSC	Health and Safety Commission (U.K.)
HSE	Health & Safety Executive (U.K.)
ICC	Interstate Commerce Commission
IChemE	Institute of Chemical Engineers (Great Britain)
IDLH	Immediately Dangerous to Life and Health
IGE	Institute of Gas Engineers
IHI	Individual Hazard Index
IMDG	International Maritime Dangerous Goods
IMO	International Maritime Organization
IRR	Institute for Risk Research
ISBN	International Standard Book Number
ISO	International Standards Organization
LANDSAT	Land Satellite
LC	Lethal Concentration
LCL	Lower Confidence Limit
LD	Lethal Dose
LEPC	Local Emergency Planning Committee
LFL	Lower Flammable Limit

LNG	Liquefied Natural Gas
LPG	Liquefied Petroleum Gas
MA	Metropolitan Area
MARDATA	Maritime Data
MAWP	Maximum Allowable Working Pressure
MC	Motor Carrier
MIN	Marine Investigations
MRI	Midwest Research Institute
MSDS	Material Safety Data Sheet
MSIS	Marine Safety Information System
MSRC	Marine Spill Response Corporation
NAS/NRC	National Academy of Sciences/National Research Council
NCC	National Climatic Center
NDT	Nondestructive Testing
NFPA	National Fire Protection Association
NGL	Natural Gas Liquid
NOAA	National Oceanic and Atmospheric Administration
NRC	National Research Council
NRDA	Natural Resource Damage Assessment
NTIS	National Technical Information Service
NTSB	National Transportation Safety Board
OBO	Oil/Bulk/Ore
OMB	U.S. Office of Management and Budget
OPA	Oil Pollution Act
OSHA	Occupational Safety and Health Administration
OTA	Office of Technology Assessment
PIH	Poison Inhalation Hazard
PIRS	Pollution Incident Reporting System
PRA	Probabilistic Risk Assessment
RIPS	Relative Index of Pipeline Safety
RMP	Risk Management Program (USEPA)
ROD	Average Rate of Death
ROW	Right-of-Way
RPI	Railway Progress Institute
RQ	Reportable Quantity
RR	Railroad
RSPA	Research and Special Programs Administration
SAM	Station of Atmospheric Measurements
SEPC	State Emergency Planning Commission
SIC	Standard Industrial Classification
SOCMA	Synthetic Organic Chemical Manufacturers Association

SPOT	Satellite pour Observation de la Terre (Earth Observation Satellite)
SRD	Safety and Reliability Directorate (U.K. Atomic Energy Authority, Warrington, England)
SRS	System Reliability Service
STAR	Stability Array Program Data
STARS	State Traffic Accident Record System
STCC	Standard Transportation Commodity Code
STS	Summary Tapes
TASAS	(State) Traffic Accident Surveillance and Analysis System
TIGER	Topologically Integrated Geographic Encoding and Referencing System
TIUS	Truck Inventory and Use Survey
TNT	Trinitrotoluene
TNO	Netherland's Organization for Applied Scientific Research
TTN	Technology Transfer Network
TRA	Transportation Risk Analysis
UCL	Upper Confidence Limit
UFL	Upper Flammable Limit
UKAEA	United Kingdom Atomic Energy Agency
USCG	United States Coast Guard
USGS	United States Geological Survey
VCE	Vapor Cloud Explosion

Introduction

Management Overview

Risk analysis has been used as a management tool to understand and control the risks of selected transportation movements for over twenty years. However, despite the similarities to risk analysis for fixed facilities and the number of techniques which are common between both types of evaluations, the use of transportation risk analysis (TRA) is quite limited. Historically, this was due in part to the manner in which transportation activities were managed in many organizations. Purchasing and distribution were generally housed together and kept quite separate from the manufacturing or production operations, and transportation safety was restricted to DOT compliance issues. In addition, the ownership of and liability for many transportation movements was shared among many parties.

Over the past five to ten years, some companies began to integrate their fixed facility and transportation risk management activities. This led to an increase in TRA studies, both qualitative and quantitative. However, the regulatory requirements for fixed facility hazard (and risk) evaluations have increased significantly since about 1992. This has drawn both attention and resources away from TRA.

Given that transportation activities occur in areas beyond a facility's control and that other parties may be involved (such as railroads and other carriers), attention to TRA is critical. To facilitate increased use of TRA, the Center for Chemical Process Safety of AIChE developed this book as a guidance manual. This book is directed toward the assessment of episodic, short-term hazards, rather than chronic health hazards.

Together with the *Guidelines for Chemical Process Quantitative Risk Analysis* (CCPS, 1989), this book will allow a process engineer or similarly qualified individual to conduct basic analyses and effectively manage more sophisticated analyses. Such studies may be qualitative, semiquantitative, or quantitative in nature.

TRA can be used to manage and control transportation risks by identifying the parameters with the greatest influence on a given movement, or to identify and evaluate risk reduction strategies. Such strategies might involve changes in mode of transportation, shipment size, route, container, etc. TRA can be used early on when a fixed site is being selected, when evaluating alternate suppliers, to develop emergency response plans, or on existing movements. Using TRA in such situations, provides a more objective set of inputs into the decision-making process. It also helps allocate resources for risk reduction more consistently across both fixed facilities and transportation activities.

The philosophy behind this volume is to provide an introduction to TRA methodology in sufficient depth so that an engineer with some practice can undertake basic TRA studies with minimal outside assistance. The engineer should be able to:

1. respond to a request for a TRA;
2. convert the request into a definable study objective;
3. develop a scope of work;
4. understand the types of data and sources of information required;
5. estimate the time and costs for the study;
6. calculate the results;
7. analyze the results for reasonableness;
8. present the results in a useful format.

The careful definition of scope and depth of study in the application of TRA is crucial to success because more quantitative approaches are cost and resource intensive. Detailed, quantitative TRA should be used sparingly and only to that depth of study necessary to achieve a study's goals and objectives. If not properly controlled, even a simple quantitative TRA can generate an unmanageable calculation burden.

Careful study of the material in these guidelines can produce only a basic level of competence. Supplementary review of important references and training is essential. Discussion of regulatory issues, public risk perception, risk criteria, and tolerable risk are excluded.

Organization of the Guidelines

This book provides an introduction to the concerns and techniques of TRA, so that process engineers and others can undertake basic TRA studies or manage more sophisticated analyses. Used in conjunction with the *CPQRA Guidelines* (CCPS, 1989) for details of certain techniques common to both fixed facility and transportation risk analysis, the worked examples and case studies illustrate the various approaches and techniques for TRA. This book discusses both qualitative and quantitative TRAs, but the emphasis of the examples is on quantitative TRAs.

Chapter 1 describes the general framework of TRA, key definitions, similarities to and differences from fixed facility evaluations, basic techniques and current practices.

Chapter 2 discusses hazard identification and frequency analysis for pipelines (Section 2.1), rail (Section 2.2), road transport (Section 2.3), barges (Section 2.4), ocean-going vessels (Section 2.5), and intermodal containers (Section 2.6). General techniques for frequency analysis are also given (Section 2.7). Both accident- and non–accident-initiated releases are considered.

Chapter 3 covers a range of special topics, including selection of release scenarios (Section 3.1), population densities (Section 3.2), breaking a route into segments with common characteristics (Section 3.3), ignition algorithms (Section 3.4), marshalling yards and transloading operations (Sections 3.5 and 3.6), special consequence modelling concerns (Section 3.7), alternate failure frequency models (Section 3.8), environmental risks (Section 3.9), and available software (Section 3.10).

Chapter 4 discusses sources of TRA data and reviews over 40 data sources. In addition, data enhancements and adjustments (Section 4.3) and data normalization (Section 4.6) are addressed.

Chapter 5 reviews the measurement, calculation, and presentation of risk estimates (Sections 5.1 and 5.2). The concepts of risk uncertainty, sensitivity and importance are also discussed (Section 5.3).

Chapter 6 addresses the application and utilization of TRA for three representative types of studies: a qualitative evaluation focussing on consequences (Section 6.1); a semiquantitative study concerned with frequency (Section 6.2); and a quantitative analysis addressing risk (Section 6.3).

Chapter 7 presents five case examples—for a pipeline (Section 7.1), a tank truck (Section 7.2), a railcar (Section 7.3), a barge (Section 7.4) and a tanker (Section 7.5). Each of these examples focusses on several different aspects of quantitative TRA.

Chapter 8 discusses future needed or anticipated developments in TRA and related areas, ranging from improvements in various types of data (Sections 8.2–8.4) to more advanced/comprehensive techniques (Sections 8.1, 8.5, and 8.8) to risk management issues (Sections 8.6, 8.7, and 8.9).

A number of appendices are also given for assistance in conducting TRAs.

A comprehensive bibliography provides references for further details and more advanced topics.

1

Transportation Risk Analysis

Transportation risk analysis (TRA) methodologies have existed for about the same time period as have chemical process quantitative risk analysis (CPQRA) methodologies, yet they are far less widely used and understood. TRA draws on many of the same tools and techniques as does CPQRA, but distribution activities are often housed in a separate part of an organization and may not be aware of all the internal resources available in risk analysis. Unlike the regulatory demands for some level of hazard analysis as part of OSHA 1910.119 or individual state requirements, transportation risks have primarily been managed through carrier and equipment selection programs and ongoing inspection, maintenance, and training programs. Transportation activities do not have an equivalent to OSHA regulations, and those regulations that exist under the Department of Transportation focus on hardware and procedures.

The appropriate application of TRA can allow risk management decisions to be made about:

- alternate modes of transport
- appropriate routes
- travel restrictions (e.g., speed, weather, time of day)
- shipment size (total volume per shipment)
- shipping conditions (pressure and temperature)
- use of more protective containers
- unit size (e.g., bulk versus drums)

These and many more examples will be discussed later. Without the benefits of a TRA, judgments about such issues are made subjectively and the status quo is often accepted.

There have been very few disasters associated with the transportation of hazardous materials (HAZMAT), especially when one considers the volumes carried per container, the number of HAZMAT shipments each year, and the direct proximity of the public. Table 1-1 summarizes just a few of the larger incidents over the last thirty or so years. Coupled with events where thousands of people had to be evacuated, the potential severity of transportation incidents can be seen. Historically, however, only a limited number of incidents have had any fatalities. Data gathered by the DOT Research and Special Programs Administration indicate that 0.3 percent of HAZMAT truck incidents result in any deaths and 2.2 percent result in any injuries (Harwood, 1989).

The total number of releases is also relatively low compared to activity levels. For instance, the Office of Technology Assessment (1986) estimates

TABLE 1-1
Selected Major Historical Transportation Accidents

Date of Incident	*Mode*	*Area/Location*	*Substance and Quantity Released*	*Nature of Initiating Event; Outcome*	*Fatal-ities*
6/28/59	Rail	Picnic area/ Georgia, USA	LPG: 18 tons	Derailment; car ruptured by coupler; explosions and fire	23
7/31/81	Rail	Urban/San Luis Potosi, Mexico	Chlorine: 90–150 tons	Derailment; tank breached; toxic cloud	14–20
5/11/76	Road	Urban/ Houston, TX	Ammonia: 19 tons	Vehicle crashed and fell 30 feet; tank rupture; toxic cloud	6
7/25/62	Road	Urban/ New York	Propane: 13 tons	Tank failure from jack knifing after collision with tree; fire	10
7/11/78	Road	Campsite/San Carlos, Spain	Propylene: 22 tons	Rupture of overfilled tank; fireball	200
1/8/79	Marine	Terminal/Bantry Bay, Ireland	Crude oil: ?	Incorrect ballasting caused corroded hull to buckle; fire and explosion	50
8/30/79	Marine	Mississippi River/Louisiana	Butane: 600 tons	Freighter collided with barge and ruptured tanks; explosion and fireball	12
2/25/84	Pipeline	Urban/ Cubatos, Brazil	Gasoline: 700 tons	Failure due to corrosion; ignition	89
6/4/89	Pipeline	Urban/ Asha-Ufa, USSR	LPG: ?	Leaked for four hours before ignition from passing train; explosion and fire	>500

Source: Advisory Committee, 1991.

that there were 784 billion ton-miles of hazardous material transported by rail, truck, air, and marine in 1982. Using an annual average of roughly 20,000 hazardous material incidents (releases), based on an estimated 178,683 incidents from 1976 to 1984, gives a rate of roughly 25 incidents per billion ton-miles. This includes a number of loading and unloading incidents. Many of the reported incidents posed no risk to the public; however, they are of concern because they illustrate the potential for a major incident in an uncontrolled environment, whether or not people are present.

The Chemical Manufacturers Association (CMA) has a Distribution Code of Management Practices as part of its Responsible Care®[1] program (CMA, 1990). There are 21 management practices that address risk management (3), compliance review/training (4), carrier safety (2), handling and storage (7), and emergency preparedness (5).

The goal of this program is to provide good management practices around which individual companies can develop their own specific systems and practices, and to promote the following of these practices by member companies and others. This in turn helps control and reduce transportation risks.

An increasing concern over the level of risk associated with transportation activities led the U.K. Advisory Committee on Dangerous Substances, Health and Safety Commission to undertake a nationwide study on the risks of moving hazardous materials (including explosives) by road, rail, and marine. Entitled *Major Hazard Aspects of the Transport of Dangerous Substances* and published in late 1991, the report discusses the TRA methodology used only in general terms and focuses on the results. This is typical of many of the TRA studies conducted to date. This particular study reached many conclusions, including that none of the risks studied were deemed intolerable. The HSC concluded that, therefore, cost–benefit trade offs must be addressed. The report also strongly endorsed the need for good risk management standards and practices.

The purpose of this book is to give guidance on conducting, overseeing, and/or managing TRA studies, not on specific results or on reduction strategies. As demonstrated by the list of events in Table 1-1, transportation risks may be caused by accident-initiated events such as derailments and collisions, or may result from non–accident-initiated causes such as a prior overfilling of a tank causing a release while enroute. This book emphasizes accident-initiated releases, but also discusses how non–accident-initiated releases can be incorporated into a TRA. It presents the issue of domino

1 Responsible Care® is a registered service mark of the Chemical Manufacturers Association

type (or secondary) events and how they can be handled, but focuses on initiating events. Loading and unloading operations are not included, as they are best handled by a CPQRA approach.

TRA can be conducted on a qualitative or quantitative basis. Qualitative approaches include risk screening methodologies, which are generally unique for each company. This book discusses a number of the screening approaches used today, but presents more information on quantitative techniques since so little guidance is available elsewhere. Many companies will find that qualitative techniques are sufficient for their goals and needs. Other qualitative approaches include carrier screening programs, route and container selection, and driver training and selection programs.

TABLE 1-2
TRA Hazards, Initiating Events, and Potential Consequences

	Initiating Events		
TRA Hazards	*Accidents*	*Non-Accidents*	*Potential Consequences*
Hazardous material releases from: Pipelines Rail tank cars Road tankers Barges Ships ISO-containers Drums Cylinders	Derailments Collisions Groundings Rammings Overturnings Grade crossings Allisions Fires and explosions Flooding Capsizing Breakaways Material failure Container drops These may be due to: equipment defects human factors track and infrastructure defects external events crane failures forklift failures navigational system failures control system failures incorrect ballasting/balancing	Corrosion Metallurgical failure Overpressure Construction defects Third-party Anchor dragging Operator error (pipelines) Valve failures Overfilling Rupture disk failures Contamination Overheating Freezing Fitting failures Loss of refrigeration	To public along route: Toxic effects Thermal effects Overpressure effects Radioactive Environmental Building damage Equipment damage

Table 1-2 illustrates the scope of this book in terms of the hazards of concern, potential initiating events and their causes, and consequences. Small packages and air transportation are not addressed. The emphasis is on acute harm to humans (e.g., fatalities, injuries, and exposures) from HAZMAT releases. However, the analysis of release frequencies addresses other potential outcomes (i.e., building damage, environmental damage, radioactive materials) equally well, and the overall approach can often be extended to these other outcomes by modifying the consequence models and selected impact criteria. This book does not go into this extension explicitly.

Many of the concepts presented in this chapter are totally consistent with their use in the area of CPQRA. CCPS has published the *CPQRA Guidelines* (CCPS, 1989) which provide further details on CPQRA tools, techniques, and concepts. The reader is encouraged to use both books together for the greatest understanding of risk analysis and its applications to fixed facilities and transportation activities.

The remainder of this chapter outlines:

- the definitions of TRA terminology (Section 1.1),
- the general framework of a TRA (Section 1.2),
- typical reasons for conducting a TRA (Section 1.3),
- the similarities of and differences between TRA and CPQRA (Section 1.4),
- the scope and depth of TRA studies (Section 1.5),
- screening techniques (Section 1.6),
- the limitations of TRA (Section 1.7),
- the current practice in TRA (Section 1.8),
- critical TRA studies (Section 1.9),
- the utilization of TRA results (Section 1.10),
- project management (Section 1.11), and
- the maintenance of study results (Section 1.12).

1.1. TRA Definitions

Table 1-3 and the Glossary define terms as they are used in this volume. Other tabulations of terms have been compiled (e.g., IChemE, 1985) and may need to be consulted as there is no single authoritative source of nomenclature and definitions. Specific regulations also have particular definitions associated with them. The definitions applied in this document are consistent with those found in the *CPQRA Guidelines* (CCPS, 1989).

Readers should recognize the relationship that exists among an incident, an incident outcome, and an incident outcome case as these terms

TABLE 1-3

Selected Definitions for TRA

Accident-Initiated Event: An event (or the first event in an event sequence) that is caused by a movement related transportation accident (e.g., a train derailment or a barge grounding).

Non–accident-initiated Event: An event due to causes unassociated with the movement related aspects of transportation (e.g., a pressure build-up due to contamination).

Hazard: A chemical or physical condition that has the potential for causing damage to people, property, or the environment (e.g., a tank truck carrying 20 tons of ammonia).

Incident: The loss of containment of material or energy (e.g., a leak of 10 lb/sec of ammonia from a fitting on a railcar); not all events propagate into incidents.

Mode or Modal: the specific type of transportation of concern (i.e., truck, rail, barge, pipeline, etc.).

Transportation: Activities that occur after a container is filled with a product and before it is unloaded; usually restricted to off-site activities.

Risk Analysis or Transportation Risk Analysis: The development of a qualitative or quantitative estimate of (transportation) risk based on engineering evaluations and possibly mathematical techniques for combining estimates of incident consequences and frequencies.

Incident Outcome: The physical manifestation of the incident; for toxic materials, the incident outcome is a toxic release, while for flammable materials, the incident outcome could be a boiling liquid expanding vapor explosion (BLEVE), flash fire, vapor cloud explosion (VCE), etc. (e.g., for a 10 lb/sec leak of ammonia, the incident outcome is a toxic release).

Incident Outcome Case: The quantitative definition of a single result of an incident outcome through specification of sufficient parameters to allow distinction of this case from all others for the same incident outcomes (e.g., a concentration of 3333 ppm (v) of ammonia 2000 ft downwind from a 10 lb/sec ammonia leak at location *X* is estimated assuming a 1.4 mph wind, and Stability Class D).

Consequence: A measure of the expected effects of an incident outcome case (e.g., an ammonia cloud from a 10 lb/sec leak under Stability Class D weather conditions, and a 1.4 mph wind traveling in a northerly direction from location *X* will injure 50 people).

Effect Zone: For an incident that produces an incident outcome of toxic release, the area over which the airborne concentration equals or exceeds some level of concern. The area of the effect zone will be different for each incident outcome case [e.g., given an IDLH for ammonia of 500 ppm (v), an effect zone of 4.6 square miles is estimated for a 10 lb/sec ammonia leak]. For a flammable vapor release, the area over which a particular incident outcome case produces an effect based on a specified overpressure criterion [e.g., an effect zone from an unconfined vapor cloud explosion of 28,000 kg of hexane assuming 1 percent yield is 0.18 km^2 if an overpressure criterion of 3 psig is established]. For a loss of containment incident producing thermal radiation effects, the area over which a particular incident outcome case produces an effect based on a specified thermal damage criterion [e.g., a circular effect zone surrounding a pool fire resulting from a flammable liquid spill, whose boundary is defined by the radial distance at which the radiative heat flux from the pool fire has decreased to 5 kW/m^2 (approximately 1600 Btu/hr-ft^2)].

TABLE 1-3
Selected Definitions for TRA (*continued*)

Likelihood: A measure of the expected probability or frequency of occurrence of an event. This may be expressed as a frequency (e.g., events/year), a probability of occurrence during some time interval, or a conditional probability (i.e., probability of occurrence given that a precursor event has occurred, e.g., the frequency of a stress corrosion hole in a pipeline of size sufficient to cause a 10 lb/sec ammonia leak might be 1×10^{-3} per year; the probability that ammonia will be flowing in the pipeline over a period of 1 year might be estimated to be 0.1; and the conditional probability that the wind blows toward a populated area following the ammonia release might be 0.1, yielding an overall likelihood of 1×10^{-5} per year).

Risk: A measure of potential economic loss, human injury, or environmental damage in terms of both the incident likelihood and the magnitude of the loss, injury, or damage.

Risk Assessment: The process by which the results of a risk analysis are used to make decisions, either through a relative ranking of risk reduction strategies or through comparison with risk targets (e.g., the risk of injuring 50 people at a frequency of 1×10^{-5} events per year from an ammonia incident might be judged intolerable, and remedial measures required).

are used throughout this book. An incident is defined in Table 1-3 as "the loss of containment of material or energy," while an incident outcome is "the physical manifestation of an incident." A single incident may have several outcomes. For example, a leak of flammable and toxic gas could result in:

- a jet fire (immediate ignition)
- a vapor cloud explosion (delayed ignition)
- a vapor cloud fire (delayed ignition)
- a toxic cloud (no ignition).

The third and often confusing term used in describing incidents is the incident outcome case. As indicated by its definition in Table 1-3, the incident outcome case specifies values for all of the parameters needed to uniquely distinguish one incident outcome from another. For example, since certain incident outcomes are dependent on weather conditions (wind direction, speed, and atmospheric stability class), more than one incident outcome case could be developed to describe the dispersion of a dense gas.

1.2. General Framework of a TRA

The general steps of a comprehensive TRA are depicted in Figure 1-1. This figure focuses on the TRA, and does not explicitly address the broader

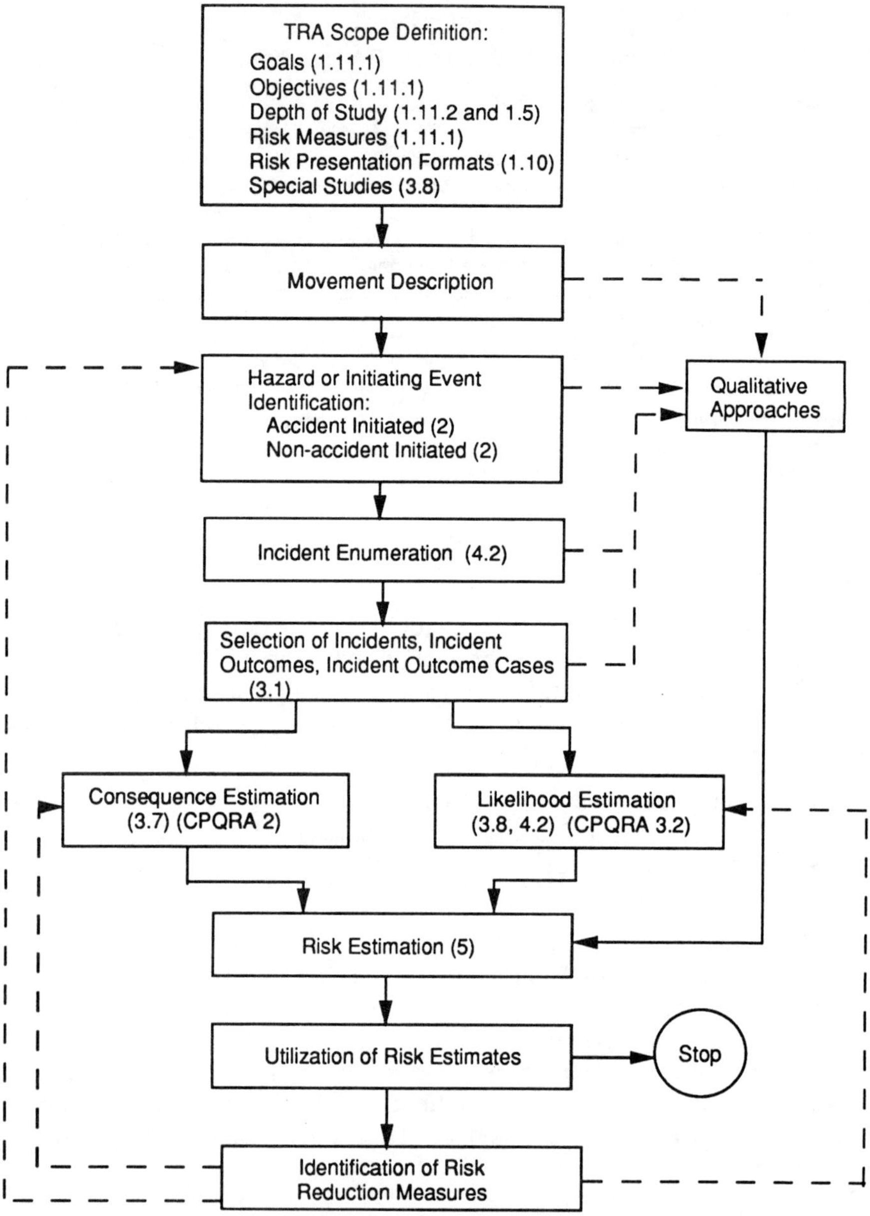

FIGURE 1.1. Framework for comprehensive TRA. *Note:* Relevant section numbers are indicated in parentheses. Section numbers preceded by "CPQRA" refer to a section in the *CPQRA Guidelines* (CCPS, 1989).

scope of a risk assessment. The relationship between risk analysis and risk assessment is illustrated in Figure 1.2 of the *CPQRA Guidelines* (CCPS, 1989).

Figure 1-1 not only shows the steps of a comprehensive TRA, it also gives cross-references in parentheses to the various chapters and sections of this book that address each of these steps in further detail. Each of the basic steps is briefly described below.

- **TRA Scope Definition** converts user requirements into study goals and objectives (Section 1.11.1). Risk measures (Section 1.11.1) and risk presentation formats (Section 1.10) are chosen in finalizing a scope of work for the TRA. A depth of study (Sections 1.11.2 and 1.5) is then selected based on the specific objectives defined and the resources available. The need for special studies (e.g., additional failure analyses) is also considered (Section 3.8). TRA definition ends with the definition of study specific information requirements.
- **Movement Description** is the compilation of the transportation activity information needed for the risk analysis. For example, mode, container specification, weather data, number of trips, volume per container, material, shipping conditions, route or origin and destination, and population data. This data set is then used throughout the TRA.
- **Hazard or Initiating Event Identification** is a critical step in TRA. The accident-initiated events of concern can generally be identified based on historical data. Non–accident-initiated events may be identified through hazard identification techniques described in *CPQRA Guidelines* (CCPS, 1989) or *HEP Guidelines* (CCPS, 1992) if they relate to overfilling or contamination, or based on experience, checklists, equipment failure experience, etc. Typical non–accident-initiated hazards are listed in Table 1-2. With pipelines, more traditional hazard identification techniques may be used.
- **Incident Enumeration** is the identification and tabulation of all incidents without regard to importance or initiating event. In TRA, the incident enumeration may be closely related to the available data (Section 4.2).
- **Selection** is the process by which one or more significant incidents are chosen to represent all identified incidents, incident outcomes are identified, and incident outcome cases are developed (Section 3.1).
- **Consequence Estimation** is the methodology used to determine the potential for damage or injury from specific incidents. A single incident (e.g., rupture of a pressurized liquid tank) can have many

distinct incident outcomes [e.g., Vapor Cloud Explosion (VCE), Boiling Liquid Expanding Vapor Explosion (BLEVE), flash fire]. These outcomes are analyzed using source and dispersion models (*CPQRA Guidelines* Section 2.1) and explosion and fire models (*CPQRA Guidelines* Section 2.2). Effects models are then used to determine the consequences to people or structures (*CPQRA Guidelines* Section 2.2).

- **Likelihood Estimation** is the methodology used to estimate the frequency or probability of occurrence of an incident. Estimates may be obtained from historical incident data on failure frequencies (Section 4.2), from failure sequence models such as fault trees and event trees (*CPQRA Guidelines* Section 3.2), or from special failure models. (Section 3.8).
- **Risk Estimation** combines the consequences and likelihood of all incident outcomes from all selected incidents to provide a measure of risk (Chapter 5). The risks of all selected incidents are individually estimated and summed to give an overall measure of risk. The sensitivity and uncertainty of risk estimates and the importance of the various contributing incidents (including various release locations) to estimates are discussed in Section 5.3.
- **Utilization of Risk Estimates** is the process by which the results of a risk analysis are used to make decisions, either through relative ranking of risk reduction strategies or through comparison with specific risk targets. This step is actually what takes a risk analysis to a risk assessment and is not discussed in detail in this volume.

It may be appropriate to stop an analysis at any one of a number of points. This may be a result of having obtained enough information to distinguish among options, or because the level of maximum consequences is low enough that a frequency estimate is not required. The purpose of the study may also be to identify potential hazards—not to evaluate risks.

In a screening or other qualitative analysis many of these steps are not carried out explicitly. Risk screens may be developed to implicitly take one or more of these steps into account. A screening level analysis can even be conducted by comparing the data set for the movement(s) of concern with the data set and results for another movement for which a quantitative TRA has been conducted, and simply determining if the movement of concern poses more or less risk than the previously evaluated movement. Even in quantitative risk analyses it may be possible to take the results of other studies and use them as the basis of one or more of the steps in the TRA. These other results may first need to be scaled or adjusted before using them.

In the **Selection** step, the number of incidents, incident outcomes, and incident outcome cases need not be more than the minimum necessary to meet the study objectives and should be consistent with the data available. For instance, if one is comparing worst case consequences, it may be adequate to look only at a catastrophic failure and worst case meteorological conditions at the point of highest population density.

1.3. Typical Reasons for Conducting a TRA

TRA can help a decision-maker in:

- choosing a site for a facility or process relocation or expansion, by taking both fixed site and transportation risks into account
- selecting appropriate risk reduction alternatives by demonstrating the effectiveness of various alternatives
- choosing among alternate routes by providing information on the relative risks associated with each route
- selecting a mode of transportation when more than one is feasible
- selecting the most effective container or knowing if additional protective measures are warranted
- developing emergency response plans
- understanding the influence of material state on risks, such as shipping HCl in solution or as anhydrous HCl
- examining the trade-off between larger/less frequent and smaller/more frequent shipments
- controlling transportation risks
- making judgments about the tolerability of existing or increased movement levels
- considering the benefits of trades, swaps, or toll operations
- meeting regulatory requirements for determining a preferential route

Depending on both the information available and the purpose of the study, different levels of analysis are appropriate. These evaluations may be based on a relative (or comparative) or an absolute estimation of risks. For instance, making decisions about the tolerability of the risks associated with a set of movements might be as simple as a qualitative examination of the characteristics of the movements of concern with those for some other "tolerated" movement—this is a relative evaluation. Alternatively, if the company has quantitative risk targets, a fairly detailed quantitative study may be required—this generally requires an absolute risk estimate to compare to these targets.

Evaluating the effectiveness of protective measures, particularly those for which historical data are unavailable or insufficient, often justifies the use of a quantitative risk analysis. While simply reviewing the effects on the frequency of release may be enough, a more comprehensive risk analysis offers several benefits. A protective measure that reduces the likelihood of an accident by 30 percent might sound very promising. However, if the risks are very high, this 30 percent reduction may be inadequate, and if the risks are very low, the reduction may be unnecessary—especially if the funds are then available to reduce other (larger) risks. In many instances, a change of 30 percent is within the overall uncertainty of the estimates.

A risk reduction alternative may effect more than the frequency of an accident, in that it may also change the quantity potentially released. Reducing the speed of a train can affect the likelihood of an accident, the chance a car derails, the probability of a spill *and* the expected size of a release. A relatively detailed evaluation would be necessary to understand the full effects of this alternative.

Other risk reduction alternatives may have no impact on the frequency of an accident or release, but might alter the consequences. One such alternative would be changing material temperature to limit flashing upon release.

Later sections in this chapter will provide further discussion on the process of selecting the appropriate type and depth of study for various situations.

1.4. Transportation versus Fixed Facility Risk Evaluations

There are both similarities and differences between transportation and fixed facility risk evaluations. Both groups are discussed below.

1.4.1. General Similarities to Fixed Facility Evaluations

The general process of a quantitative TRA involves:

- defining the scope of the analysis
- describing the system or movement
- identifying hazards or initiating events (accident and non-accident)
- enumerating incidents
- selecting incidents, incident outcomes, and incident outcome cases
- estimating consequences
- estimating frequencies
- combining frequencies and consequences to estimate risk
- evaluating risk reduction alternatives

These steps are virtually the same as in a CPQRA, but there are subtleties within each step. Table 1-4 lists some of these. Both TRA and fixed facility risk evaluations can be qualitative or semiquantitative as opposed to quantitative. In these cases, many of these steps are handled on a comparative basis or addressed in much less detail.

The modeling of release consequences is largely independent of the cause of the release and is therefore basically directly transferable from the techniques of CPQRA. Section 3.7 discusses some of the unique consequence modeling issues for TRA; otherwise the CPQRA guidance should be followed.

Risk measures are also common to the two types of quantitative risk analyses, for both individual and societal risk. The only real differences are in the details of how the risks are estimated and the units chosen. For example, in transportation one may wish to measure risk per trip or risk per year (or both), while in CPQRA risks are generally measured on a per year basis or on a comparative basis with other results. Table 1-5 lists many of the risk measures available; a number of these are illustrated in the worked examples provided in Section 7.

Qualitative or semiquantitative estimates of risk generally take into account experience, judgment, good practices, training, procedures, inspection and maintenance, codes and standards, past performance, etc., whether one is dealing with fixed facilities or transportation movements.

1.4.2. Differences from Fixed Facility Evaluations

The most fundamental difference between CPQRA and TRA is that TRA deals with a linear source of risk, versus a relatively discrete point source for a CPQRA. This linear source may be static as in the case of pipelines or may be a moving source for other modes of transport. In transportation, a release can occur anywhere along a route between the origin and destination. The unpredictability of the exact release location often requires the use of generalized approaches to limit the data needs and number of incident outcome cases. These generalized approaches may relate to one or more of the following:

- *identification and selection of initiating events*—may utilize an aggregate accident rate or a limited breakdown, such as derailments and collisions, rather than a detailed breakout by failure mode.
- *selection of incidents and incident outcomes*—particularly release sizes and rates, release orientation, material temperature and pressure at time of release. While a release in a facility is reasonably predictable in terms of the material conditions, these could change

TABLE 1-4
Similarities and Differences between Fixed Facility and Transportation Risk Evaluations

Topic	*Similarities*	*Differences*
Basic quantitative approach	Same basic steps and consequence models	More likely to include spills on water or without containment in TRA; more reliance on data bases and less on fault trees; TRA less common
Data	Chemical information	Less situation specific data in TRA; may include external events; more applicability and availability of nationwide data in TRA
Risk measures	Measures and their calculation	Units of frequency measure—may use per trip or per ton-mile in TRA; can use more than one frequency base in analysis
Risk reduction alternatives	May address frequency, consequence, or both	Fewer options in TRA and less control over infrastructure; limited amounts of reduction available in TRA
Qualitative approaches	Consider same factors or types of factors	More value for TRA to use these approaches due to commonality in operations
Scope of study	Flexibility in defining	More likely to include entire route in TRA than full site in CPQRA
Nature of risk	Generally computed on annualized basis	Point or small area for facilities vs. extended linear for transportation; constant vs. intermittent risk to community
Role of human error	Significant	Even greater for TRA yet harder to control as outside groups or third parties involved
Location of release	None	Unknown for TRA; necessitates use of generalized approaches for TRA
Release size and orientation	Significant to result	Harder to predict for TRA for accident-initiated events; orientation may vary such that relief valves contact liquid or are blocked

with the seasons for many HAZMAT movements, and even from one end of a route to another.

- *meteorological conditions for modeling*—wind roses and stability class distributions vary from location to location, as does the ambient temperature and humidity.
- *ignition probabilities*—the number, type, and proximity of ignition sources vary along a route, and it may be very difficult to get route specific data.

TABLE 1-5
TRA Quantitative Risk Measures

Risk Measure	*Perceived Benefits*
Average Individual Risk	Matches generalized assumptions about ignition source location, population distribution, etc.; comparable to background risk statistics; reflects transient nature of risk, depending on vehicle location.
Individual Risk Profiles	Provides greater detail about changes in risk with distance from route; comparable to background risk statistics; accounts for noncontinuous presence of transportation risk.
Societal Risk per Year	Provides details on overall levels of risk from activity; gives information on number of people potentially involved in various events.
Societal Risk per Trip	Similar to societal risk per year, but removes the multiplier associated with number of trips per year; some believe to be a better measure of the risk of a given material/mode.
Societal Risk per Ton-Mile	Similar to societal risk per year, but removes multipliers for both length and number of trips; while insight on overall risk levels may be diminished, more direct comparisons with other results are possible.
Societal Risk per Community (10–20 mile segment)	Maintains number of trips multiplier, yet reduces length to a finite segment; can allow comparison with CPQRA results and use of common risk targets.
Comparative Measures	Allows risks of one movement to be compared to those of another movement or material where data are available; allows judgments to be made on a relative basis; may be very helpful in prioritizing large numbers of movements for minimal effort.

- *population distribution*—not only is there usually very little, if any, buffer distance between a route and the general population, but the population distribution (density) constantly changes along the route.

Generalized approaches help address these variabilities in a consistent and usually conservative manner. Several approaches are discussed in Chapter 3. The ability to eliminate or prevent risks is much greater for fixed facilities. In transportation, risk reduction is generally all that can be obtained. The Office of Technology Assessment (1986) estimated that 62 percent of hazardous material incidents are due to human error. The degree of variability and influence of human performance is often cited as being much greater for some modes of transportation than others and for transportation in general in comparison to fixed sites. This is particularly true for road transportation where the route taken can vary from one trip to another. The role of outside forces such as weather and other drivers or operators can also be quite significant. For instance, extensive training of truck drivers does not change the performance of other drivers on the road. Thus, the ability to control the overall impact of human errors is diminished in transportation.

The need to consider other drivers is also an example of the reduced degree of control of many of the parameters that influence accident rates and/or release probabilities in transportation. Other examples point out the limited potential to reduce the basic accident rates or make system-wide changes. This is particularly true for a chemical company that does not have responsibility for the transportation system itself, as it does for its chemical plants. For instance, changing rail cars does not improve track quality and pipeline failure rates are often dominated by the potential for third-party damage. Where there are potential risk reduction alternatives available, they are often quite costly and time consuming, particularly if they involve new equipment. A fixed site may have a recommendation to add a valve or alarm that can be done easily, but a new fleet of railcars or barges may be suggested as risk reduction alternatives for transportation and may be much more difficult to fund and implement. Questions of availability come up in such circumstances—the desired equipment must often be built, and even retrofits can take a considerable amount of time.

Comparative or relative risks (versus absolute risks) are more commonly estimated and used in transportation risk assessments. This leads to a greater use of qualitative and semiquantitative approaches to estimate risks for transportation, whereas, fixed facilities may stop at the identification of potential hazards in their qualitative studies.

When estimating the frequency of accidents, historical data are often available and adequate, unlike the case of CPQRA. This is a direct result of many accidents not being equipment dependent. However, the likelihood or chance of a release is very equipment-specific, and is often less easily estimated via other techniques such as fault trees. This can increase the role of judgment and extrapolation in TRA.

The nature of TRA data can be different from CPQRA data. They are often expressed as a function of distance traveled or per trip, transit, or visit. External event causes of accidents are generally included in the data–including such items as vandalism for rail transport, adverse weather for marine transport, and third-party damage for pipelines.

The actual ownership of and liability for a material in transport is also less clear cut. Legal agreements may specify when a customer "owns" the product, but any incident that occurs is going to adversely affect the shipper, customer, and carrier, regardless of product ownership. (Industry as a whole is also adversely affected whenever an accident occurs—through image, regulation, insurance costs, etc.) All three parties therefore have a vested interest in the product, and may even have conflicting requirements for product handling. Truck drivers may prefer driving at night when the roads are less congested and (they believe) less likely to produce an accident. However, the shipper and customer may be concerned about nighttime travel due to the increased speed that may be possible, the greater number of people in their homes along residential portions of a route, and possible difficulties in emergency response.

One last difference between transportation and fixed facility risk evaluations is the nature of the risk reduction and mitigation alternatives available. By virtue of the unknown location of a transportation release (prior to its occurrence), it is much more difficult to identify and implement effective mitigative (postrelease) strategies. Secondary containment via diking, selective grading, water sprays, foams, evacuations, etc. are either not feasible, or can only be initiated some significant amount of time after the release has occurred. Given the rapid dispersion of many large releases, such mitigation measures may be totally infeasible or untimely.

1.5. Scope and Depth of TRA Studies

The scope and depth of study is especially important in TRA, since the available alternatives range from very straightforward and simple qualitative efforts to very detailed quantitative analyses. As such, the associated cost, time, data, and other resource requirements can also vary significantly. The initial scope must also be borne in mind throughout the analysis in order to control the progress of the effort and to avoid going into undesired detail where it is not appropriate. This section introduces the concept of a study cube (Figure 1-2) to integrate scope, study goals, and level of effort. Since the identification and selection of release events are two of the processes with the greatest influence on the level of effort and the potential usefulness of a study to meet certain goals, these items are also discussed.

1.5.1. The Study Cube

The three primary factors driving the overall scope of a TRA are: the TRA focus, the nature or complexity of the study, and the number of incident outcome cases that will be evaluated. These items are reasonably similar to their CPQRA counterparts. While the study cube has three representative divisions for each axis, this division is arbitrary and is only meant

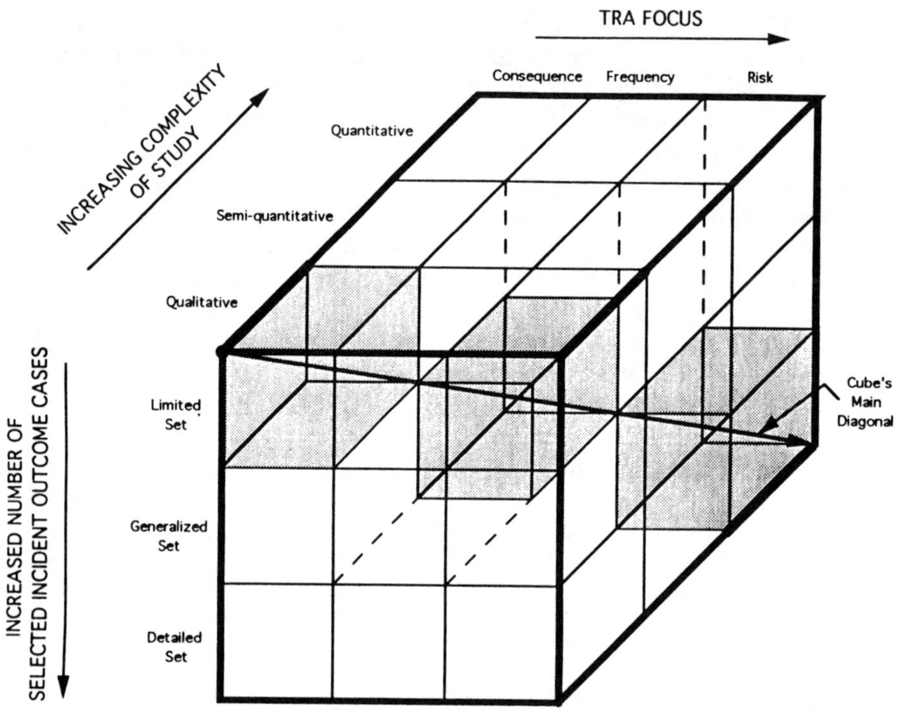

FIGURE 1.2. TRA Study Cube.

for illustrative purposes. In reality there is a continuum of possible choices, with very few constraints on one's choices. The actual likelihood of someone pursuing a qualitative risk evaluation of many different incident outcome cases may be somewhat remote, however. In general, certain diagonals or slices through the cube will represent the most typical scopes of TRAs.

TRA Focus

Some TRA studies focus on the potential consequences of releases. These studies may involve consequence modeling in various levels of detail, comparisons with consequence results obtained for other materials or release quantities, or even a review of the demographics along a route to minimize potential impacts from a release. At the simplest level, a review of the types of potential consequences may be all that is addressed.

Other TRA studies focus directly or indirectly on frequencies. Carrier inspection or selection programs, equipment selection and inspection/maintenance programs, and driver training programs are all geared toward improving safety through the reduction of accidents and releases. More recognizable frequency evaluations include studies to determine accident rates and spill probabilities using historical data or other techniques discussed elsewhere in this document.

TRA studies that emphasize risk bring together both the frequency and consequence elements. In a qualitative evaluation, the total effort required may not be much more than the total for the two elements. In a quantitative evaluation, the risk calculations themselves can require considerable additional efforts.

Nature or Complexity of Study

Whereas a CPQRA study is quantitative by definition, TRA studies range from totally qualitative to semiquantitative to quantitative in nature. (There are equivalent types of studies available for fixed facilities, but they are not within the scope of CPQRA.) Within each of these categories, there is also an increasing amount of detail that can be considered. This level of detail is generally a reflection of the number of accident-initiated events and non–accident-initiated events that will be considered and how they are identified.

The least detailed case might involve using only accident-initiated events, as determined from historical data. An increasing degree of sophistication comes from trying to use more situation-specific data, or drawing on various analytical tools to handle situations not covered by historical data. At another level, non–accident-initiated events may also be addressed. These may be identified through standard hazard identification techniques, checklists, historical data, etc.

Number of Incident Outcome Cases

Unlike CPQRA where this axis focused on the number of incidents, in TRA the concern is on the number of incident outcome cases. As discussed in Section 1.3, there are many generalizations that can be made to reduce the number of incident outcome cases. In the absence of this, the number of cases gets extremely large.

A screening level analysis might only consider one limited situation or might consider a worse case release with one or two sets of meteorological conditions and only a few hypothetical release locations. Some techniques do not even identify the situation or scenarios upon which they are based. More detailed studies would be expected to increase the number of releases, the number of meteorological conditions, the number of ignition assumptions, and the number of different release locations (i.e., different population densities), but to still maintain certain generalized assumptions. A very detailed (and numerous) set of incident outcome cases would represent the extreme position on this axis.

1.5.2. Range of Incidents

Both the complexity of study and the number of incident outcome cases are affected by the range of initiating events and incidents covered. This not only reflects the inclusion of accident and/or non–accident-initiated events, but also the size of those events. For instance, studies may evaluate one or more of the following:

- catastrophic failure of container (very rapid total release)
- large hole (large continuous release)
- smaller holes (continuous release)
- leaks at fittings or valves (small continuous leaks)
- "popping" of relief valves (short duration limited release)

The latter two of these might occur in transit or while stationary after an accident or parked at a rest stop.

In general, quantitative studies do not include very small continuous releases or short duration limited releases if past experience or preliminary consequence modeling shows that such releases do not contribute to the overall risk levels. Similarly, small releases while in motion are so diluted and distributed so as to have minimal acute human health impacts. Environmental effects could be more significant.

Qualitative studies and inspections or reviews may actually focus on these smaller releases and releases while moving, in large part due to their generally increased likelihood of occurrence, and to the greater potential to reduce or even eliminate them.

1.5.3. Selection of Initiating Events and Incidents

The selection of initiating events and incidents should take into account the goals or objectives of the study and the data requirements. Table 1-6 gives some of the purposes of conducting a TRA study from Section 1.3, along with the types of initiating events and incidents that might be appropriate for each purpose.

The data requirements increase significantly when non–accident-initiated events are included and when the number of release sizes increases. While the potential range of release sizes is tremendous, groupings are both appropriate and necessitated by data restrictions (i.e., on the likelihood of various release sizes). The main reasons for including release sizes other than catastrophic are to reduce the conservatism in an analysis and to better understand the relative contributions to risk of small versus large releases.

The information in Table 1-6 assumes that consequences are being evaluated. There may be circumstances where the TRA only looks at

TABLE 1-6

TRA Purposes and Incident Selection

	Initiating Events		*Type of Release*		
Purpose	*Accident*	*Non-accident*	*Rapid Loss of Contents*	*Continuous*	*While Moving*
Consider transportation in site selection	●	**O**	●	●*	○
Demonstrate effectiveness of various alternatives	●	**O**	●	**O**	○
Choose among alternate routes on relative basis	●	○	●	**O**	○
Select mode of transportation	●	**O**	●	●*	○
Control transportation risks	●	●	**O**	**O**	●
Select container and/or additional protective features	●	●	●	●*	**O**
Develop emergency response plans	●	**O**	●	**O**	○
Understand influence of material state	●	●	●	**O**	**O**
Examine trade-offs in container size	●	○	●	○	○
Make risk tolerability judgments	●	●	●	●*	○
Consider benefits of trades, swaps, or toll operations	●	**O**	●	**O**	○
Meet regulatory requirements for route determination	●	○	●	**O**	○

● Very common to include in study.
O Sometimes included in study.
○ Uncommon to include in study.
* Medium to large releases.

frequency—either of an accident or of a release or of a catastrophic release. This consideration of frequency need not be quantitative. Table 1-7 takes the same study purposes and indicates whether a frequency, consequence or risk evaluation is generally conducted.

Both Tables 1-6 and 1-7 are meant to supply guidance to the selection of the appropriate scope and depth of study—they do not prohibit alternate scopes, where the situation demands or suggests them.

TABLE 1-7

TRA Purposes and Selection of TRA Focus

Purpose	*Consequence*	*Frequency*	*Risk*
Consider transportation in site selection	⭘	⭘	●
Demonstrate effectiveness of various alternatives	○	●	⭘
Choose among alternate routes on relative basis	○	⭘	●
Select mode of transportation	○	○	●
Control transportation risks	⭘	●	⭘
Select container and/or additional protective features	○	●	⭘
Develop emergency response plans	●	○	○
Understand influence of material state	●	⭘	⭘
Examine trade-offs in container size	○	○	●
Make risk tolerability judgments	⭘	○	●
Consider benefits of trades, swaps, or toll operations	○	○	●
Meet regulatory requirements for route determination	⭘	●	⭘

● Very commonly the focus.
⭘ Sometimes the focus.
○ Uncommonly the focus.

1.6. Screening Techniques

TRA screening techniques range from simple checklists based on DOT regulatory compliance issues to risk-based screens that consider both frequency and consequence and produce a numerical or categorical score. In addition, consequence-based screens using simplified algorithms are common. A stated or unstated goal of the application of screening techniques is to stop at this point whenever possible, in order to make the most effective use of available resources. In circumstances where data availability is limited or impractical—such as a review of a large number of different movements—screening analyses are most applicable.

Some screens are designed for qualifying carriers. Such screens review:

- insurance coverage
- years of experience
- operating authority
- safety rating

- accident rate
- financial data
- hazardous materials experience
- emergency response planning and communication
- equipment type and ownership
- driver certification and training
- quality assurance program
- refrigeration or heating controls

These parameters influence accident rates, and may influence release probabilities and consequences. Some companies also review vehicle maintenance and inspection records and programs.

The Dow Chemical Company has a Distribution Ranking Index (DRI) that provides a method for rating the inherent hazards of a material as they pertain to potential transportation incidents. Generally, the DRI represents the acute hazards to humans and/or the environment potentially affected by a single release of a material in a transportation accident.

Seven specific indices are considered for each material; namely,

- Fire and Explosion
- Chemical Exposure
- Inhalation Toxicity
- Aquatic Toxicity
- Skin Absorption Toxicity
- Skin/Eye Corrosivity
- "Other Effects"

Each of these indices is rated on a scale from low (1) to high (4), except for skin/eye corrosivity which can only receive a maximum ranking of 3, and the "other effects" index which has only one ranking of 4. This latter index is intended to identify those materials which would not normally achieve a high DRI hazard rating by the other indices shown above, but which, because of their other health or environmental effects, are believed to be especially harmful. The overall DRI for a material is given by the highest value achieved on any one of the seven individual indices.

An indexing system for pipelines has also been developed by Dow Chemical. The RIPS (relative index of pipeline safety) system provides a logical method for computing the relative risk of a pipeline, based on four indexes that take into account the pipeline attributes and preventative actions regarding:

- third-party damage
- corrosion
- design
- incorrect operations

Each index receives between 0 and 100 points and the overall sum is adjusted by a leak impact factor that reflects the proximity to populated areas and inherent material hazards. Described briefly in Muhlbauer (1991), the details of this screening tool are given in Muhlbauer (1992).

Texaco Marine Services screens the vessels used for transportation in a three-stage process. The stages are a one-hour preliminary vetting, vessel inspection, and a final vetting that updates and enhances the screening conducted as a part of the preliminary vetting. No vessel is rejected if it meets Texaco's minimum standards, and conditional approvals can be given for a single voyage. The process is very methodical and considers such factors as:

- flag state registry
- Lloyd's MARDATA Casualty History
- vessel cargo experience factor
- corrosion control program
- insurance coverage
- credit rating
- vessel's certification status

A total of 29 factors are scored and weighted.

Other screening tools (see Section 3.10) include ADLTRS™, which is a risk-based screening computer program that considers worst-case consequences as well as the likelihood of a release. HazTrans® has a geographic information system (GIS) based routing risk screen that may also be used in more detailed assessments. As part of an Inter Industry Rail Safety Task Force project, a rail-based risk screening model has been developed. This Task Force includes representatives of the Chemical Manufacturers Association (CMA), the Association of American Railroads (AAR), and the Railway Progress Institute (RPI). The screening model includes data on 125 hazardous materials moved by rail in tank cars.

Screens are often useful when dealing with just one mode of transportation. In order to provide maximum value when dealing with multiple modes of transport, the screen must appropriately reflect some of the trade-offs involved among modes. These might include: fewer trips at the expense of increased container volumes; different likelihoods of large spills relative to small spills; different accident rates; etc.

1.7. Limitations

For qualitative TRAs, the greatest limitations relate to the applicability of other results when they are used for comparison or scaling. Not all comparisons are obvious—one material may be less flammable than another, but might also be toxic, posing additional hazards. The applicability of general nationwide trends to specific situations may also be questioned. In general, the limitations of qualitative approaches are in terms of whether or not they can accurately discriminate among alternatives.

The data needs in a detailed quantitative TRA can be very extensive. Even if large amounts of data are obtained, they still may not cover the specifics of a given movement. This is particularly true for the container being used and the selection of release sizes and rates. The quality and level of detail of the available data should be used to guide the selection of the TRA techniques.

A summary of technical and management limitations, their implications, and possible means for reducing their impact are given in Tables 1-8 and 1-9. An extensive discussion of risk analysis limitations is given in the *CPQRA Guidelines* (CCPS, 1989).

As in CPQRA, the major limitations of any TRA are related to uncertainty. Uncertainty should decrease in the future as models become standardized, data collection is more thorough and consistent, risk analysis expertise is more available, and human consequence effect data (e.g., toxicity) are more complete. The uncertainty over event location is a basic characteristic of a TRA and will not change.

Quantitative TRAs, like the CPQRAs, have uncertainties that can span one or more orders of magnitude. This often has a greater effect when evaluating multiple modes than when working within a single transport mode, when working with multiple materials rather than just one material, and when working with absolute risk estimates instead of relative ones. When uncertainties cannot be estimated, the sensitivity of the results to variations in key parameters or assumptions can be evaluated and used as a guide.

1.8. Current Practice

TRA is a "state-of-the-art" risk management tool. While the Distribution Code of Responsible Care® calls for some level of hazard and risk-based screening to be done, no other systematic requirements for TRA currently exist in the U.S. for the chemical process industry. Various communities are using TRA approaches to understand risk levels or help with route restrictions, but as yet there are no equivalents to the risk management program requirements enacted by many states for fixed facilities.

As was found in the survey conducted as part of the *CPQRA Guidelines* (CCPS, 1989) preparation, a number of companies do conduct quantitative risk analyses and a few have clear corporate risk policies and targets.

TABLE 1-8

Technical Limitations of TRA and Means to Address Them

Technical Causes of Limitation	*Implications to TRA*	*Remedies*
Incomplete or inadequate enumeration of initiating events	• Underestimate risk of accident or non–accident-initiated events • Leads to improper selection	• Interface with CPQRA • Involve experienced TRA practitioners • Peer review/quality control • Review by distribution and operations personnel
Improper selection of initiating events	• Underestimate (or overestimate) risk of accident or non–accident-initiated events	• Involve experienced TRA practitioners • Sensitivity testing • Interface with data availability • Peer review/quality control • Review by distribution and operations personnel
Incomplete or inadequate enumeration of incidents	• Underestimate risk of incident • Leads to improper selection	• Interface with CPQRA • Involve experienced TRA practitioners • Peer review/quality control • Review by distribution and operations personnel
Improper selection of incidents	• Underestimate (or overestimate) risk for all incident groupings • Overly simplistic, generalized assumptions	• Involve experienced TRA practitioners • Interface with data availability • Sensitivity testing • Peer review/quality control • Review by distribution and operations personnel
Unavailability of data required for selected scope	• Possibility of systematic bias • Uncertainty in consequences, frequencies, or risk estimates • Incorrect prioritization of major risk contributors	• Secure additional resources • Expert review/judgment • Use benchmark/landmark study results • Check results against other models or historical incident records • Sensitivity testing • Use different level of detail
Consequence or frequency model assumptions/validity	• Similar in effect to data limitations	• Peer review • Ensure models are used within design range • Sensitivity testing • Comparisons with historical data • Test with other models

TABLE 1-9
Management Limitations of TRA and Means to Address Them

Management Causes of Limitation	*Implications to TRA*	*Remedies*
Resource limitations (personnel, time, models)	• Mismatch of time and depth of study • Insufficient depth of study • Inadequate quality of study	• Extend schedule • Defer study until resources available; use benchmark studies; use consultants • Identify major risk contributors and emphasize these
Skills unavailable	• Incorrect preparation and analysis • Improper interpretation of results	• Amend scope of work • Provide training, acquire new staff, use consultants

Many or most of these companies have policies for quantitative TRA as well, recognizing that distribution activities may actually dominate their overall risk portfolio. Many more companies have qualitative programs.

Current practice in TRA ranges from no activity; to screening by judgment; to screening by consequence; to screening by risk; to detailed quantitative risk analyses, with decreasing numbers of practitioners at each step of the sequence. Sharing of screening techniques is becoming more common (see Section 1.6), but to date only a few very large detailed studies have been published (see Section 1.9). A fair number of small (one or two movements) detailed studies are conducted each year by individual companies, but there is very little sharing of results. CMA, for example, through Responsible Care®, is promoting the role of TRA in risk management and calling for more sharing across industry of tools, experience, and expertise.

Few companies have integrated their TRA and CPQRA efforts. Such integration is particularly valuable in site selection problems where both transportation and fixed facility risks should be prominent concerns.

Where TRA is applied quantitatively, there is no consensus at this time over the appropriate risk measure or measures to be used. This limits the potential to compare results if the measures are not readily convertible. Some of the most common measures and their perceived benefits are outlined in Table 1-5.

1.9. The Use of Existing TRA Studies

Existing studies include both pivotal published studies such as the recent UK Advisory Committee (1991) report and the International Consensus

Conference corridor application (Saccomanno, 1993) and companies own evaluations of their past or present movements. These latter studies can serve as benchmarks or comparisons for other company movements; internal studies often document past risk decisions and therefore provide guidance on risk targets as well. The interim or final results of both types of studies can be used to evaluate other situations, through direct application or with necessary scaling.

Table 1-10 lists some of the data and techniques used in several recent landmark studies and their suitability for use in other situations.

In general, accident rates and release probabilities can be taken from other studies when the same or similar routes are used and the same type of equipment is used. If a movement has better equipment than that used in another study, then the values from that second study could be used as upper bounds or as a starting point for further adjustments in the absence of specific data.

Consequence results, in terms of hazard modeling areas, can also be used directly for the same or very similar materials if the same incident outcomes are being used in another study. The incident outcome cases and final consequences will vary however. If one toxic material is known to be less hazardous than another toxic material, it may be possible to use previous results as upper bounds on the estimated risk levels. However, in the case of flammable releases, one must use caution as ignition and explosion probabilities may also change from one material to another.

TABLE 1-10
Information Provided in Recent Landmark Studies

Study	*Modes*	*Type of Information*	*Comments*
International Consensus Conference corridor application (Saccomanno, 1993)	Rail; Road	Accident rates	Accident rates from many different countries were used; however, some applications just used hypothetical rates.
		Methodology	Vague problem definition led to varied approaches and results.
Advisory Committee on Dangerous Substances, Health and Safety Commission (1991)	Rail; Road; Marine	Accident rates	Very specific to UK rail, road, and port characteristics.
		Release probabilities	Equipment types may not be directly applicable to US.
		Overview of methodology	Limited details on methodology; hard to apply some techniques.

In some selected cases, it may be possible to use the risk results of one study to estimate the risks in another study. When doing such a comparative estimation, one needs to review each of the following parameters:

- number of trips
- type of container
- quantity of material per container
- route length
- route characteristics (population density, weather conditions, etc.)
- mitigation measures
- mode

Which of these are critical will depend on the risk measure being used. In general, the same mode and preferably the same material should be addressed in the two studies. If very few parameters differ and the effects of those differences are well understood, then a comparative estimation of risk may be developed. The result should be treated as a rough approximation of absolute risk or as a relative measure of risk.

1.10. Utilization of TRA Results

As indicated in Section 1.3 and Tables 1-6 and 1-7, there are many different uses for TRA results for specific decisions. There are also uses of TRA that are better classified as monitoring of the risk. In these cases, qualitative or screening approaches may be routinely applied to large numbers of movements in order to detect (and act upon) changes in risk. The triggers for such detection might be increases in carrier accident rates, more problems identified during inspections, violations of management practices, increases in numbers of annual movements, new customers, changes in routes, etc. Further study may be required in the event that one of these changes is noted.

The outcome of a TRA is generally some form of written report. A TRA study is useful only when its information is incorporated into an action plan. Action plans may address changes in container design, enhanced driver training, changes in route, switching to another mode, ceasing a movement, or relocating a process to eliminate the need for transportation. In order to determine the need for and to develop an action plan, it is necessary to make judgments about the output of the TRA. These may be relative or absolute judgments.

Relative uses of results include a comparison and ranking of various risk reduction schemes based on their competitive effectiveness in reducing risk. (There are generally fewer options and less reduction available in transportation than are found for fixed facilities.) A table of risk benefits versus costs is often prepared. Given the potential amount of time involved to make certain changes, it should be addressed explicitly in both the risk benefits and the costs. This type of judgment can be made based on qualitative or semiquantitative results and is generally less affected by errors in quantitative analyses than are absolute comparisons of risk estimates with specified targets. The overall role of uncertainty is very important to keep in mind, however, regardless of whether the TRA was qualitative or quantitative.

Another relative application is a comparison of the results of several evaluations in terms of the features or parameters that had the greatest contribution to the differences in the results. It may be possible to target risk reduction efforts based on which parameter is most critical, and to limit future evaluations to only the selected parameter(s).

The use of TRA to develop absolute risk estimates forces some level of quantitative analysis to be performed. If a maximum release frequency per trip is established, only a frequency estimate may be needed. Alternatively, a target may be expressed as a maximum number of potential consequences, and no frequency estimation may be performed. In general, though, absolute risk targets usually require quantitative risk estimates. It may still be plausible to start with a simplified level of detail and only increase the study complexity if the established target is not met, and if this is believed to be a result of simplifications or conservative judgments in the analysis.

Risk targets have been established by several governmental agencies and companies for in-plant risks; however, there are very few such targets for transportation risks. The work in the United Kingdom (Advisory Committee, 1991) presents some criteria that have been *proposed* for use in the UK, but discussions with some of the authors of the report make it clear that these have been set forth for the purpose of generating public reaction. The nature of the public–regulatory interface in the UK and the much shorter maximum and average route lengths in the UK as compared with the United States also constrain the applicability of these criteria to other situations.

A few companies are starting to work with targets for transportation risks. Where targets are employed, initial risk estimates are compared to the targets. If the target is not met or exceeded, the potential for risk reduction measures to reduce the risk estimate to (or below) the targeted level is determined. This can be difficult in transportation. In CPQRA additional risk reduction is often considered if the associated cost is reasonable. Such options are limited at times for transportation.

1.11. Project Management

As with many other aspects of TRA, the project management process is very similar to that used for a CPQRA study. The primary differences are at a detailed level—such as determining the study objectives or the actual risk measures to be used. However, this same degree of variation can be found between two CPQRA studies! As a result, most of the discussion found in the *CPQRA Guidelines* (CCPS, 1989) is relevant and is only summarized here. Figure 1-3 presents a logic diagram outlining the TRA project management process.

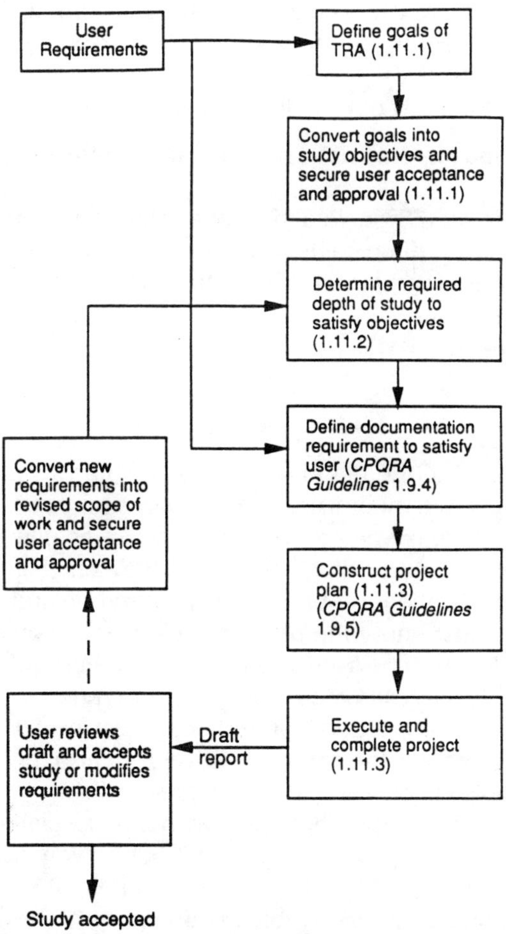

FIGURE 1-3. TRA Project Management Process.

1.11.1. Study Objectives

Typical study goals are presented in Table 1-6. These goals may originate from internal or external sources, including senior management, supplier requests, regulatory agencies and internal risk management initiatives.

Regardless of how the goals are established, they need to be turned into study objectives. The study objectives help to select appropriate risk measures and to establish the depth of study necessary and appropriate. The objectives also establish/influence time and resource requirements. Given the role that the objectives play in directing all further efforts, it is vital that they be communicated to and agreed to by senior management and other potential users of the study and its results. Should any of the management team or users change assignments during the course of the study, it is often very helpful to get the new personnel familiar with the existing study objectives.

Potential TRA study objectives include:

- determination of societal risk from transportation of a set of chemicals
- determination of societal risk for a particular chemical movement
- identification of cost-effective risk reduction measures to achieve a target risk level for an existing transportation movement
- evaluation and ranking of transportation alternatives based on individual risk levels.

1.11.2. Depth of Study

The goal is to define the depth of study in such a way as to achieve the study objective while minimizing the resources required. The calculation workload can increase by orders of magnitude as one moves outward from the study cube origin along any one of the axes (Figure 1-2). Thus, in order to determine the level of effort, calendar time and other resources required, one must know the depth of study. The project manager will generally need assistance from a risk analyst to determine the appropriate depth of study and to estimate resource requirements.

The most appropriate depth of study for a given situation is determined by considering not only the study objectives, but also the quality and type of data available. If only limited data are available, a high degree of differentiation between two options may not be available quantitatively, and a very detailed depth of study would not be warranted. Similarly, if selected data are considered to be very uncertain, it may not be worthwhile to be as precise as is possible in other parts of the analysis.

The expected future uses of the study should also be considered. For instance, if the present study objective involves an evaluation and comparison of two routes the project manager might determine that it is unnecessary to have a detailed evauation of the accident rate and release probabilities, if these variables will be common to the two routes. However, if it is already known (or it it is considered likely) that a subsequent study will be conducted to select the type of container, then a more rigorous estimation of the accident rate and release probabilities for the container used in the initial study might save time and other resources when the second study is performed.

Selecting the depth of study involves choosing the cell within the study cube, that is, specifying:

- the TRA focus (consequence, frequency, or risk)
- the complexity of the study (qualitative, semiquantitative, or quantitative)
- the number of incident outcome cases of concern (limited, generalized, or detailed set)

Section 1.5 discusses each of these points and provides initial guidance on selecting the appropriate depth of study. Chapters 5 and 6 will also provide assistance.

1.11.3. Project Plan and Schedule

A written project plan should be prepared and circulated, and should address the project objectives, responsibilities, schedule, resource requirements, special documentation requirements, quality assurance, training needs, and other items typical to any project plan. This plan should be updated as necessary.

The project plan, particularly the schedule, will be largely defined through the resource requirements in conjunction with resource availability. If the study objectives have any particular deadlines associated with them, these deadlines may have a significant impact on the project plan.

TRA is even more variable than CPQRA in terms of potential labor requirements. This variability is due to both the potential to have one study cover a single chemical movement or many such movements and to the difference between a qualitative TRA and a quantitative TRA. In addition, every time an alternate mode of transportation is evaluated, it is comparable to evaluating a whole new processing unit in a fixed site, not just the addition of some protective features. Thus, the differences in risk reduction strategies can play a significant role in determining resource requirements for quantitative studies, especially those with a risk focus.

Table 1-11 gives estimates for several different depths of study for a TRA where the three depths of study described are the cells on the main diagonal of the study cube given in Figure 1-2. The ranges are only approximations, and can be influenced significantly by the need to acquire substantial amounts of data. The assumed scope includes only one chemical and a few routes.

With regard to project scheduling, not all of the tasks need to be conducted in sequence as some efforts can be done in parallel, such as some of the frequency and consequence estimation. However, there may also be occasions when there are interruptions in the project flow due to staff unavailability or while waiting for critical data. As a coarse rule, the calendar time required will generally be 60 to 100 percent of the total labor estimate.

Section 1.9.5 of the *CPQRA Guidelines* (CCPS, 1989) provides further details on project scheduling (1.9.5.2), a thorough discussion of quality assurance issues (1.9.5.3), and brief sections on project cost control (1.9.5.5) and project execution (1.9.5.6). A discussion of training requirements is also provided (1.9.5.4), but is less directly applicable to TRA. As for CPQRA, experience and the judgment that comes with experience is critical. Less experienced practitioners will need greater scrutiny of both their work products and their time estimates.

There are very few textbooks or training courses specific to TRA, especially at an overview level. References are given in subsequent chapters for documents on specific topics, and a general understanding of CPQRA and its affiliated techniques provides a good introductory foundation for TRA. Thus, many of the other publications in the CCPS Guidelines series (see Preface) are helpful, as is the topical bibliography on TRA at the end of this book.

1.11.4. Review and Approvals

TRA documents should be reviewed and approved, particularly when the study objective includes support of a decision about the tolerability of the risks. Generally such reviews are conducted within the organization requesting the study, unless an outside opinion is desired due to:

- the magnitude of the risk
- difficulty in identifying risk reduction strategies
- regulatory requirements
- unavailability of key data

In these circumstances, a peer review may be requested or even an independent analysis conducted for confirmation of the results.

Any copies of a TRA report that are distributed should be uniquely identified and tracked, so that updates and revisions can be forwarded to all parties.

TABLE 1-11
Representative Labor Requirements for Different Depths of Study for TRA

Depth of Study (from Figure 1-2)	*Activity*	*Person-Weeks*
Qualitative/Limited Set of Incident Outcomes/Consequence	Data compilation	0.1–0.5
	Incident identification and selection	0.1–0.2
	Consequence estimation	0.1–0.2
	Frequency estimation	—
	Risk estimation	—
	Documentation and management	0.1–0.2
	Total	**0.4–1.1**
Semiquantitative/Generalized Set of Incident Outcomes/Frequency	Data compilation	0.5–1.5
	Incident identification and selection	0.5–1.0
	Consequence estimation	0–0.2
	Frequency estimation	0.1–0.5
	Risk estimation	—
	Documentation and management	0.2–1.0
	Total	**1.3–4.2**
Quantitative/Detailed Set of Incident Outcomes/Risk —same mode and routes for risk reduction	Data compilation	2–4
	Incident identification and selection	1–3
	Consequence estimation	1–2
	Frequency estimation	1–2
	Risk estimation	1–2
	Documentation and management	3–5
	Total	**7–18**
—same mode but alternate routes for risk reduction	Data compilation	2.5–5
	Incident identification and selection	1.5–4
	Consequence estimation	1.5–3
	Frequency estimation	1.5–3
	Risk estimation	1.5–3
	Documentation and management	4–6
	Total	**14–24**
—different mode for risk reduction	Data compilation	4–8
	Incident identification and selection	2–6
	Consequence estimation	2–4
	Frequency estimation	2–4
	Risk estimation	2–3
	Documentation and management	5–8
	Total	**17–33**

1.12. Maintenance of Study Results

TRA results should be maintained after the completion of a study as part of the overall risk management program. While the overall program should be kept up to date, individual TRAs may or may not be updated. For critical studies (i.e., those supporting key risk reduction decisions or decisions not to reduce risk further at the present time) the need for keeping the studies up to date is much greater. These studies should be modified whenever a key parameter changes significantly. These might include new carriers/routes, new containers, additional high volume destinations, changes in shipping data, changes in toxicity information, etc. It may be sufficient to simply test the sensitivity to the change, rather than reevaluate any results.

For noncritical studies, such as broad-based screens of all hazardous material movements, the studies may be routinely and periodically repeated—at intervals ranging from one to five years. This is a reflection of the small amount of change that is expected from one version to the next, particularly if a system of checks and balances is built in to catch major changes as they occur.

Given the nature of the distribution function at many companies, it is possible for large fluctuations in the number and location of customers or their required shipments to take place on an annual basis—without those with risk management responsibilities being aware of the changes. This requires that some form of interactive communication be established.

Documenting the approach and assumptions used in a TRA not only helps future readers use the results, but makes it much easier to periodically assess and revisit the validity of the results in light of the current situation. It also serves as an explicit statement of what the situation was at the time the risk was estimated. Otherwise one is faced with such questions as:

- Did we remember that shipments to Customer A are made in a different container?
- Was that before or after we switched carriers?
- Did we consider material-specific hazards or just accident-initiated events?

It is important to monitor and control the distribution of copies of TRA reports, so that corrections and updates can be sent as appropriate. This reduces the potential for decisions being made based on outdated or incorrect information. Other controls on distributed copies may be required as part of a company's own internal record retention program.

In addition to controlling the distribution of copies and keeping the study current, it is also important to have a known location for a master copy and for key back up files. Supporting documentation makes it much easier to answer questions that may arise, consider additional mitigation measures as they become available, update studies, and to use interim results in other studies. It is very unusual for a detailed quantitative TRA to be fully documented in one report.

1.13. References

This section lists the references cited in this chapter. Additional references that may be of interest to the reader can be found in the topical bibliography.

Advisory Committee on Dangerous Substances. *Health and Safety Commission; Major Hazard Aspects of the Transport of Dangerous Goods*. Report and Appendices; HMSO, 1991.

Center for Chemical Process Safety. *Guidelines for Chemical Process Quantitative Risk Analysis*. American Institute of Chemical Engineers. New York, NY, 1989.

Center for Chemical Process Safety. *Guidelines for Hazard Evaluation Procedures, Second Edition with Worked Examples*. American Institute of Chemical Engineers. New York 1992.

Chemical Manufacturers Association. *Responsible Care®: Distribution Code of Management Practices*. Chemical Manufacturers Association, Washington, DC, 1990.

Harwood, D.W., E.R. Russell, and J.G. Viner. "Characteristics of Accidents and Incidents in Highway Transportation of Hazardous Materials," preprint from 1989 Transportation Research Board Annual Meeting, January 1989.

IChemE. *Nomenclature of Hazard and Risk Assessment in the Process Industries*. Institution of Chemical Engineers, U.K., 1985; ISBN 0-85295-184-1.

Muhlbauer, W.K. *Pipeline Risk Management Manual*. Gulf Publishing Company, Houston, TX, 1992; ISBN 0-88415-035-6.

Muhlbauer, W.K. "RIPS—A Pipeline Safety Evaluation System," *Proceedings of the 1991 Pipeline Risk Assessment and Repair Conference*. Gulf Publishing Company; Houston, TX, 1991.

Office of Technology Assessment. *Transportation of Hazardous Materials*. OTA-SET-304, 1986.

Saccomanno, F., K.Cassidy. *Transportation of Dangerous Goods: Assessing the Risks*. Institute for Risk Research. Ontario, Canada, 1993.

2

TRA Frequency Analysis

TRA focuses on the loss of containment of a hazardous material (HAZMAT) due to either an accident breaching a transport container or a failure of associated equipment such as relief valves or fittings while in transit. Accident-initiated releases tend to dominate the risk of HAZMAT transportation, but, for certain products, reactivity, contamination, extreme temperatures, or other material-specific hazards, can cause non–accident-initiated releases to become more significant or even dominate the risk. Both types of releases can be addressed in a TRA and the analyst should take care to assure that all significant hazards are identified.

The chance of a release is generally a function of the distance traveled. Thus, the frequency of an accident is often expressed as an accident rate per mile. Contributions from non–accident-initiated events are typically expressed on a frequency per hour or per year basis. Thus, the duration of the HAZMAT movement is a key parameter. Figure 2-1 illustrates the basic calculation sequence for one trip or movement. If multiple trips are made, the total risk is equal to the number of trips times the risk per trip.

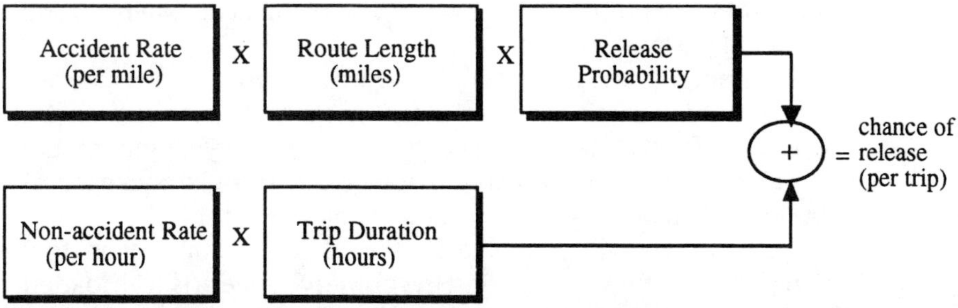

FIGURE 2-1. Calculating the Chance of Release.

For accident initiated events, the chance of a release given an accident must be included. For non–accident-initiated events, the data is expressed as the chance of a release and needs no further adjustment for release probability. The parameters identified in Figure 2-1 can also be reviewed in a qualitative or semiquantitative study on a relative basis.

This basic calculation sequence will have minor variations for each mode of transport, and can be broken down into greater detail as needed. Increased detail might include different accident rates and segment lengths for each segment of a route or might explicitly address the accident rates and release probabilities for different accident causes. An increased level of detail requires more data, and also increases the number of calculations.

This chapter summarizes the key considerations in conducting a frequency analysis using and applying historical data for a TRA. Another approach that uses predictive methods (also known as mechanical engineering methods) is described in Section 3.8. The emphasis in this chapter is on qualitative descriptions of the major issues with examples of data. Quantitative data sources are presented and described in Chapter 4. The sections that follow discuss typical operations, hazards, failure modes, key parameters influencing accident rates and release probabilities, type of containers, mode-specific issues, and general calculation procedures for a number of modes of HAZMAT transport. Specifically, these modes are:

- *pipelines*—may be defined as starting at a storage tank, a pump, or a facility boundary. TRAs generally include all valve and pumping stations between the defined start and a similarly defined end point.
- *rail*—rail TRAs address the movement of (loaded) railcars from the supplier to the customer. This generally involves visits to several classification yards where the train may be broken down and reconfigured. Other yard stops may be made to change crews, but are generally not critical as the cars are not handled or moved. "Empty" cars containing residual material can be studied as well. While trucks may contain one hundred pounds or less of material when they are empty, an empty rail car might still contain several thousand pounds of material.
- *truck*—generally movements of loaded trucks from gate to gate are analyzed. As with rail, the movement of empty vehicles can also be reviewed. Trucks may be tank trucks, trailers carrying intermodal containers, or others carrying various types of bulk packages of HAZMAT. Truck stops may need to be considered.
- *barge*—may be self-propelled or towed, individually or in tow strings. As with the other modes, the primary concern is with loaded barges. Overnight moorings may take place and need to be included in the TRA.

- *ocean-going vessels*—vessels may be dedicated tankers, container ships or general cargo vessels. Due to the proximity to people, generally harbor and coastal areas are analyzed, but not open ocean waters. Interim storage in port may need to be included.
- *intermodal*—intermodal containers may be moved via only one transport mode, or may be transferred among modes. The transfer activities, interim storage, and actual transport legs are all included in a typical TRA.

Risks arising from maintenance activities or vessel inspections or cleanings are not typically addressed. Loading and unloading risks are generally estimated using fixed site CPQRA techniques, and may be reviewed as part of a facility evaluation.

The discussion of pipelines includes numerous comments related to design practices to minimize risks, because this is the one mode of transport where plant engineers are heavily involved in design issues. Due to its fixed location and greater control of all operations, this mode may also have more opportunities for risk reduction, but not without interrupting service.

2.1. Pipelines

A pipeline is actually a system of equipment designed to allow material to flow (continuously or intermittently) from one location to another. The receiving site could be another facility a few thousand feet away, or one that is 100 or more miles away. Gas transmission pipelines can be several feet in diameter, and may cross through both rural and heavily populated areas. The pipeline that failed on March 24, 1994 in Edison, New Jersey is an example of the type of pipeline system that is often forgotten about until an incident occurs—similar to gasoline transportation by truck.

A pipeline system contains many or all of the following elements:

- pumps or compressors
- buried pipelines
- above-ground pipelines
- booster pumps or compressor stations
- isolation valves—manually, remotely, or automatically activated
- relief valves—thermal or other types
- pipebridges or other supports
- cased sleeves under roadways or rail lines
- control systems
- leak detection systems
- river crossings

- pig launchers/receivers
- storage tanks, wells, and caverns

Thus, pipeline failures can involve more than just pipe failures. Table 2-1 presents a detailed listing of potential incidents for a pipeline system. While these incidents have been grouped into human errors, equipment failures, failures of systems and procedures, and external events, these categories do not necessarily correspond to the way in which failure rate data are collected (see Section 2.1.1). Rather, they refer to general categories of hazards that should be considered for qualitative hazard identification, and as further input to a quantitative study.

In particular, the hazard categories and their contents can be used as a checklist at a design stage, or as a list of items for comparison between two pipelines or between one pipeline and a "standard" that serves as a benchmark for risk evaluations.

2.1.1 Failure Modes

For many years pipeline failures have been reported either by law and made public, as in the United States or by law but under conditions of

TABLE 2-1
Pipeline Hazards—Initiating and Contributing Causes

Human errors	*Equipment failures*	*System or procedural failures*	*External events*
Pigging operations Hot tapping Slug operation Repair/replacement of a section of line Startup Changing operating conditions Shutdown Preparation for maintenance Valve operation	Thermal expansion Internal corrosion External corrosion Bad welds Fatigue Failure of cathodic protection Cyclic stress Galvanic corrosion Internal erosion Control system failure Brittle fracture Support failure Construction defects Overpressure Polymerization Plugging or fouling	Inspection Operation Startup/shutdown Communication Maintenance Leak detection Emergency temporary repair Material specification and testing Modifications	Accidental excavation Post digging Wind (above-ground lines) Earthquake Subsidence Avalanche Flood/scouring Lightning Fire Vandalism/sabotage Freezing/thawing Rail/road crossing (pipebridge or sleeved casing) Railroad derailment Mining Anchor dragging Mooring pole

confidentiality, as in some European countries. The U.S. Department of Transportation regularly publishes statistics of oil and gas pipeline failures, which record the type of failure, type of release, and other data. In Europe, the major gas companies of Belgium, Denmark, France, Germany, Italy, the Netherlands, the United Kingdom, and Spain have gathered their national statistics and published failure rate data under the European Gas Pipeline Incident Data Group. The European oil companies also publish annual failure rate data under the organization CONCAWE (the oil companies' study group for **con**servation of **c**lean **a**ir and **w**ater—**E**urope).

The databases published by these organizations include a large number of pipelines and incorporate many years of operating experience. Therefore, the general quality of data is good and it can be used with confidence in predicting the likelihood of pipeline failure in risk assessment.

When considering all these databases together one broad conclusion comes out of the statistics, despite some variation caused by dissimilarity in the type of data collected—failures occur in roughly equal proportions in three broad categories:

- Failures caused by external mechanical interference,
- Failures caused by corrosion defects,
- Failures caused by miscellaneous factors such as pipe material defects, natural hazards or operator error.

Some of these failures are time-independent occurrences—such as external mechanical interference, earthquakes, or overpressure, while others are time dependent—such as corrosion and fatigue failures. Table 2-2 provides a comparison of pipeline failure rates from two sources. Cross-references to the appropriate transportation data source sheets in Chapter 4 are also given.

TABLE 2-2
Gas Pipeline Failure Rates

	Rate Per 1000 Mile-Years	
Cause	*US*[1]	*Europe*[2]
External mechanical interference	0.68	0.67
Corrosion	0.20	0.17
Material failure/construction defect	0.26	0.20
Other	0.10	0.08
Total	**1.25**	**1.12**

[1] Jones, 1986 (excluding 1983 data); Transportation Data Source 4-9
[2] European Gas Pipeline Incident Data Group, 1988; Transportation Data Source 4-14

2.1.1.1. Failure by External Mechanical Interference

Failures caused by third party external mechanical interference include such causes as:

- being damaged by excavators or other equipment in use by other utility or construction companies,
- damage following derailments on railroads,
- hot tap in error by other utilities,
- damage during deep plowing by farmers,
- damage during construction of land drainage,
- damage at river crossings by dragging of ships anchors, or
- damage during the construction of barge pole moorings.

The type of failure generally caused by third party mechanical interference is a puncture or split of the pipe or a gouge severely reducing the wall thickness of the pipe. Failure can be immediate or may occur some time later by fatigue. This type of incident is likely to have severe consequences and historically some of the most serious pipeline accidents resulting in ruptures have been caused by such incidents. Many pipelines are laid adjacent to railroads and delayed pipeline failures have resulted from rail accidents when the damage to the pipelines was not recognized at the time of the derailment. When flammable substances are released as a result of immediate failure due to external mechanical interference, a high probability of ignition is likely as a result of the presence of ignition sources on the equipment causing the incident.

2.1.1.2. Failure from Corrosion Defects

Pipeline failures by corrosion can be due to internal corrosion or external corrosion. External corrosion failures are due to moisture in the ground and aggressive soils and can take two forms—small pinhole failures caused by pitting and more generalized corrosion leading to a reduction in pipe wall thickness over a plane area. External pitting corrosion leads to small leaks that are often difficult to detect but that gradually grow in size over a period of time. External area or plane defects cause a generalized reduction of wall thickness, that can eventually fail catastrophically under pressure, leading to a large scale release. Section 2.1.2.4 discusses some of the mechanisms for preventing or reducing external corrosion.

Internal corrosion leads to similar failure mechanisms as external corrosion but is caused by the corrosive nature of the products being transported. Thus, each substance to be conveyed by pipeline should be assessed for its corrosion potential at the design stage. Failures are generally caused by the formation of acids resulting from the presence of water, carbon dioxide or sulphur compounds in the substances being transported. Crude oil pipelines often exhibit corrosion caused by sulphur compounds or by a failure pattern in the bottom half of the pipe caused by the presence of water. Gas pipelines often exhibit corrosion caused by hydrogen sulphide or carbon dioxide in the gas. Internal corrosion can often be prevented by the use of corrosion inhibitors continually injected into the pipeline.

2.1.1.3. Failure by Miscellaneous Causes

Pipelines can also fail for a variety of other causes. Typical causes are construction defects, pipe material defects, operator error, equipment failure, failure due to internal erosion and failure due to ground slip, flood ground erosion, earthquake, or mining.

Unidentified construction defects are most frequently the type in which mechanical damage to the pipe is caused by machinery during the backfilling or land restoration activity, and are often discovered shortly after pressurization. However, some defects, such as inadequate welding or joint corrosion protection and corrosion coating damage during laying will only be identified some time later.

Pipe material defects usually result from inadequate specification and testing of materials. Failures can be of any type such as ruptures following brittle fracture or leaks due to longitudinal seam defects in pipe.

Failures initiated by operator error may be much less likely on pipelines where control system designs are more fault tolerant. However, accidents frequently occur during operation in abnormal conditions such as following equipment failure or during maintenance activity. Recent accidents have also been reported where operators have ignored alarms such as high rate of loss of pressure or tank high level alarms.

Pipeline associated equipment failures can occur at any time and pipelines should be designed with protection systems (see Sections 2.1.2.3–2.1.2.8) to reduce hazards to a minimum under such circumstances. Equipment failure is sudden and unexpected, and the lack of protective systems (for example on older pipelines) can often result in a lack of control of the consequences such as the inability to isolate sections of line by remotely operated valves.

The ground forces imparted to pipelines during ground slips, mining activities or earthquakes are very high and sufficient to cause pipeline rupture. In earthquakes two types of event can lead to failure. These are ground movement leading to overstressing of the pipeline and ground liquefaction causing the pipeline to lose support. Both result in failures by bending. Mining activities can also lead to ground movement over a much longer time period than earthquakes, but with similar effects.

Pipelines can also fail by internal erosion. The causes are either very high gas velocities in gas pipelines or the presence of abrasive particulate matter in slurry pipelines. Failures will result from a gradual reduction in wall thickness and are likely to be catastrophic in nature.

2.1.2. Parameters Influencing Accident Rates

2.1.2.1. Age of Pipelines

Pipeline failure rate databases are beginning to show an increase in incidents involving older pipelines. There are several factors influencing these failures. Older pipelines were built when standards of design, quality, and construction practices were inferior to those of today. In the 1950's and earlier, the codes of practice did not require cathodic protection.These older pipelines were often laid with inadequate corrosion protection, or even none at all. Welding standards were less advanced and materials that could resist, for example, corrosion from hydrogen sulfide were unavailable. In their early years the maintenance was often poor and the design made the possibility of inspection by internal inspection devices difficult without extensive modification. The control of construction activities was not rigorous and the depth of cover and precise location of the pipelines as built was less than certain. Finally, pipelines that have operated over a wide range of cyclic pressure variation may be approaching the end of their expected life due to fatigue. If data sources include a wide range of pipeline ages, they may not be totally applicable to a newly designed pipeline.

2.1.2.2 Third Party Activities

Failures due to third party interference can occur because the third party is unaware of the presence of the pipeline. Alternatively, the pipeline protection at the scene of a repair, excavation, or other activity may be so inadequate that an incident occurs.

Pipeline failures by third party activity can be reduced by ensuring that members of the public, landowners, and utilities are always aware of the location of the pipeline and by providing adequate protection to the pipeline in areas where the risks appear high. Many companies fly or walk their pipeline routes at frequent intervals to identify any construction activities that have the potential to cause pipeline damage or to identify small leaks whose effects are particularly visible. Generally, it would be expected that the rate per mile of third party incidents would be higher in populated areas. In practice, analysis of the failure rate database does not demonstrate this to be the case. This is probably explained by the extensive notification procedures put in place by the operators in populated areas, public awareness campaigns, and the increased wall thickness that some pipeline standards require in populated areas.

Pipelines crossing other utility service lines (e.g., water/gas mains, phone lines, etc.) have an increased potential for third party incidents. Pipelines should always have adequate protection (i.e., physical monitoring, shielding) in these cases so that work on the services will not cause damage to the pipeline. Pipelines crossing rivers that carry shipping traffic should be protected from anchor and mooring pole damage by strict work enforcement procedures in the vicinity of the pipeline and by rock or concrete protection laid above the pipe.

The depth of agricultural plowing has increased in recent years and it is expected that an increase in incidents of third party damage caused by agricultural machinery may occur. Older pipelines were not buried to the same depth of cover as they are today, making them more vulnerable than modern pipelines; many of today's pipelines are buried to depths of three to five feet. The installation and maintenance of drainage systems may pose similar hazards.

2.1.2.3. Pipeline Designs and Standards

The ANSI B31.8 Code for Gas Transmission and Distribution Piping Systems (1986), in common with new European Standards, recognizes the increased risk from pipelines in populated areas and recommends improved safety requirements. These take the form of progressive increases in pipe wall thickness with both increased population density and types of road and railroad crossing. In Europe, the latest British Standard BS 8010 Code of Practice for Pipelines (1991) similarly increases wall thickness in areas of increased population density. It also recommends minimum safety distances to occupied buildings to reduce individual risk levels for pipelines carrying gas, volatile liquids and chemicals. These minimum safety distances were derived from risk criteria and vary according to the substance transported.

Where pipelines are transporting cryogenic substances or where a blowdown operation could result in low temperatures the pipe steel could fall to a temperature that results in catastrophic failure due to embrittlement. To reduce the probability of failure, pipeline designers should ensure that the steel toughness is adequate for such operation or that the line is adequately protected by low temperature trip systems, such as described in 49 CFR Part 193.

2.1.2.4. Corrosion

In the early years steel pipelines were laid without any form of external corrosion protection and, where they were buried in wet and aggressive soil, failure by corrosion was very rapid. To reduce these failures external coatings such as bitumen and coal tar enamel were developed and these had considerable effects in reducing failure rates. Later, cathodic protection systems were applied in which electrical potential differences are set up between the pipe and surrounding earth to prevent the formation of corrosion cells where the pipe external coating had failed locally. Over recent years epoxy and polyethylene coating systems have been developed that have further enhanced the quality of external corrosion coatings.

Nevertheless, corrosion defects are a growing concern for the operators of older pipelines and the problems appear to be more severe in hazardous liquids pipelines than in gas pipelines. The reason could be that liquids pipelines comprise a larger proportion of the older pipelines in operation today—some of these systems have been in operation upward of sixty years. They were regarded as being less hazardous in terms of public safety than gas or chemicals pipelines when originally constructed. Further, the impact of leaks from these lines on the environment was not regarded with the same degree of public concern as it is today.

Cathodic protection systems are designed to reduce the likelihood of external corrosion. To work effectively, regular readings of pipe to soil potentials should be carried out to ensure that rapid corrosion is not taking place locally. Particular areas prone to corrosion are where other metallic services pass close to a pipeline and could interfere with the workings of the cathodic protection system, leading to accelerated corrosion. In these areas local bonding should be undertaken and the conditions monitored. Other areas prone to corrosion are in areas where the pipeline is shielded from the cathodic protection currents. Cases with corrosion where pipelines pass within steel sleeves (at crossings) have been so severe that some international pipeline codes now preclude the use of steel sleeves. (See Section 2.1.4.2.)

2.1.2.5. Maintenance

High maintenance standards are vital in ensuring that a pipeline operates safely. The as-built condition of a pipeline should be able to be maintained almost indefinitely by high quality maintenance. Regular inspection of the cathodic protection system to ensure that high currents are not being drawn should be undertaken so that early indication of the failure of the corrosion coatings is known. Inspection devices, known as "smart" or "intelligent" pigs, are available that travel internally along the pipeline and measure the condition of the pipe wall. These devices are unable to pass around sharp radius bends or reduced bore valves or fittings and are therefore sometimes difficult to use on old pipelines unless the pipeline is modified. However, all new pipelines are generally designed to accept such devices.

2.1.2.6. Fatigue Loading

Providing pipelines are designed, built, and maintained to the latest international standards their life should ultimately be determined by the fatigue life of the pipe material. Studies have therefore been undertaken to estimate the factors that determine the fatigue life of pipelines. Since pipelines can operate in a highly stressed condition, generally up to 72 percent of the specified minimum yield strength of the steel, it follows that regular wide cyclic variations in pressure could lead to a failure by fatigue after many years of such conditions (IGE, 1984). Such pressure swings are seen in gas distribution systems where the pressure within the pipeline system varies over each day, falling in times of high demand and rising in times of low demand. Such variations can exceed 50 percent of the design pressure. To respond to this concern about fatigue life many companies have aggressive monitoring, inspection, and replacement programs.

2.1.2.7. Other Factors

In densely populated areas pipelines often run in pipeline corridors. The effect of this is to provide a protected zone where the likelihood of failure by external mechanical incidents is reduced. It is nevertheless important that such pipelines are laid sufficiently far apart to allow access for maintenance and that the access for emergency vehicles is not hindered by valve pits or short above-ground sections.

The main parameters influencing accident rates for pipelines laid in railway rights of way and road crossings is the increased risk of damage due to external mechanical interference. In the case of railroads the risk is from derailments and in the case of roads from the work associated with the maintenance work of other utilities.

Above-ground sections of pipework, including pipe bridges, have the advantage that inspection and maintenance against corrosion are easier. While above-ground sections are clearly visible to third parties they are also vulnerable to damage from secondary accidents, for example, those involving road vehicles, particularly where the pipeline runs along a road or crosses a road by pipebridge.

2.1.2.8. Pipeline Wall Thickness

The European Gas Pipeline Incident Data Group (1988) has published statistics showing the relationship between pipe wall thickness and the likelihood of incidents involving a gas release due to external interference. The relationship is shown in Figure 2-2. This shows that the likelihood of pipeline failure reduces as the pipe wall thickness increases, for the range of thickness typically in service. Since most incidents are related to either puncture owing to third party interference or corrosion it is expected that puncture is less likely and corrosion defects will reveal themselves less quickly with thicker wall pipe.

Some researchers have speculated a link with pipeline diameter and failure rate, with large diameter lines having lower failure rates. Although

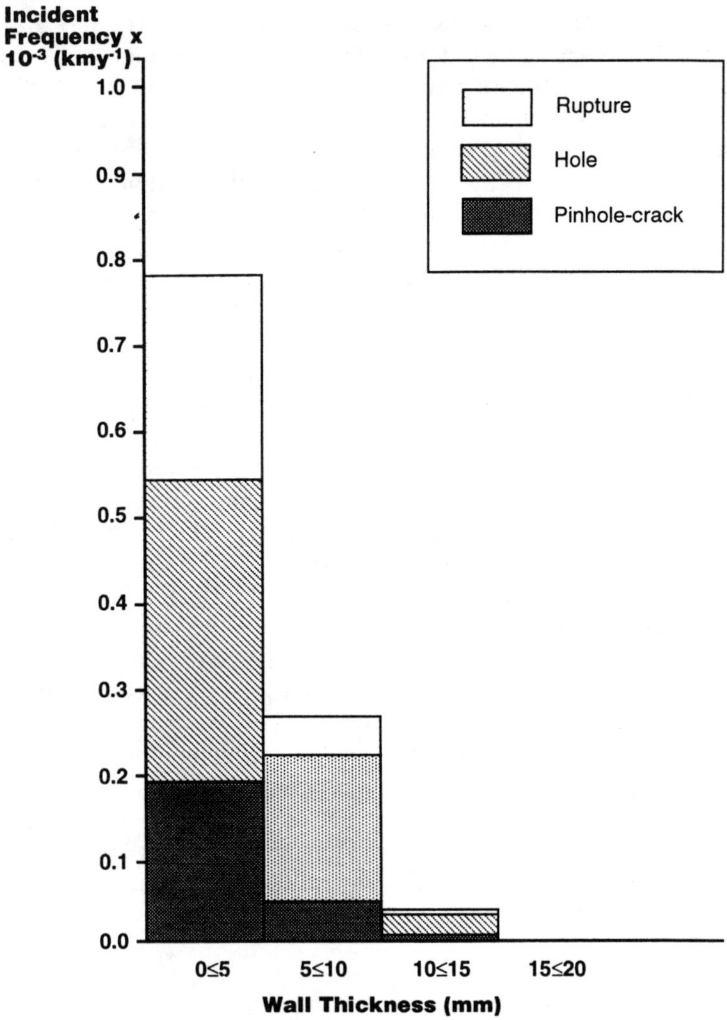

FIGURE 2-2. External Interference as a Function of Wall Thickness. (European Gas Pipeline Incident Data Group, 1988.)

larger diameter pipelines are generally newer and more modern in design and as a result are probably built to higher standards, there is also an increase in the required wall thickness with increases in diameter for similar operating pressures and steel grade. Other researchers believe that this increased wall thickness is having a greater influence on pipeline failures than diameter.

2.1.3. Parameters Influencing Release Probabilities

2.1.3.1. Relationship between Leaks and Ruptures

The failure modes for liquids, volatile liquids and gas pipelines are significantly different. In liquids pipelines the pressure will fall rapidly following puncture owing to the relative incompressibility of the liquid. The initial high rate of release will therefore reduce rapidly and the drain down of the pipeline will be driven initially by the hydraulic head of the system and finally by the static head, until equilibrium conditions are reached. The release duration will depend on the operating pressure and the time taken to detect the leak or rupture before pump station shutdown and block valve closure have been achieved.

For volatile liquids pipelines (i.e., those transporting LPG, NGL, ammonia, etc.), the material is conveyed as liquid under pressure, but following loss of pressure due to a leak or rupture the material will "flash" to a mixture of cold vapor and liquid. This regime initially results in very unstable release rates as "slugs" of liquid and gas are alternatively released. The release conditions produce a heavier than air flammable or toxic vapor cloud. Ultimately all the liquid is evaporated and a simple gaseous release takes place.

For gas pipelines and pipelines containing liquids that evolve gas upon decompression, the highly compressible nature of gas results in a high level of energy remaining in the vicinity of the puncture for a sufficient period of time to allow small holes to propagate to full bore ruptures under certain circumstances. Gas will continue to be discharged from the pipe and, for high pressure releases, the relatively long duration will be self regulated by gas leaving the pipelines at sonic velocity. However, decompressive cooling under certain circumstances can result in pipeline embrittlement and sudden failure which causes a small leak to turn into a large leak.

British Gas (Townsend and Fearnehough, 1986) have published research showing the relationship between defects propagating to full bore ruptures or remaining as leaks. The graph in Figure 2-3 shows the range over which leaks will propagate to full bore ruptures for different stress levels within the pipe wall against the size of initial leak. Following this research British Gas concluded that leaks would not propagate to full bore ruptures if the ratio of operating stress to yield stress was less than about 0.3. (The y-axis in Figure 2-3 represents this ratio while the x-axis represents defect length.)

Because of the compressible nature of gas, the relationship between leaks and rupture only applies to gas pipelines. Another factor to be considered is that certain pipeline materials are more likely to leak before they break. This should be taken into account at the design stage.

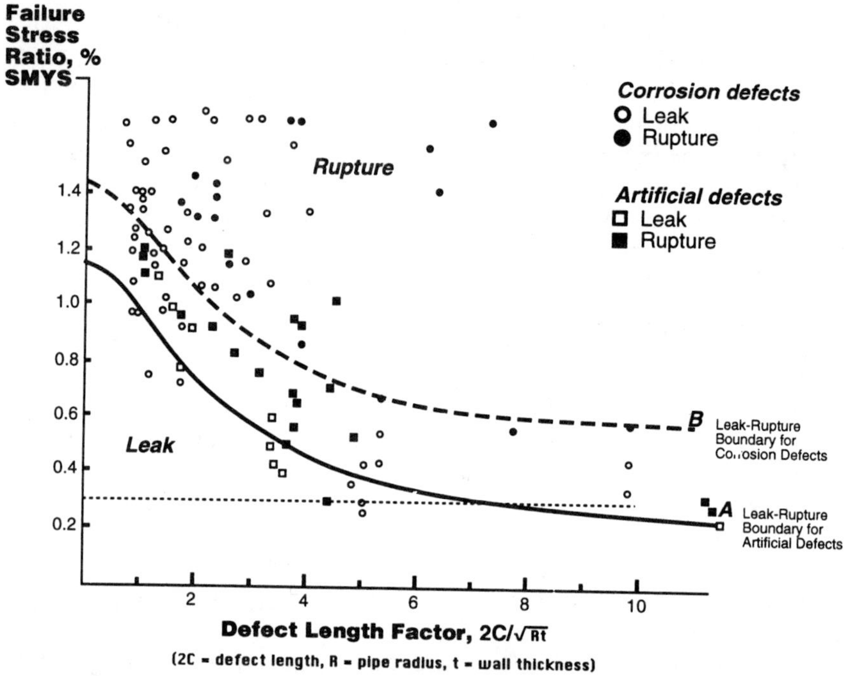

FIGURE 2-3. Leak–Rupture Behavior of Pipeline Defects. (Townsend and Fearnehough, 1986.)

2.1.3.2. Insulated Pipelines

Based on proprietary data presented at a recent industry seminar, failure rates for insulated pipelines have been greater historically than for uninsulated pipelines. Because of the shielding effect of the insulation, cathodic protection systems are difficult to apply adequately to insulated pipelines. Most insulated pipelines have failed following a breakdown in the bonding between the insulation, anticorrosion coating, and the pipe, allowing water to enter the insulation and provide a corrosive medium. This breakdown is invariably caused by relative movement between the pipe and the insulation during the temperature cycles resulting in crushing of the insulation at bends.

Special attention should therefore be given to the design of insulated pipelines and consideration given to preventing relative movement of the pipe to the outer coating of insulation. This could be achieved by anchoring the pipe to the ground at major changes of direction or by the development of an insulation bond to the pipe sufficiently strong to transmit the stresses between the pipe and soil caused by thermal movement without causing damage to the insulation. Alternatively, temperature cycling of the line could be avoided.

Release probabilities are higher at mechanical joints such as flanges than for welded joints and this is reflected in the pipeline design codes that recommend the minimization of flanged connections below ground. However, some equipment, such as valves, will be generally difficult to seal positively owing to the presence of stem packing. Sufficient attention should be given to small leaks from such equipment, as these leaks can grow rapidly and could even lead to catastrophic failure of the valve as a result of brittle fracture in extreme cases.

Isolating (or insulating) joints, which are installed in pipelines to separate cathodic protection sections, have also been reported as sources of leaks. Great care should be taken in the installation of such joints to ensure that undue stresses are not introduced across the insulation faces thereby precipitating early failure of the joint and subsequent leakage.

2.1.4. Containment Systems

2.1.4.1. Pipe Body

Pipelines are generally manufactured as seamless pipe, electric resistance/induction welded (ERW) pipe or as submerged arc welded pipe. The manufacturing methodology chosen is largely related to the pipe size. Seamless pipe usually covers smaller diameters, ERW pipe a mid range of diameters and submerged arc pipe the largest diameters.

Failures of seamless pipe have been associated with laminations introduced by the manufacturing process. They are a particular problem at the ends of the pipe where the girth weld quality can be affected by the presence of a lamination close to the end of the pipe. Most pipe codes now specify extensive nondestructive testing (NDT) of pipe ends of seamless pipe (ANSI/ASME B31.8, 1986). Seamless pipe can also be damaged during manufacture by the rolling in of loose material, which becomes subsequently detached resulting in a local, possibly severe, reduction in wall thickness. Modern NDT methods should be able to detect such defects.

ERW pipe failures are often associated with defects in the longitudinal ERW weld resulting in lack of fusion and pinhole leaks. Again, modern NDT systems can identify such defects and one would expect their occurrence to decrease over time.

Submerged arc welded pipe has had a good performance record to date and is not prone to specific types of failure.

2.1.4.2 Pipe Sleeves

Some pipelines are constructed within steel sleeves, usually at crossings, to protect against impact damage or stressing as heavy vehicles pass over the line. Unfortunately, the placing of one steel pipe inside another, and allowing ground water to enter in the gap between the pipes, results in high levels of corrosion of the carrier pipe and pipeline failures are often reported at sleeved crossings. The reason for this is that the pipe within the sleeve cannot be protected by cathodic protection owing to the shielding effect of the sleeve and failure often occurs at the point of exit of the carrier pipe from the sleeve. Designs have been put forward that involve sealing the space between the pipe and sleeve and filling it with an inert gas. This has met with partial success but all too often the sealing device fails and the inert gas escapes. Many international codes now strongly advise against the installation of pipelines within steel sleeves (BS8010, 1991) and other forms of impact protection are suggested such as increased wall thickness carrier pipe, concrete slabbing or nonmetallic sleeves filled with sand or earth, which allow the passage of cathodic protection currents.

2.1.4.3. Total Pipeline Containment

Many large-scale hot water heating systems associated with cogeneration plants are laid with one steel pipe within another steel pipe along the whole pipeline length. Short lengths of heated oil and natural gas pipelines as well as very toxic material pipelines have been laid with such double containment systems to reduce the risk of releases or operational difficulties. Such pipelines must be insulated and have severe thermal stresses imposed on them by the expansion and contraction of the steel under the effects of the hot water. Such pipelines are pre-stressed during laying by filling with hot water and, at regular intervals, welding the outer and inner pipelines sections together while being clamped in a stressed condition. This will ensure that the insulation material is not crushed during the expansion caused by the introduction of hot water during the commissioning process. Such pipelines will also be entirely sealed against corrosion apart from the outer pipe, which can be treated as a conventional pipeline.

2.1.5. Accident Trends with Time

Pipeline failure rates have not fallen over recent years in the same way as railroad accidents (see Section 2.2.5) and it has been suggested that, although the standards of new pipelines are improving, there is an increased likelihood of older pipelines failing. The failure mode speculated in this case is that of failure by corrosion. CONCAWE have recently tested the hypothesis and evaluated the failure rates of hazardous liquids pipelines in Europe by age. The results of this confidential study showed a significant correlation of reported incidents involving loss of product from pipelines with increasing age of the pipeline.

Many operating companies are now examining their pipelines with smart pigs and the early identification and repair of corrosion defects is expected to result in a reduction in failure rates in the long term. However, many older lines cannot be inspected with smart pigs—other forms of monitoring and additional instrumentation should be considered for these lines.

2.1.6. Confidence in Data

The transportation data source sheets given in Chapter 4 present a variety of pipeline failure rate data sources, covering oil and gas pipelines in both the United States and Europe. In total these pipelines represent many thousands of mile years of operating experience and a statistically significant number of failures per year.

2.1.7. Mode-Specific Issues

Many pipelines contain more than one substance and a pipeline failure may result in more than one hazard in the atmosphere upon release. In this case it will be necessary to consider each hazard in turn and evaluate their harmful effects to the environment. Typical of such pipelines are sour gas pipelines that contain both methane (which is flammable) and hydrogen sulphide (which is toxic and flammable). Similarly, unstabilized crude oils will contain volatile flammable liquids dissolved in the crude oil which, when released, will boil off as flammable gases. Also, NGL comprises volatile flammable liquids that similarly behave as gases when released to the atmosphere but that are pumped as a liquid in the pipeline.

In some cases, where the risk increases when the proportion of the various components within a pipeline changes over time, authorities have been known to request the operator to control the flowrate in the pipeline to maintain the risk profile below an agreed maximum level.

Multiproduct pipelines transport different substances in batches along the pipeline and long pipelines can contain several different hazardous substances at any one time traveling along the pipeline. Such pipelines present a changing risk profile to the community as the batches progress along the pipeline if the changes in materials also represent major changes in hazard properties or characteristics. For instance, take a multiproduct pipeline carrying ethane (55 percent of the time), propane (33 percent) and ethylene (12 percent), which is over one thousand miles long. Individuals living near this pipeline could be exposed to a release of any one of these products, depending on the specific product carried at the time of the release. The hazards of the three materials are similar, but the ethylene releases were found to have slightly greater hazard distances in this case(about 25 percent), increasing the number of people potentially at risk.

2.1.8. Releases Associated with Supporting Equipment

The primary releases of concern here are from valve stations, compressor stations, pumping stations, etc. along the pipeline. These locations can be evaluated as mini-CPQRAs, and their results added to the general pipeline frequency analysis.

2.1.9. General Calculation Procedure

The general calculation procedure for pipelines involves multiplying a pipeline failure rate per mile-year by the pipeline length by a release probability, which might be the chance of a significant release. To this can be added the point sources of risk such as valve stations. Thus, the overall frequency of a release might be given by:

$$\begin{aligned}\text{Release Frequency} = {} & [\text{Failure rate} \times \text{length} \times \text{release probability}] \\ & + \text{Frequency of release from associated equipment} \qquad (2.1)\end{aligned}$$

Pipelines can be in many different operating states, including fully operational, closed in and circulating, closed in and holding, or empty. When they are in a less than continuous service, a reduced failure rate may be appropriate. However, if the line is idle and unmonitored while not in-service, then failures might go undetected and a release might be experienced the next time the line is put into service. In such circumstances, it is usually not appropriate to apply the reduced usage multiplier. (If a pipeline is only circulating or is in a holding position, the release rate and quantity will also be reduced and this should be addressed in the analysis.) Also, such cycling can increase the stress on the line, leading to early failure.

2.2. Rail

Rail transportation can involve several transfers of the railcars from one train to another, or even one railroad to another. Figure 2-4 illustrates some of the typical operations involved in a rail movement. Some steps, such as mainline travel and switching in yards where cars are moved from one train to another, may occur several times along a route, while others may not occur at all on certain movements. For instance, siding visits may not take place if the movement uses only heavily traveled corridors where the railroads have installed several tracks to eliminate the need for passings. Classification yards are used to configure trains. A useful overview of railroad operations is provided in *The Railroad—What It Is, What It Does* (Armstrong, 1978).

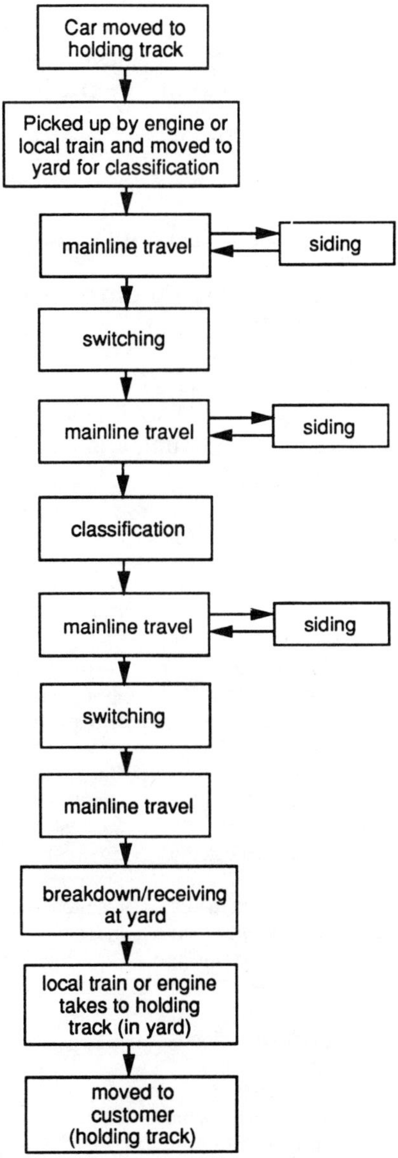

FIGURE 2-4. Typical Rail Transportation Operations.

All of these transfers introduce opportunities for human errors and control system failures in addition to the other hazards of concern. There will also be hazards attributed to the specific product that is carried. Table 2-3 provides a detailed listing of potential incident causes. As for pipelines, this tabulation is for the purpose of hazard identification, qualitative and quantitative risk evaluations, or relative risk comparisons. The way in which data are collected for use in determining accident rates is discussed in the following sections.

2.2.1. Failure Modes

A release of a hazardous material from a railroad tankcar or other type of car can occur as a result of either a railroad accident, or as a result of

TABLE 2-3
Rail Incidents—Initiating and Contributing Causes

Human errors	*Equipment failures*	*System or Procedural Failures*	*External events*
Humping	Nondedicated car	Routing	Vandalism/sabotage
Switching	Crossing guard failure	Control —yards —mainline —in-plant	At-grade crossing
Coupling	Overpressure	Interim storage at holding track	Flood/washout
Transloading	Leaking valve	Car tracking	Earthquake
Inerting	Roller bearing failure	Container specification	Rockslide/landslide
Contamination	Coupling failure	Emergency response training	Avalanche
Heating and cooling	Broken rail	Maintenance	Fire on rail siding
Overfilling	Brake failure	Inspection	Fog/blizzard
Speeding	Roadbed failure	Circuitous routing	Bridge failure
Ignoring closed block signals	Protective coating/ insulation/thermal protection failure		
Driver impairment	Relief device failure		
	Track sensor failure		
	Switchgear failure		
	Signal failure		
	Communications system failure		
	Broken wheel		
	Suspension failure		
	Fitting defect		
	Corrosion		
	Material defect		
	Bad weld		

damage or mishandling of car outlets or equipment failure, not associated with an accident. This second type of event is covered in Section 2.2.8.

Railroad accidents are a result either of an error on the part of railroad operating personnel, or a technical failure in track, rolling stock or signalling and train control equipment. (Railroad at-grade crossing accidents generally cause minimal damage to the train involved, but can be quite serious for the highway vehicles.) Such errors or failures can either lead to a collision between trains, a collision between a train and another obstruction on the track, or a derailment where a train loses the support and guidance provided by the track. It is customary in the analysis of railroad accidents to group failure modes or causes into four categories that correspond to those used in the Federal Railroad Administration's (FRA) railroad accident reporting system. The four categories are:

- *Human Factors.* These accidents generally result from a failure to obey railroad operating rules and procedures by train crews and dispatchers. When such errors lead to an accident it is most commonly a collision, but could be a train derailment if, for example, a train is operated at excessive speed, or encounters a wrongly set track switch.
- *Track and Infrastructure Defects.* These accidents are caused by a structural failure in the track. Leading causes include broken rails, inadequate fastenings between rails and ties, and track buckling. The latter can occur when continuously welded rail is under compressive stress due to high temperature, and is inadequately retained laterally by the ballast. Track defects almost always cause derailments, as opposed to collisions, if they are not detected and repaired.

  This category also includes failures in signal, train control and communications equipment that lead to false or misleading information conveyed to a train crew. Such events are rare, but can lead to serious collisions.
- *Rolling Stock Defects.* These are accidents caused by a technical failure of a component of a freight car or locomotive. A leading cause is roller bearing failure leading to excessive heating and failure of the axle. Other common causes include broken wheels, failures in trucks and suspension components, and brake defects.
- *Miscellaneous Causes.* These are accident causes that do not fit into the other three categories, either because they have multiple causes or because they are substantially outside the control of the railroad. Significant individual causes include grade crossing collisions (usually caused by an error of judgment on the part of the highway user), acts of vandalism or sabotage, natural disasters, and

track–train interaction derailments. Train–track interaction derailments result either from situations where there are related problems in both the track and the rolling stock, or instances where the failure cannot be clearly attributed to one of these categories. Bridge failures would also fall into this category.

It should be noted that railroad accidents have a large number of very different causes. No one cause or failure mode dominates, and even the most common types of failure, such as broken rails or failed roller bearings, cause less than 10 percent of the total.

It is not generally necessary to disaggregate railroad accident rates in more detail than by the four cause groups described above, and between derailments and collisions when performing analyses of transportation risk. The exception is when analyzing the benefits of changes in railroad practice. In these cases, it is necessary to focus on the particular accident causes affected by the proposed change.

2.2.2. Parameters Influencing Accident Rates

The accident rate of interest is the frequency with which a loaded freight car will be derailed in a railroad accident (collision or derailment), and thus be exposed to physical damage and the possibility that this will result in a release of hazardous material. For mainline railroad operations this rate is the combination of the frequency with which trains are involved in accidents, combined with the conditional probability that a car in such a train will be derailed. Usually, mainline accident rates are expressed as a function of train and car-miles operated.

For railroad operations in yards and on sidings, where movements involve single cars and small groups of cars rather than trains, the accident rate is expressed as the frequency of car derailments rather than train derailments. Usually yard accident rates are expressed as a function of yard visits (by loaded cars).

2.2.2.1 Frequency of Train Accidents

The primary factors affecting train accident (derailment) frequency are track quality, train speed, and train size. Track quality is usually quantified by track class as defined in the FRA Track Safety Standards (49 CFR Part 213). There are six FRA classes. The regulations specify minimum acceptable quality standards for each class (geometrical deviation, proportion of defective components, etc.) and a maximum permitted operating speed as follows. These are the legal operating speeds. Most railroads limit the speed of general merchandise trains to about 60 mph, regardless of higher legal limits.

FRA Track Class	1	2	3	4	5	6
Maximum Speed (mph)	10	25	40	60	80	110

Derailment rates are high on low class track and reduce rapidly as track quality is improved. As would be expected, this effect is particularly strong for accidents caused by track defects. Table 2-4 gives derailment rates by failure mode and track class. The overall rates are given for both train-miles and car-miles, and the car-mile rates are also given by failure mode.

Because train operating speeds are normally at close to the maximum speed permitted for a particular track class except for classes 5 and 6, accident rates appear to decline as speed increases. However, if track quality is held constant, there is little evidence to suggest a strong relationship between speed and accident rate.

Train size has a direct effect on accident rate. With more cars in a train there is a corresponding increase in the likelihood that a defective car will

TABLE 2-4

Estimated Derailment and Accident Frequencies for Loaded cars on Class I Railroad Main Tracks (Average per Year, 1985–1988)[a]

	Track Class			
Cause Group	*2*	*3*	*4 and Above*	*All*
Human factors—Number of cars derailed /year	149	199	334	805
Cars derailed per 10^6 car-miles	0.266	0.062	0.031	0.055
Track Defects—Number of cars derailed/year	549	628	647	2212
Cars derailed per 10^6 car-miles	0.980	0.194	0.060	0.151
Rolling Stock Defects—Number of cars derailed/year	161	809	825	1866
Cars derailed per 10^6 car-miles	0.288	0.250	0.076	0.128
Miscellaneous Causes—Number of cars derailed/year	173	203	321	801
Cars derailed per 10^6 car-miles	0.309	0.063	0.030	0.055
All Causes—Number of cars derailed/year	1031	1611	2127	5449
Cars derailed per 10^6 car-miles	1.841	0.499	0.197	0.372
All Causes—Number of accidents/year	152	244	270	771
Accidents per 10^6 train-miles	4.0	2.4	1.0	1.9
Exposure—Millions loaded car-miles/year	560	3230	10810	14632
Millions train-miles/year	38	102	270	414

[a]FRA (1992); Transportation Data Source 4-18.

cause an accident, or that a track component will fail under repeated loading. Also long, heavy trains can suffer severe longitudinal shocks (due to bunching or stretching of the cars resulting from accelerating, braking, or gradient changes) if not handled carefully by the engineer, leading to derailment.

There are also a number of secondary parameters that have a limited influence on the frequency of train accidents. These include the following:

- *Type of signal system installed.* Signal system capabilities can directly affect the frequency of a human error accident.
- *Inspection type and frequency.* More frequent or comprehensive inspections of track or cars increase the likelihood that defects will be found and corrected before they cause an accident. This includes use of automated inspection systems such as hot bearing detectors.
- *Types and density of railroad operation.* High density operations (very frequent trains) or operations that involve a great deal of switching activity may be more vulnerable to accidents than lower density operations, and those with little switching activity. Indeed, switching activity differences may account for some of the accident rate differences by track class.
- *Railroad type.* Whether the railroad is a major "Class I" railroad with broad technical and financial resources, or a small railroad with more limited resources may affect the safety performance of a railroad. There is also variability by individual railroad from year to year.

It should be recognized that it is difficult to estimate the magnitude of these secondary effects empirically from available accident data. It is possible to develop estimates from engineering analysis of the failures that cause accidents, but the application of expert judgment is involved, and there will inevitably be uncertainty regarding the validity of the results. Weather is a very minor contributor to the overall accident rate. Such accidents are generally a result of major events like avalanches or flash floods, which are quite infrequent.

2.2.2.2. Probability That a Car Will Be Derailed Given a Train Accident Given that the train is involved in an accident, the conditional probability that a specific car will be derailed is a function of train speed, train size, type of accident, and position of the car in the train.

Figure 2-5 illustrates the strong relationship between the number of cars derailed in an accident and train speed for all mainline freight train accidents in which at least one car is derailed. Similar relationships can be developed by accident cause; for example, track failure-caused accidents tend to be more destructive than those due to car defects at the same speed. However, in most analyses it will only be necessary to use the aggregated relationship. More detail is important only in evaluating the

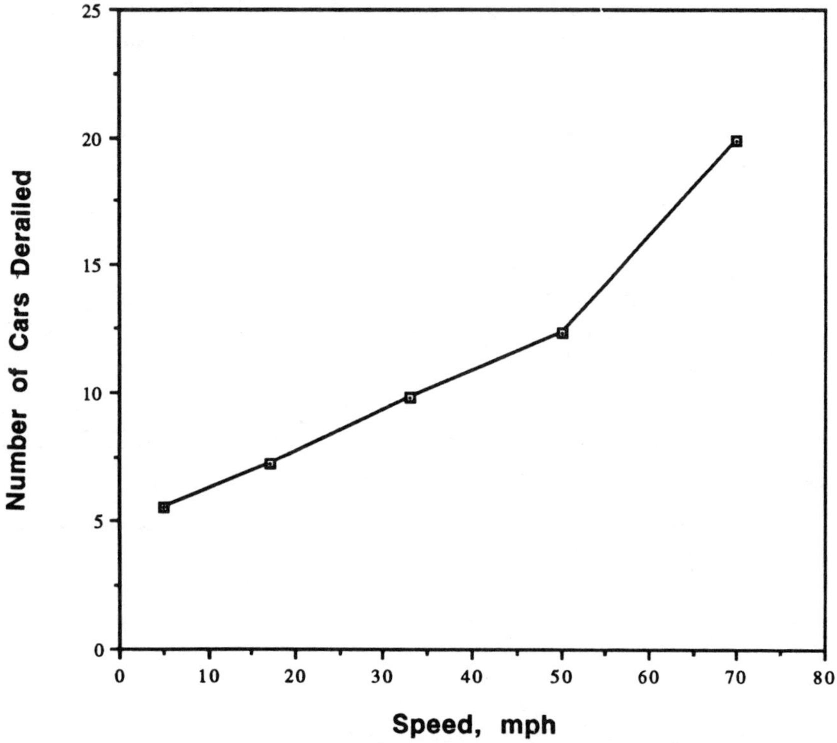

FIGURE 2.5. Number of Cars Derailed by Speed. All mainline freight train accidents in which at least one car was derailed. (Transportation Data Source 4-18; data from 1989 to 1991.)

benefits from improvement measures that reduce the frequency of specific accident causes.

The fraction of mainline freight train accidents in which at least one freight car is derailed varies by the type of accident (Transportation Data Source 4-18 in Chapter 4, 1989–1991):

Derailments	0.97
Collisions	0.28

The average number of cars derailed is about eight per mainline accident, based on data from the FRA accident database (Transportation Data Source 4-18 in Chapter 4).

Train size also has an effect. There is a slightly lower conditional probability of a car derailing in a longer train than a shorter train. However, the effect is relatively small, and is not normally included in risk analyses.

The variation in car derailment risk by car position in the train is indicated in Figure 2-6. (Each position represents 20 percent of the cars;

hence, Position 1 in a 70-car train would be the first 14 cars.) The lowest risk position is toward the rear of the train. However, it is questionable whether any practical use can be made of this effect. Railroad operational efficiency requires that blocks of cars be positioned in the train so as to minimize in-train longitudinal forces, which can cause accidents. Adding loaded cars near the rear of the train could increase such forces, and thus the incidence of some types of accidents.

2.2.2.3. Frequency of Car Derailments in Yard Operations
Yard operations are characterized by train disassembly and reconfiguration, and by numerous switching activities. The predominant types of accident causes in yards are:

- *Track failures*—because these operations typically take place at low speed on lower quality track, that is, Classes 1 and 2 from the FRA breakdown given earlier in this section.

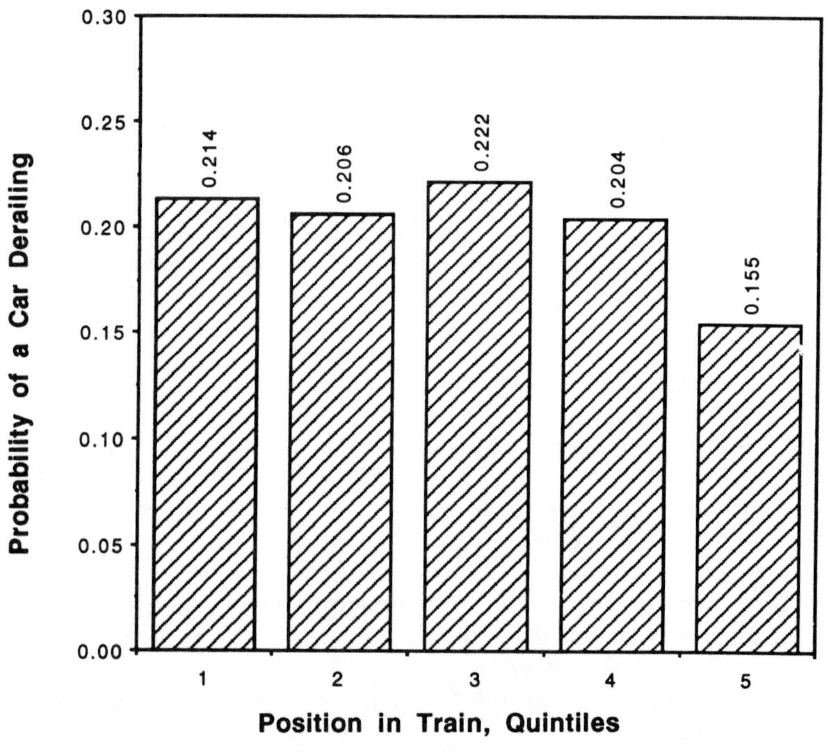

FIGURE 2-6. Distribution of Derailed Cars by Position in Train. All mainline freight train accidents in which at least one car was derailed. (Transportation Data Source 4-18; data from 1985 to 1990.)

- *Human factor accidents*—especially minor collisions and derailments due to incorrectly aligned turnouts.

Track quality (as reflected by FRA Class) is the primary parameter influencing derailment rates in yard accidents. The care taken in operations is also clearly important, but difficult to quantify in any meaningful way. The maximum allowable speed may also influence accident rates, and speed limits for handling HAZMAT cars in yards are applied in the railroad industry. However, data with which to quantify the effect of speed on derailment rates are lacking.

2.2.3. Parameters Influencing Release Probabilities

The conditional probability of release given that a car is derailed is a function of speed and car design features. The likelihood of a release increases as speed increases, and speed is the only significant railroad operating parameter affecting release probability. Using data from the FRA railroad accident reports and the Research and Special Programs Administration (RSPA) HAZMAT release reports, estimates can be made of both the overall probability of release for the two principal types of tank cars—pressure cars and nonpressure cars, and the effect of train speed on the release probability (Transportation Data Sources 4-3 and 4-18; using data from 1985 to 1988, and Transportation Data Source 4-16; R-709).

Overall release probability in mainline accidents

Nonpressure cars	0.29
Pressure cars	0.08

The adjustment factor for train speed has been estimated based on a regression analysis and is shown in Figure 2-7.

Release probability in yard accident (all speeds)

Nonpressure	0.13
Pressure	0.02

Car design features also have a very strong effect on release probability. There are four mechanisms by which releases occur given an accident:

- Head puncture
- Shell puncture
- Bottom fitting damage
- Top fitting damage

Resistance to head and shell puncture in an impact is a function of shell thickness and material of construction, and also whether the car is equipped with jacketed insulation. This insulation typically consists of four

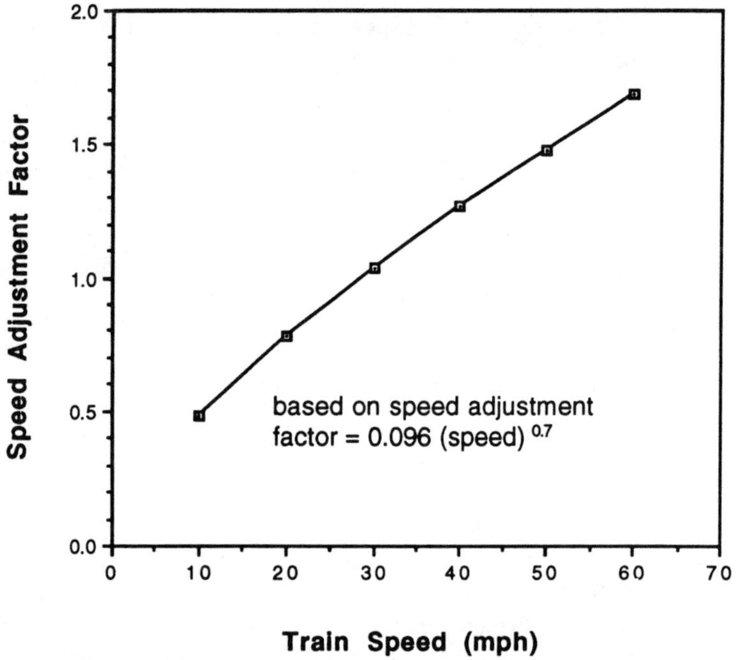

FIGURE 2-7. Adjustment Factors for Release Probabilities as a Function of Train Speed. (Transportation Data Sources 4-3 and 4-18; data from 1985 to 1988.)

to six inches of glass wool contained inside a ⅛-inch steel jacket. Distortion of the jacket absorbs impact energy, and reduces the severity of the impact on the containment shell. Tank car heads can be protected by a head shield, either half or full height, which also reduces the risk of head puncture. Head puncture risk is also reduced by the use of shelf couplers, which lessen the risk of couplers disengaging in an accident and striking the tank car head or shell. All HAZMAT cars in the U.S. are now equipped with shelf couplers.

Depending on car type, bottom fittings are either not used, or are protected from damage by skids. Top fittings can be protected by using low-profile fittings, which are less vulnerable to damage, or by protecting the fitting with skids.

2.2.4. Container Types

Details of tank car specifications by USDOT class are given in Appendix A. Brief descriptions of the principal classes are given below:

- *Nonpressure, noninsulated tank cars (DOT 111A).* These cars normally have a 7⁄16-inch-thick tank and head, and are not equipped with head shields. Various outlet configurations are used, but bottom outlets are prohibited on cars used for corrosive commodities such as acids.
- *Nonpressure, insulated tank cars (DOT 111A).* These are similar to noninsulated cars with the addition of jacketed insulation. The insulation is applied to maintain the contents at the desired temperature: the improved damage resistance is a side benefit.
- *Pressure, noninsulated cars (DOT 105 and 112).* These cars typically have a tank and head thickness of between 9⁄16 inch and nearly 1 inch, depending on the pressure rating. Most pressure cars are equipped with head shields and shelf couplers. Noninsulated cars used for flammable commodities also have thermal protection applied to the outside of the tank. This is intended to reduce the risk of an explosion (BLEVE) if the car is exposed to a postaccident fire.
- *Pressure, insulated cars (DOT 105 and 112).* These have three to six inches of glass wool insulation contained within a steel jacket. Jacket thickness around the shell is typically 1/8 inch, increasing to ½ inch at the head to provide protection equivalent to a head shield.

2.2.5. Accident Trends with Time

As illustrated in Figure 2-8, the number of railroad accidents reported to the FRA has decreased by a factor of three between 1979 and 1991. Railroad traffic (measured in ton-miles) increased by 5 percent over the same period. The decrease in track-defect-caused accidents has been the most notable. The number of accident-related releases from HAZMAT cars has also declined, due both to the reduction in accidents and the application of protective measures such as head-shields and shelf-couplers in the period 1978–1984. The variability of the total number of accidents in recent years is largely a result of variation in train miles operated as shown below (Transportation Data Source 4-17 in Chapter 4):

	Accidents	*Train Miles (Millions)*
1986	2761	567
1987	2647	581
1988	3051	609
1989	3080	621
1990	3045	609
1991	2814	577

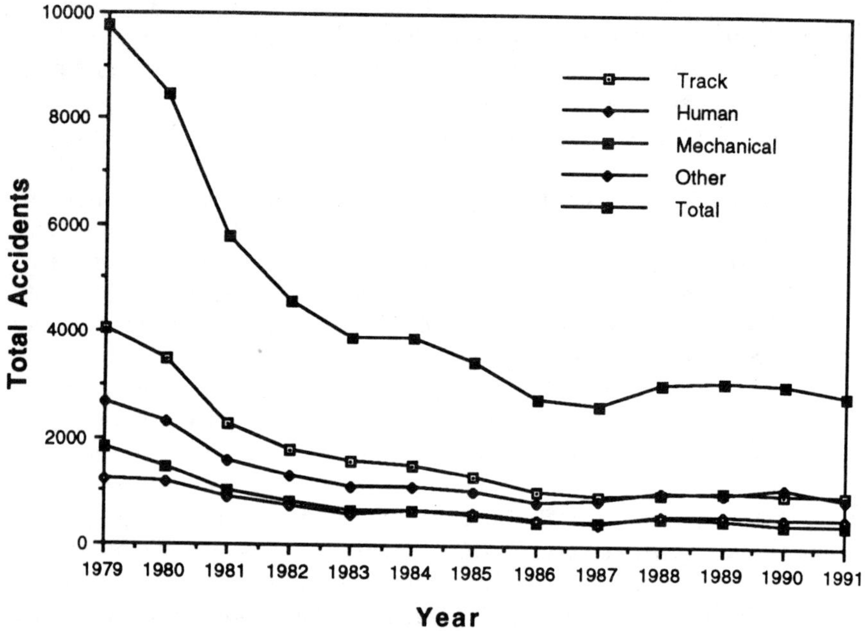

FIGURE 2-8. Trends in Railroad Accident Causes (numbers of reported accidents). (Transportation Data Sources 4-17 and 4-18.)

These trends have been noted across all track classes and all railroads. Thus, there need not be any concern beyond identifying the appropriate track class and railroad type for your movement.

2.2.6. Confidence in Data

In order to obtain accident frequencies and release probabilities for a risk analysis, it is necessary to obtain data on both the incidence of accidents, disaggregated as necessary by accident type or by the type of HAZMAT car involved in the accident, and also to quantify the potential exposure to accident risk, usually quantified by train or car-miles and separated by causal parameters such as track class or car type. These provide the numerator and the denominator, respectively, for determining the accident rate.

The FRA railroad accident reporting system provides accurate data regarding accident incidence by cause. These data include counts of HAZMAT cars involved in accidents and cars releasing, but there is no disaggregation by HAZMAT car type. The Research and Special Programs Administration (RSPA) of the USDOT collects data on HAZMAT releases by commodity, from which car type can be inferred, but these data lack information on accident circumstances (speed, train size, etc.). Efforts to match RSPA and FRA HAZMAT data have been only partially successful.

A large tank car damage database has been assembled by the Railway Progress Institute—Association of American Railroads Railroad Tank Car Safety Research and Test Project. This database is currently being analyzed to provide improved data on the conditional probability of release from a tank car as a function of tank car features, and also release size distribution. There is little information on the rate of release.

With regard to exposure data, there is a high degree of uncertainty regarding the breakdown of railroad traffic volume (train and car-miles) by track class. Therefore, accident rates by track class are highly uncertain. Data collection efforts are in progress in this area, and it is hoped that improved data will become available. There is also uncertainty regarding the exposure (shipments and car-miles) by HAZMAT car type. This is because there is not a direct one-to-one correspondence between car type and commodity carried, and also there are some difficulties in interpreting commodity shipment data contained on freight waybills.

2.2.7. Mode-Specific Issues

The mode-specific issues arise from the nature of railroad operations. For a shipper, the shipment is normally a single carload or a small number of car loads. The railroad procedures in transporting these cars to their destination are as described below, leading to a number of mode-specific issues that have to be reflected in the risk analysis.

- The car will be picked up by a way-freight serving the originating point, and taken to a nearby classification yard. Often this movement will be over a branch line having a low track quality.
- The car will be switched in the yard and put into a mainline train traveling in the direction of, but not necessarily to, the final destination. This train may typically have about 100 cars, and may include HAZMAT cars containing different types of commodity.
- The routing of the car depends on the railroad's operating plans, and may change from shipment to shipment, or be circuitous.
- The cars may pass through one or more classification yards en-route, and may also be interchanged among different railroad companies. Reclassification typically occurs every 500–1000 miles, but can be much more frequent.

- The car is finally reclassified at a yard near the destination, and is put into a local way-freight for final movement to the destination.

The principal exception to the above procedure is when large multiple-car shipments are moved in a unit train. A unit train moves directly from origin to destination without intermediate switching.

These operating procedures can mean that risks may have to be calculated for each stage in the movement using accident frequencies and conditional probability, which have been estimated separately. This includes the local branch-line movements at the origin and destination, the number of classification yard visits, and the characteristics of different main line segments of the movement. When analyzing accident consequences, the probability that other HAZMAT cars and commodities are involved has to be considered, as does the potential for multiple car spills of the same commodity when several cars are moved together.

2.2.8. Non–Accident-Initiated Releases

These are releases, usually through tank car outlets and fittings, resulting from a malfunction or mishandling of the fitting. By definition they are not caused by any kind of railroad movement accident. There are several ways in which such a release can occur, including:

- operator error—such as valves not closed and secured properly
- via an outlet that has become defective as a result of misuse or general wear and tear
- deliberate sabotage or vandalism.

Such releases usually involve the slow loss of car contents through a defective or inadequately secured opening. The consequences are usually minor due to the small quantity of material lost. While the car is in motion, this loss is usually distributed along many miles of railroad, with minimal consequences. Consequences can be more serious in yards, since the car stands in one place for a long time and local concentrations can be high. There is also a hazard to railroad personnel working on or near the defective car who may suffer injury from breathing fumes or direct contact with a harmful commodity.

Another possible source of non–accident-initiated releases is loss of a commodity through a relief valve. This can occur if the car is overfilled, or if the car is left standing in unusually hot weather leading to expansion of the commodity and release through the valve. Product contamination and reactivity can lead to releases through the relief valve or more catastrophic failures of a car.

The data on all of these releases are very limited and situation-specific. Where there is a particular concern about non–accident-initiated releases, they should be analyzed separately through other techniques, such as fault tree analysis.

2.2.9. General Calculation Procedure

At the greatest level of detail, the overall chance of release is calculated separately for the various segments of the railroad movement and for yard visits, and summed to yield the overall release risk for a specified movement. A movement is specified by origin, destination, and railroad routing, which should include speeds, track class, and the length of each track segment over which the car passes, as well as the locations and number of yards visited.

The risk of car derailment during classification yard visits is simply the product of the accident rate (loaded car derailments per yard visit) and the aggregate number of yard visits. This must be multiplied by the release probability to get the overall chance of a release.

Calculation of the chance of the car being derailed during a mainline movement is a function of several parameters. Accident rates must reflect both train-miles and car-miles operated. This is because the causality of some accident types (e.g., collisions caused by operator error) are dependent only on train miles and are independent of the size of the train. Other accident types (e.g., caused by a bearing failure on a freight car) are a function of the number of cars in the train, as well as the train-miles operated. Therefore, railroad mainline train accident rates, derived from historical accident data, are expressed as follows:

$$\begin{aligned}&\text{Overall frequency of train accidents per train mile}\\&\quad = \text{Accident frequency per train mile}\\&\qquad\quad {\scriptstyle\text{for accident causes where frequency is a function of train miles}}\\&\qquad + \text{Accident frequency per car mile} \times \text{number of cars in train}\\&\qquad\quad {\scriptstyle\text{for accident causes where frequency is a function of car miles}} \qquad (2.2)\end{aligned}$$

Table 2-5 illustrates the format of the frequency data needed to calculate the overall accident frequency for a specific train.

If available data do not support this level of breakdown or if the analysis does not require this level of detail, it is possible to combine track classes and collisions and derailments directly. "Per train-mile" and "Per car-mile" accident rates should not be combined without the proper conversion to a common set of units.

Data of the form presented in Figure 2-6 are used to determine the probability that a particular car will be derailed given that the train it is in

TABLE 2-5
Format for Railroad Accident Frequencies

Track Class	*Collision Frequency*		*Derailment Frequency*	
	per train-mile	*per car-mile*	*per train-mile*	*per car-mile*
1				
2				
3				
4				
5				
6				

is involved in an accident, and those from Figure 2-7 are used to determine release probability as a function of speed and car type.

Thus, the risk of release during movements over a specific track segment is calculated as follows:

$$\begin{aligned}\text{Release Frequency} = &\ \text{train accident rate}\\ &\times \text{length of segment}\\ &\times \text{number of cars of concern shipped per train}\\ &\times \text{derailment probability}\\ &\times \text{release probability} \qquad (2\text{-}3)\end{aligned}$$

Release frequencies can ideally be calculated separately for each segment of the main line rail movement, including yards, and summed to give overall risk. However, data restrictions/limitations and the purpose of a particular risk analysis may make this impractical if not impossible.

2.3. Road Transport

Transportation of hazardous materials by truck involves many fewer operations than a rail movement. The main activity is driving, interrupted by rest stops and weigh station or company required inspections. Figure 2-9 illustrates one possible sequence of operations on a relatively long route. The overnight parking may or may not be in a secured area.

There may also be many iterations of the same steps, particularly if it is a multiple day journey. The types of hazards of concern are identified in Table 2-6. As for the previous modes, these are subdivided into human

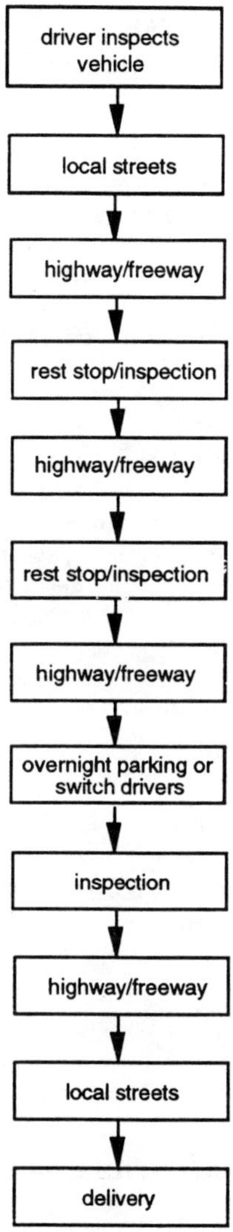

FIGURE 2-9. Typical Truck Transport Operation.

errors, equipment failures, failures of systems and procedures, and external events. The detailed listing of hazards can be used for overall hazard identification, identification of potential non–accident-initiated events, qualitative risk evaluation or reduction, etc. The grouping of hazards for the purpose of estimating accident rates is discussed below.

TABLE 2-6
Truck Incidents—Initiating and Contributing Causes

Human errors	*Equipment failures*	*System or procedural failures*	*External events*
Driver impairment	Nondedicated trailer	Driver incentives	Vandalism/sabotage
Speeding	RR crossing guard failure	Driver training	Snow
Driver overtired	Leaking valve	Carrier selection	Rain
Enroute inspection	Leaking fitting	Container specification	Ice
Contamination	Brake failure	Route selection	Fog
Heating and cooling	Insulation/thermal protection failure	Emergency response training	Wind
Overfilling	Relief device failure	Speed enforcement	Flood/washout
Inerting	Tire failure	Driver rest periods	Rockslide/landslide
Other vehicle's driver	Soft shoulder	Maintenance	Fire at rest areas/parking areas
Taking tight turns/ramps too quickly (overturns)	Overpressure	Inspection	Hurricane
Unsecured load	Material defect	Time of day restrictions	Tornado
	Vacuum		Earthquake
	Steering failure		Existing accident
	Sloshing		
	High center of gravity		
	Corrosion		
	Bad weld		
	Excessive grade		
	Poor intersection design		
	Suspension system		
	Fifth wheel failure		

2.3.1 Failure Modes

Although a variety of mechanisms may cause a truck accident and cargo release, several general categories are of greatest relevance with respect to risk analysis. Typically, these are relatively broad categories for which useful data are available and which share common characteristics with respect to releases. These are:

- Vehicular collisions,
- Collisions with fixed objects,
- Vehicle overturnings,
- Railroad grade crossing accidents, and
- Non–accident-initiated releases.

Vehicular collisions involve a collision between two vehicles. Due to the involvement of two vehicles, the potential for substantial damage and large releases is present. Collisions with fixed objects can also represent relatively energetic impact accidents with the potential for significant damage and/or cargo release. Overturned vehicles are most likely during trucking operations where, for some truck designs and cargoes, the vehicle center of gravity is high, especially on tight curves such as ramps. Railroad grade crossings present an additional hazard, but, due to their limited number, typically do not contribute substantially to overall accident risk on a long route. Non–accident-initiated releases are characterized by equipment failures not associated with accidents such as leaks of pipes and fittings or failures of relief valves and rupture disks. They are described in Section 2.3.8.

These represent the major failure modes that require consideration during a truck risk analysis. Typically, efforts to refine the list of modes are limited by the availability of data to quantify risks and consequences. Furthermore, these classes of failure are fairly broad in the sense that the majority of accidents may be characterized as belonging to one of the above general classes. Unlike rail, the classification of accidents is not uniform nationwide. The minimum damage level making an accident reportable is $50,000 as of December 31, 1991 (49 CFR 171.15).

2.3.2. Parameters Influencing Accident Rates

In most instances, enroute accident rates are the most important components of a truck risk analysis. The rate is, in general, affected by numerous parameters describing the nature of the trucking operation, environmental, and road conditions, and specific risk reduction measures that may have been implemented. In particular, considerations such as urban versus rural routes and divided versus undivided highway have a direct influence on the likelihood of an accident and its severity. For example, the high traffic density associated with urban operation significantly increases the chance of a collision, but the typically lower speeds reduce the potential for a severe consequence due to a collision. Conversely, a divided limited access highway significantly lowers the likelihood of an accident, but the high speed operation presents a much higher potential for a severe accident and release. Table 2-7 illustrates some accident rates and HAZMAT release probabilities for different types of roads.

Location-specific conditions such as excessive grade, obstructions to vision, poorly designed intersections, etc. can have a significant localized influence on accident rates, but are difficult to quantify in general. Additionally, such local circumstances are seldom considered a significant influence on overall risk since their site-specific nature applies to a rela-

TABLE 2-7
Truck Accident Rates for California, Illinois, and Michigan (Combined)[a]

Highway Class		*Truck Accident Rate (per 10^6 vehicle miles)*	*Release Probability*
Area	*Roadway*		
Rural	Two-lane	2.19	0.086
Rural	Multilane, undivided	4.49	0.081
Rural	Multilane, divided	2.15	0.082
Rural	Freeway (limited access)	0.64	0.090
Urban	Two-lane	8.66	0.069
Urban	Multilane, undivided	13.92	0.055
Urban	Multilane, divided	12.47	0.062
Urban	One-way street	9.70	0.056
Urban	Freeway (limited access)	2.18	0.062

[a]Harwood and Russell (1992); Transportation Data Source 4-27.

tively short distance compared to the longer route segments associated with the overall route. Weather conditions such as fog, storms, or icing conditions can also influence accident rates but typically are incorporated into a risk analysis only in fairly general terms based on the general climate of the transport region. Only if one of these issues or affected areas is of particular concern is it considered explicitly in the analysis, provided that there are sufficient data.

Driver training programs, fleet maintenance, speed monitoring, and other characteristics specific to an individual carrier can also influence accident rates. To date, the effectiveness of such programs have been determined only on a carrier by carrier basis, based on changes in reportable accident rates.

In summary, truck risk analyses are normally based on accident rates characteristic of broad classes of route types (e.g., urban versus rural, divided versus undivided highway) for which useful data are available. The route type incorporates several parameters of concern, namely traffic density, speed, and potential for head-on collisions. Location- specific hazards (such as poorly designed intersections or access ramps) may be noteworthy and are generally the subject of qualitative comment, but it is uncommon for such hazards to have a significant quantitative effect on the overall risk estimate unless there are particularly severe weather conditions such as fog or ice. Carrier-specific variations in accident rates can also be considered if available for the carrier of concern or one with similar programs and reputation.

2.3.3. Parameters Influencing Release Probabilities

Given the occurrence of an accident involving a truck carrying hazardous material, the conditional probability of release is a function of vehicle characteristics and the nature of the accident. This conditional release probability has a significant influence on overall risk since it typically addresses the relative likelihoods of different sizes of releases having substantially different potential consequences. The probability of various sizes of releases is also affected by shipping conditions, for example, pressurized or nonpressurized cargo. When moving smaller packages (i.e., drums and cylinders) improper fastening or securing of the containers can allow one to impact another in the event of an accident, sudden acceleration, or rapid stop.

With respect to the nature of the incident, the most significant factor is the general accident type. In truck risk analyses, the following four accident classes are typically used to characterize the type of accident:

- High speed collision,
- Low speed collision,
- Overturning, and
- Non–accident-initiated release.

By considering these general categories, one can then analyze the available data to attempt to characterize the probability of various release sizes. In general, the data available are sufficiently limited that at present analyses use aggregated release probabilities. Harwood and Russell (1992) present one set of aggregate numbers by road type, but do not reflect differences due to container design.

Vehicle characteristics influence the likelihood of release via the strength and integrity aspects of the container as well as specific mitigation factors such as antisloshing devices and the height of the center of gravity of the loaded vehicle. Strength and integrity considerations include the material and wall thickness of the container, the presence or absence of a double wall, compartmentalization of cargo and the presence of any protective shielding devices. These measures either affect the potential for container rupture or puncture (e.g., strength, double walls or shielding) or, in the event of container failure, the quantity of material released (e.g., compartmentalization).

A secondary effect on release probability and a more important effect on overall release size results from the characteristics of the transported material and the conditions of transport. Relevant considerations include the material phase (liquid or gas), material temperature and pressure. The effects of these material property considerations are relatively straightforward, with pressurized and gas phase cargos more likely to result in major releases. However, the rate of release may be relatively slow if the accident caused only a small hole.

2.3.4. Container Types

Road transport of hazardous materials is conducted using a variety of truck types ranging from basic single-walled tank trucks, which may or may not have internal compartments to refrigerated, double-walled insulated tank trucks. Tube-trailers are used to transport many industrial gases. Materials are also sometimes transported using drums and cylinders and other small containers to contain the material. Descriptions of typical truck types and their DOT designations are provided in Appendix A. Small bulk containers are sometimes moved in pairs if local regulations allow double-trailers.

2.3.5. Accident Trends with Time

The overall rate of trucking accidents has been relatively constant over time, with most variation within the overall uncertainty of the estimates and the performance of individual carriers. For example, freeway or interstate truck accident rates of roughly 1 to 2 per million vehicle-miles have been reported from 1966 through the present as shown in Table 2-8.

Variations may also exist for individual routes whereas national averages have remained more constant. A more important trend over time relates to the actions of individual carriers to reduce the risk of HAZMAT trucking accidents. In particular, carriers with concern about reducing their potential exposure to HAZMAT trucking accidents have implemented mitigation measures to reduce the likelihood and/or severity of trucking accidents. In particular, enhanced safety related training programs, improved equipment inspection and maintenance and stricter enforcement of company safety policies have provided carriers with reduced incidence of HAZMAT accidents. In general, therefore, accident experience may be pooled over relatively long periods of time to provide better estimates and appropriate adjustments to base rates can be used to reflect the effects of specific mitigation measures such as training and maintenance.

2.3.6. Confidence in Data

The degree of confidence in accident data relates directly to the size and quality of the database used to estimate rates and release size probability distributions. In general, statistics describing accident rates exhibit higher confidence than statistics with respect to release sizes since the total number of accidents is significantly higher than the number of accidents with HAZMAT releases in general, let alone by specific size or type of truck.

TABLE 2-8

A Comparison of National, State, City, and Route-Specific Truck Accident Rates

Source	*Accident Rates (per million vehicle-miles)*		Comments
	All vehicles	*Trucks*	
Survey of California Highways (ADL, 1990)	0.44–5.67	0.49–4.49	• Based on Caltrans (California D.O.T.) data • Accident data for 1985–1988 • Traffic volume data for 1986
Toronto Study (Saccomanno and Chan, 1985)		0.85–4.38 (Expressways) 0.95–13.25 (Collectors, ramps, intersections)	• Data Year 1981 • Accidents in and around a city (Toronto) • Variations due to type of road, visibility, and pavement (wet/dry)
University of Michigan Study (Ervin et al., 1980)	1.45 (Freeway) 5.0 (U.S. and Michigan highways) 8.25 (city streets and roads)		• Compiled from data from Michigan State Police • Data years 1974–1978
Battelle Study on Propane (Geffen, 1980)		2.5	• Data years 1966–1970 • Reporting limit $250 or more • Transportation data sources 4-1 and 4-2
Rensselaer Report (Abkowitz et al., 1984)		0.65 (Interstate) 2.26 (U.S. & State highways) 3.65 (Highways with intersections) 1.4 (Average)	• Data years 1980–1982 • California, Texas, and New Jersey data
Insurance Institute for Highway Safety Report (1985)		*Double Trailers* 2.28 (rural freeway) 4.68 (rural nonfreeway) 3.88 (urban freeway) 4.28 (urban nonfreeway) *Single Trailer* 1.10 (rural freeway) 0.99 (rural nonfreeway) 2.14 (urban freeway) 0.93 (urban nonfreeway)	• Based on a FHWA report (1981)
National Safety Council (1987)	7.55	3.52 (petroleum trucks) 8.84 (All trucks)	• Data year 1986 • Only participating fleets • No minimum damage • Transportation data source 4-22
Caltrans Report on Hazmat Transportation (1986)	2.4 (excluding trucks)	4.6	• 1985 data

Since the same data sources are generally applied to estimate rates under a variety of conditions, the confidence level tends to decrease as the conditions are made more restrictive. For example, the overall accident rate estimates tend to exhibit relatively high confidence since they are based on a large number of incidents. As analysis is restricted to specific types of trucks or specific commodities, the number of relevant incidents in the databases is reduced with an associated reduction in confidence in the estimate. Similarly, the restriction to specific types of roadways, weather conditions and time of day results in a smaller quantity of relevant data and thus, lower confidence in the resulting estimates. The practical ability to estimate quantitative confidence levels is often restricted by the nature of reporting and database access. Where feasible, however, a detailed analysis should include a quantitative assessment of the confidence in the estimates.

2.3.7. Mode-Specific Issues

In addition to general accident rate considerations, a number of specific considerations relating to truck transportation may significantly alter the estimated rates and release probabilities. Many of these measures are alternatives that are considered as potential risk mitigation measures for trucking operations. Most are not clear reductions and have associated tradeoffs that must be evaluated to provide an overall indication of the effectiveness of the reduction measures. Key operational measures include the following:

- Route modification to avoid populated areas,
- Modification of shipment time to avoid periods of high traffic density,
- Overnight travel to avoid traffic and outdoor populations,
- Rerouting or trip interruption to avoid adverse weather conditions, and
- Use of interim storage and parking areas.

As mentioned above, most of these measures must be examined under the specific circumstances of the trucking operation being considered to understand the relative tradeoffs. For example, route modification to avoid populated areas or localized adverse conditions may result in more circuitous routing that causes some increase in accident risk due to the longer distance traveled. This may still decrease the risk to the public, however, it may increase the environmental risk. Adjustment of shipping time to avoid high traffic and outdoor activities may result in night-time travel where limited visibility and fatigue may play a role in increasing accident rate. Trip interruption and interim storage may avoid some adverse operating conditions such as weather and fatigue, but exposure to some non–accident-initiated releases or accidents from other moving vehicles may be increased in exchange.

The role of the driver and the incentives/disincentives offered to the driver are also unique in this mode of transport. Scheduling practices, mandatory breaks, etc. can affect driver alertness. Payment practices may encourage speeding and minimal breaks in order to get in more trips or hauls, and may also encourage driving under adverse weather conditions to meet a delivery schedule.

2.3.8. Non–Accident-Initiated Releases

As noted previously, a variety of potential release mechanisms exist during truck shipping operations that do not relate to vehicular accidents. In general, these mechanisms result in relatively small quantities of cargo being released, but in extreme circumstances can result in a significant spill. Typical non–accident-initiated releases include leaks that develop at valves or fittings. Tanks may be overfilled resulting in a release of material via relief devices or even failure of the tank. Relief valves and rupture disks may fail, resulting in a release of material when operating conditions are normal.

In general, these types of releases are related more to the duration of the time period over which specific equipment has been in operation, rather than the number of miles traveled during a particular HAZMAT shipment. These releases are therefore accommodated in an analysis by using failure rates normalized to time or to the number of operations rather than to the miles traveled. Standard failure rate sources for mechanical and/or electrical components are a common source of data representing the failure rates of these components and some even include appropriate environmental factors (such as corrosion, vibration, ...). For overfilling or product contamination scenarios, the frequencies can be estimated using other techniques such as fault tree analysis and then distributed over the full route length, or that portion felt to be at greatest risk from the event.

2.3.9. General Calculation Procedure

As discussed above, truck accident rates are typically normalized by distance traveled while non–accident-initiated releases are better characterized by rates per demand or per unit time. The procedure used to calculate the overall accident rate is thus to combine the estimates for each component of the accident rate that is being considered. In particular, the basic accident rate per mile is multiplied by the number of miles in the route and added to the total of other events multiplied by the number of demands or operating hours as appropriate. For example, consider a

500 mile (10 hour) trucking operation over a route with an expected accident rate of one accident per million miles, using a truck that has one relief valve with a failure rate of 1.7×10^{-6} per hour. This movement would present the following estimate of the overall chance of release, assuming that 1 accident in 10 results in a release (see Table 2-7).

$$
\begin{aligned}
F_R &= [(1.0 \times 10^{-6} \text{ accidents/mile})(500 \text{ miles/trip})] \times 0.1 \text{ chance of release} \\
&\quad + (1.7 \times 10^{-6} \text{ failures/valve-hour})\ (10 \text{ hours/trip}) \\
&= 6.7 \times 10^{-5} \text{ releases/trip or one release every 15,000 trips on average.}
\end{aligned}
$$

In a detailed application, a route would be divided into a number of segments, each having uniform conditions and the above calculation procedure applied to each segment. An overall estimate for the route is then obtained by summing the estimates for the individual segments comprising the route.

2.4. Barge

Barges come in a variety of sizes and designs, but all offer the advantage of requiring relatively little draft. This allows them to reach locations tankers cannot. Barge operations may be on open water (including ocean waters) or on rivers and waterways. Individual barges may or may not be dedicated to a single product and might typically have one to ten tanks or compartments. General industry practice is to transport barges in groups or "strings" (sometimes called tows), tied together and pushed (sometimes pulled) by a tug or towboat. Strings include roughly two to eight barges. Some barges are self-propelled.

Types of barges include inland, coastal, and ocean-going. Further subdivisions include:

- tramp and liner barges
- inspected and noninspected barges
- manned and unmanned barges
- integrated, deep-notch, and other designs of barges.

Barges may be relatively small or larger than some ships. Manning restrictions may vary from those for ships and should be considered for their contribution to risk. Given the specialized knowledge needed in this area, advice and information should be obtained from internal experts.

The cargo tanks on barges may be classified as:

- *integral*: forming part of the hull
- *independent*: not a contiguous part of the hull structure

- *gravity*: design pressure not greater than 10 psig. Integral tanks are of this type.
- *pressure*: design pressure greater than 10 psig. Independent gravity tanks whose shape allows a stress analysis to be performed are classed as pressure vessel type tanks.

A barge movement may be direct to a single customer, but could also entail additional port or harbor visits for partial offloading to other customers. Depending on the nature of the operation, barges may be loaded at shore-based facilities or may involve lightering from a large tanker that may not be able to transit shallower waters. Some barges will also make visits to mooring areas—either for overnight stops or to reconfigure tow strings.

Table 2-9 provides a detailed listing of potential hazards encountered during marine operations. Most of these are equally applicable to barges and ocean-going vessels; however, the grouping of hazards into human errors, equipment failures, failures of systems and procedures, and external events does not exactly match failure rate collection categories (see Sections 2.4.1 and 2.5.1). In any event, the full listing should be reviewed in order to obtain a comprehensive identification of hazards. This listing can also serve as a basis for a qualitative or semiquantitative evaluation of risks or risk reduction measures. It provides a mechanism for identifying non–accident-initiated events and can point out where systems and procedures could be broadened in coverage.

Section 2.4.1 describes the classification of hazards for the purpose of generating accident rates.

2.4.1 Failure Modes

Barge failure modes include the following types of accidents:

- collisions with other moving vessels,
- collisions with moored vessels (allisions),
- collisions with fixed objects (rammings),
- groundings,
- non–cargo-related fires and explosions,
- material failures of the barge, and
- others such as flooding, capsizing, breakaways, and towing cables cut by other vessels.

The accident and spill rates associated with each failure mode vary from one body of water to the next. The combination of fires and explosions and other accidents, such as capsizing, flooding, and breakaways

TABLE 2-9
Marine Incidents—Initiating and Contributing Causes

Human errors	*Equipment failures*	*System or procedural failures*	*External events*
Pilot impairment	Nondedicated compartment	Loading/unloading sequences	Vandalism/sabotage
Inappropriate speed	Leaking valve	Carrier selection	Fog
Other vessel's pilot	Leaking fitting	Container specification	High sea/tide/wind/current
Contamination	Piping failure	Emergency response training	Problem initiated by other cargo
Transshipping	Navigational system failure	Maintenance	Depth less than charted
Lightering	Steering system failure	Inspection	Other vessel's failure
Heating and cooling	Relief device failure	Communication	Debris or blocks of ice
Inerting	Overpressure	Operator/crew training	Hurricane
Overfilling	Material defect	Language barriers	Lost container
Ballasting	Vacuum	Passing	
Misreading charts	Corrosion		
Local harbor pilot	Bad weld		
Conning	Engine and other fires and explosions		
Steering	Propulsion system failure		
Misunderstanding regulations	Cable failure		
	Communication system (hardware)		
	Fire protection		
	Tug failure		
	Leaking hull		

typically accounts for a small portion of total spill events (less than 10 percent of all spills).

The actual breakdown by failure mode is often very location specific. Collisions with other moving vessels may occur during passing maneuvers (overtakings), head on when passing in opposite directions (meetings), or at crossings such as river forks or dock areas. Collisions with moored vessels (allisions) typically occur during operations near dock areas such as docking maneuvers or when the waterway is narrow. Collisions with fixed objects are most often with bridges, or docks, and infrequently occur with a submerged object such as a rock or a sunken vessel. In general, collisions are the most dangerous types of failures, that is, most likely to result in a spill. This is due to potential for the barge hulls to be ruptured from the very large energy of impact involved in these events.

Groundings occur where there is inadequate water depth as compared to vessel draft, and do not often result in releases. This is due to the fact that the river beds of the most frequently traveled waterways are mostly soft mud or silts. However, there is always the potential for significant damage from rocks or debris, as well as physical distortion. Fires and explosions are fairly infrequent events. Most often fires are associated with the engine of the tug that is towing the barge string. As a result, fire damage to the actual barges is generally negligible or very limited. Material failures, as they are recorded by the Coast Guard, involve incidents in which mechanical or steering failures result in a disabled tow string. Less often, they refer to actual material failure of a barge in which the integrity of the vessel is compromised. Other types of failures include flooding of vessels, breakaway barges, and capsizing. These events are rare and cases of flooding and capsizing generally occur to the tug rather than the barges.

A breakdown of incidents by primary causes (for all waterborne incidents) is given in Table 2-10.

2.4.2. Parameters Influencing Accident Rates

Vessel speed in the waterway (including rivers) is perhaps the most important factor influencing barge accident rates. Higher vessel speeds result in greater stopping distances and turning radii. As a result, higher vessel speeds present the danger of higher accident rates in addition to an increased probability of release should a collision occur. The greater the vessel density along the waterway, the greater the potential for a collision with another vessel. The most dangerous areas in terms of collisions between vessels are in and adjacent to port areas and at crossings where there is significant traffic.

The width and depth of the waterway can also affect the potential for collisions and groundings. Narrow waterways present less room to ma-

TABLE 2-10
Primary Causes of Marine Incidents (from Commercial Vessel Casualty File, 1976–1980)[a]

Incident Type	*Percentage*
Poor weather	3
Equipment failure	8
Depth less than charted	2
Fault of personnel/other vessel	69
Other	17

[a]OTA (1986).

neuver the vessel and a smaller margin of error when maneuvers are executed. Shallow waterways may result in a higher incidence of groundings. The current in the waterway may present a hazard if it is sufficient to impede controlled maneuvers. This is more often a problem for down-river and cross-river traffic than for up-river traffic since the net speed of these vessels is greater than for a vessel traveling up-river. River speed is typically seasonal, as run-off from rains can significantly increase the rate of flow in a river. The current speeds in the Intercoastal Waterways tend to be minimal except under storm or flood conditions.

The number of bridges and other fixed objects such as docks that occur along a route determines the potential for a collision with a fixed object. The greater the number of fixed objects, the greater the overall risk.

2.4.3. Parameters Influencing Release Probabilities

The type of accident/failure mode is the most important factor when considering the release probability given an accident. Collisions are the most dangerous due to the energy imparted during the impact. Other factors include: hull construction and design, which affect the structural integrity of the vessel; position of the barge in the towed string of vessels, which affects the vulnerability of the vessel (vessels at the front of strings are more exposed to collisions and groundings, although adjacent barges can also collide); whether the vessel is being pushed or pulled (vessels being pulled are less exposed to collisions and groundings); and vessel speed due to the energy involved in collision events as described above.

2.4.4. Container Types

The design, inspection, and certain operational features of tank barges are regulated by the U.S. Coast Guard under Chapter 1 of Title 46 of the Code of Federal Regulations. Title 46 includes specifications and requirements for both the types of barge hulls and for the types of cargo containment systems incorporated into the different types of hull. Classification of barge hulls by Type (Types I, I-S, II, or III) relates to construction features associated with the barge hull's ability to resist cargo releases and to the vessel's stability in the event of marine casualties. Type III is the least stringent of the barge classification types, and can generally be satisfied with a single-hull design.

The types of cargo containment systems (Types I, II, and III) as prescribed in Subpart 153 of Title 46 are based on damage criteria from collisions and groundings, barge and tank dimensions, and on various criteria related to the location of the cargo tanks within the barge hull and their separation from various spaces and equipment. A Type I containment system must be within a Type I barge hull; a Type II containment system must be within a Type I or II barge hull; and a Type III containment system can be within a Type I, II, or III barge hull. Appendix A provides further details.

2.4.5 Accident Trends with Time

The available data are insufficient to establish trends in accident rates, due to the fact that very few incidents have occurred. However, vessel design and construction have improved over time as have communications systems. There are more requirements now for new barges to be double-hulled; double-hulled barges are less likely to rupture as a result of an impact. Thus, over the long term, accidents yielding releases should decline as these become a larger portion of the total fleet. Communications improvements should also help reduce accidents by warning tow operators of hazardous conditions and informing them of the position of other vessels and hazardous objects.

2.4.6. Confidence in Data

The best source of data is the U.S. Coast Guard's vessel casualty file for U.S. waterways, the Marine Casualty Information Reporting System (CASMAIN). These data are generally reliable for the accidents that are catalogued in the database. However, because many incidents go unreported, this database does not contain all incidents and therefore might represent a lower bound by which to establish accident rates. The criteria for reporting an incident in the CASMAIN database are subject to interpretation. By law, a marine incident is reportable if it involves a death, serious injury, structural damage greater than a specified value, loss of propulsion or steering, or any occurrence adversely affecting the seaworthiness of the vessel. Additionally, all groundings, spills, and releases are reportable. (See Transportation Data Source 4-32 in Chapter 4.)

Marine traffic statistics in the waterways and harbor systems of the United States are compiled annually by the U.S. Corps of Engineers (Corps of Engineers, 1986) and tabulated for both self-propelled and non–self-propelled vessels by draft and by direction (i.e., upbound and downbound in rivers, inbound and outbound in harbors). Self-propelled vessels are further broken down into passenger/dry cargo, tanker, or tow-boat vessels, while non–self-propelled are broken down into dry cargo or tanker vessels.

These statistics, which are published in the Waterborne Commerce of the United States (Corps of Engineers, 1986; Transportation Data Source 4-28), are compiled for segments along a particular waterway (between two key ports or connecting waterways, for example). Consequently, it is impossible to determine what percentage of these transits are full transits of the entire segment and what percentage are partial transits that begin and/or end within the defined segment. As a result, using these figures as denominators when calculating accident rates is nonconservative, as they are all presented as full transits. Some adjustments to these figures are often necessary. Adjustments can be made based on knowledge of the traffic patterns along the route. For example, the location of key waterway intersections, ports, or industrial facilities along the route might give an indication of the ratio of full versus partial transits along a segment. Reducing partial transits by the percentage of the segment not transited will provide a more accurate depiction of the actual traffic flow in terms of total miles traveled or equivalent total transists along the segment. Overall, these data are considered generally reliable and other sources of accident data tend to be incomplete and/or anecdotal.

2.4.7. Mode-Specific Issues

Mode-specific issues include releases in moving water, which makes spill containment difficult due to the currents and tides, and whether or not there is immediate access to and availability of proper equipment. These do not affect the chance of an accident or release occurring, however. The possibility of failing multiple compartments affects the probability of a total loss of contents. Single or multiple barge tows are important where the position of the barge in the tow string affects the probability of its being involved in a collision. Being inside a tow can offer up to a 50 percent reduction in chance of release.

One must also consider pushed versus pulled barges for differences in collision risk. Additionally, routes may be unavailable or accidents more likely at times due to currents, droughts, ice, etc. Natural disasters such as hurricanes, flash floods, etc. can influence both the chance of an accident and the availability of the route. In general, there is adequate warning of such events to limit the influence on accident rates.

Accident rates are often defined by segment rather than length. This is because the data are collected in a manner that makes analysis by segment easier and more meaningful (as discussed in 2.4.6). The accident rate per mile can be calculated, however, by dividing the accident rate per segment by the segment length.

2.4.8. Non–Accident-Initiated Releases

The majority of non–accident-initiated releases tend to be due to previously overfilling of the barge or knocking off small fittings. These tend to be small releases (less than a few barrels of material), especially compared to major losses of cargo spilled as the result of a ruptured or punctured tank. Reliable data on the number of these types of incidents are not available.

Scenarios that yield releases while underway, but where the cause is associated with loading errors such as overfilling or contamination, can be estimated using techniques such as fault tree analysis, and then combined with releases from collisions and other accidents.

Even though all spills are required to be reported to the Coast Guard, the Coast Guard database does not describe the circumstances behind the spills, and is also thought to be missing many small spills (OTA, 1986). The number of overfill or contamination induced incidents would also be quite low just as a result of the small population of cargoes vulnerable to such events.

2.4.9 General Calculation Procedure

The general procedure is to sum the product of the following items over all segments of the route:

- Accident probability per segment
- Probability of spill per accident (segment specific)

As for all modes, the result can be multiplied by the number of trips per year. The accident probability per segment is often obtained by taking the number of barges involved in accidents divided by the total number of barge movements, over a representative time period (typically 3-5 years) for a specific segment of waterway. These numbers may be subdivided by accident type. For example, assume that one wished to determine the accident rate for 25 transits per year over the route given in Table 2-11.

TABLE 2-11
Hypothetical Barge Data

	Segment 1	*Segment 2*
Barge movements (upstream and downstream previous 5 years)	111,244	75,760
Barge accidents (upstream and downstream previous 5 years)	97	62
Spills (upstream and downstream previous 5 years)	7	3
Movements of concern per year for your product	25	25

The accident rate is calculated in the following manner:

$$\textit{Segment 1:}\quad 97/111{,}244 = 8.7 \times 10^{-4} \text{ accidents/transit}$$

$$\textit{Segment 2:}\quad 62/75{,}760 = 8.2 \times 10^{-4} \text{ accidents/transit}$$

Assuming that one-half of the total number of tank barges moving in both directions are full and thus susceptible to spills, the conditional spill probability for loaded barges is:

$$\textit{Segment 1:}\quad 7/(97 \times 0.5) = 14.4 \text{ percent}$$

$$\textit{Segment 2:}\quad 3/(62 \times 0.5) = 9.5 \text{ percent}$$

Thus, the total rate is:

$$(25 \cdot 8.7 \times 10^{-4} \cdot 0.144) + (25 \cdot 8.2 \times 10^{-4} \cdot 0.095) = 5.1 \times 10^{-3} \text{ spills/yr}$$

If there was also a frequency of spills from non–accident-initiated causes of 1×10^{-4}/yr, this could be added to the accident initiated spills to get a total of 5.2×10^{-3} spills per year.

As seen above, the probability of spill per accident is calculated by taking the number of spills and dividing by one-half the number of accidents. It is assumed that roughly one-half of all barge movements are full and one-half are empty. Accident rates per mile are calculated in a similar fashion by dividing the total segment rates by the mileage associated with each segment. If there is more than one segment, all accidents and spills can be summed over the entire route and then divided by the number of transits and by the total mileage if a single accident rate is desired, or segment-specific results can be maintained. In some cases, accident rates and release probabilities may be calculated by accident type. However, the available data are often too sparse to make this meaningful.

2.5. Ocean-Going Vessel

For ocean-going vessels, typical operations may involve several port calls for both loading and unloading the various compartments on board. Vessel draft restrictions may also necessitate transshipping through shoreside facilities or lightering into barges or smaller tankers.

If general cargo or container vessels are used, there will almost certainly be a pre-designated schedule of port calls. This sometimes influences the choice of departure port or carrier to minimize the number of interim port calls prior to reaching the destination of interest. This both reduces the total chance of an accident in transiting the port, and the chance of dropping containers onto other containers during loading and unloading operations at the interim ports (see Section 2.6).

Table 2-9 presented the hazards of concern to marine operations and should be referred to for ocean-going vessels as well as barges. This information can be incorporated into various types of TRA studies—from qualitative to quantitative. The sections below describe how the hazards are categorized for purposes of developing accident rates and also the factors that influence accident rates. These factors can also be reviewed and used on a qualitative basis.

2.5.1. Failure Modes

Ocean-going vessel failure modes are generally the same as for barges and include the following types of accidents:

- collisions with other moving vessels,
- collisions with moored vessels (allisions),
- collisions with fixed objects (rammings),
- groundings,
- non–cargo-related fires and explosions, and
- material failure of the vessel or other equipment.

Based on historical experience such as reported in CASMAIN, collisions with other moving vessels are most likely to occur in a harbor or a port or at river or channel crossings where there is heavy traffic. Collisions at sea are far less likely. In fact, available data on tankers suggest that only about one quarter of all marine casualties occur in open seas (J. J. Henry, 1973). Collisions with moored vessels (allisions) typically occur during operations near dock areas such as docking maneuvers or when the waterway is narrow. Collisions with fixed objects are most often with bridges or docks and infrequently occur with a submerged object such as a rock or a sunken vessel. They can also involve offshore oil and gas platforms. In general, collisions are the most dangerous types of failures, i.e., most likely to result in a spill. This is due to potential for the vessel hulls to be ruptured from the very large energy of impact involved in these events.

Groundings occur in shallow water areas and vary in severity depending on whether the impact is with rock or a softer material such as mud or sand. Fires and explosions generally do not start as a result of a cargo release, although they may ultimately cause a release. Material failures refer to actual material failure of the hull in which the integrity of the vessel is compromised. Hull failure may also occur when multiple compartment tankers are loaded or unloaded unevenly. In these events, uneven loading over the length of the hull creates severe stress resulting in a failure.

Another potential for release of a hazardous material is failure of the pipework on the deck. This is of greatest concern during loading and unloading when the lines are full of material and under pressure, which would be addressed in a separate analysis. Other failures include flooding of vessels and capsizing. As was discussed for barges, the actual breakdown for each cause can vary significantly from harbor to harbor. However, Table 2-12 presents the combined results for a number of U.S. ports plus Vancouver, for the time period 1981 to 1989. The results are for port calls, which generally include transits both into and out of the port.

2.5.2. Parameters Influencing Accident Rates

Higher vessel speeds result in greater stopping distances and turning radii. As a result, higher vessel speeds present the danger of higher accident rates. The greater the vessel density of vessel traffic in the harbor, the greater the potential for a collision with another vessel. The most dangerous areas in terms of collisions between vessels are in and adjacent to port areas and at crossings where there is significant traffic. The width and depth of the harbor can affect the potential for collisions and groundings. Narrow harbors and channels present less room to maneuver the vessel and a smaller margin of error when maneuvers are executed. Shallow waterways may result in a higher incidence of groundings.

The current in the waterway may present a hazard if it is sufficient to impede controlled maneuvers. This is more often a problem for down-current and cross-current traffic than for up-river traffic since the net speed of these vessels is greater than for a vessel traveling up-current. The

TABLE 2-12
Casualty Rates in Ports (per 10,000 port calls)[a]

Casualty Type	*All Tankers[b]*	*Chemical Tankers*	*Tanker Barges*
Collisions	1.0	1.25	1.5
Groundings	4.0	4.80	1.0
Strikings (rammings)	7.0	8.75	7.5
Fires/explosions	2.0	2.20	0.5
Structural failures	0.5	0.53	1.5
Total	**14.5**	**17.5**	**12.0**

[a]Sandwell (1991)
[b]Includes crude, petroleum, chemical, and coastal tankers plus tanker barges.

number of bridges and other fixed objects such as docks that occur along a route determines the potential for a collision with a fixed object. The greater the number of fixed objects, the greater the overall risk. Many vessels and/or harbors require the use of local pilots and tugs. The effect of pilot training and experience on accident rates is difficult to quantify but it is reasonable to assume that well trained pilots with experience will commit fewer errors in judgment and react better in emergency situations than a novice pilot with little or poor quality training. Therefore, one can assume that the experienced pilots with good training will present a lower accident risk. Local pilots know the surroundings and traffic pattern in their waterways, but may be less familiar with vessel characteristics.

Weather conditions can influence vessel control as well as visibility. Heavy fog, storms, high winds and other extreme situations tend to increase the likelihood of accidents, but are generally accounted for in historical data. Their effects may be offset in some instances by reduced release probabilities due to lower travel speeds.

As vessel communications systems continue to improve, there is a greater chance that dangerous situations can be avoided. Therefore, ships and harbors with better communications systems are less likely to have accidents. Some of the busiest harbors have vessel traffic systems similar to those of airports, including radar and centralized traffic control. These ports need to have these systems in place due to the large number of vessels transiting the area and the potential for collisions. Because the high traffic harbors are at greater risk for collisions than those with less traffic, one can reasonably assume that harbors with large volumes of traffic that do not have vessel traffic systems in place are at greater risk than those with such systems. However, in less frequently visited ports, the absence of a vessel traffic system may not necessarily be a liability with regard to collisions between ships.

Vessel registry is a reflection of some of the design, construction, crewing, and operating standards for a given vessel, and can produce a variability in historical performance.

2.5.3. Parameters Influencing Release Probabilities

Table 2-13 lists overall historical spill probabilities for tankers as given in a number of data sources. The lower values given in the more recent studies are felt to be a reflection of adherence to more strict regulations.

The type of accident/failure mode along with hull construction are the most important factors that determine release probabilities. Collisions are the most likely cause of release due to the energy levels involved, as well as the shape of the striking bodies. Vessel speed is the most important factor influencing the severity of the damage, as higher speeds limit the

TABLE 2-13
Historical Conditional Spill Probabilities for Tankers

Source	*Marine Casualties with Spills*
OIW, 1974	24%
IMCO, 1978	30%
Porricelli & Keith, 1974	19%
Sandwell, 1991	16%
ECO, 1989	13%

potential to avoid accidents and increase the expected penetration depth in the event of a collision, ramming, etc.

The U.S. Coast Guard, in evaluating the effectiveness of double hull construction for tank barges, utilized collision casualty data for 1973–1978 and found that only two percent of the incidents involved penetration of both hull structures while 40 percent involved penetration of the outer hull (Office of Merchant Marine Safety, 1979). While reducing the likelihood of release from punctures, double-hulls can cause vertical stability and stress corrosion problems.

2.5.4. Container Types

Ocean-going vessels must generally meet both national and international design and construction standards. They are often in international service, hence worldwide data may be applicable in generating accident rates and release probabilities.

Ocean-going vessel types include:

- Single-hulled tankers
- Double-hulled tankers, double-bottomed tankers
- LPG/chemical tankers
- LNG tankers
- OBO ships (oil/bulk/ore)
- Container Ships
- Other than container ships (i.e., containers on noncontainer vessels)

The tankers are generally grouped as liquefied gas carriers, chemical tankers, and petroleum tankers. The liquefied gas carriers have specialized containment systems with insulation and high quality primary barrier materials, and tend to be very large. The chemical tankers have cylindrical or prismatic tanks, and may have many different products onboard at one time. Tankers can be roughly ten times larger than barges. Appendix A provides further details.

2.5.5. Accident Trends with Time

It is thought that there is a general downward trend in accident rates, as vessel design and construction has improved over time, along with communications systems. Together these serve as warnings of potentially hazardous conditions, increase vessel reaction time to avoid accidents, and provide protection from releases in the event of an accident. However, there is not a sufficient amount of data to demonstrate this at this time. As time goes on and more data are available, the rate of accidents is expected to decline. Increased use of double-hulled vessels is also expected to reduce the average likelihood of a release given an accident as a greater proportion of vessels become double-hulled, but may have a small effect on the total quantity spilled or on the likelihood of large releases.

2.5.6. Confidence in Data

Accident data are quite limited and are often anecdotal. The potential size of releases does help to bound the likelihood of large releases as these are widely publicized events. Given the nature of operations and vessels, worldwide data are often used if available. The limited population of vessels, particularly by type of vessel, restricts the amount of available data.

2.5.7. Mode-Specific Issues

Many releases are so far offshore, that public impacts are not expected. Thus, in many marine risk analyses, the ocean transit is not included in the calculations—limiting the data requirements. (The analysis of risk to an offshore oil or gas platform from passing vessels is an example of a specialized application of TRA techniques, and is not the focus of this section.) If environmental spills are in the scope of the TRA, then the ocean transit will be important. The need for pilot and/or tug assistance is unique to this mode of transport and needs to be considered when addressing the effectiveness of risk reduction measures. The general issues associated with releases in water are very similar to those for barges, but equipment availability can be an even greater issue. The variability from one harbor to another can also be much more significant than is seen in the other transport modes.

2.5.8. Non–Accident-Initiated Releases

The rate of non–accident-initiated releases is low, as are the associated rates of release. Anecdotal evidence suggests that delayed problems from overfilling may account for the majority of these with the bulk of the remaining incidents consisting of hose and pipe failures. Complete data for these types of releases are unavailable, and such releases are best handled in risk analyses of loading or unloading operations. The data may then be combined with the accident-initiated releases.

2.5.9. General Calculation Procedure

Accident rates for harbors are multiplied by the total number of harbor visits (waterborne commerce and accident data are available) and the likelihood of release. These can be port specific if enough data exist. Interim port calls need to be considered and can generally be obtained from ship's schedules. At sea accidents may yield cargo loss or crew injury, but they are too far from shore for public impacts. As such, they may not be included except in the near shore area if public impact is the study focus.

Thus, the basic formula for calculating the release frequency is:

$$\text{Release Frequency} = \text{Accident Rate} \times \text{Release Probability} \times \text{Number of Harbor Visits} \quad (2\text{-}5)$$

The accident rate may be broken down by type of accident, in which case the release probability will also be broken down. If there are non–accident-initiated releases, they can be added in as well. They will generally play a very small role due to the limited time in proximity to populated areas.

2.6. Intermodal Containers

Intermodal tanks are "portable," holding roughly 4000 to 6000 gallons, that are within a protective metal frame. While intermodal containers can be boxes, may carry solids, etc., when this book refers to an intermodal container it should generally be assumed to be an intermodal tank. Such containers may be called IMO tanks, ISO tanks, or ISO-tainers. The frame of these containers is affixed to a railcar or truck bed, or stacked with other containers on a vessel. The uses of intermodal containers fall into several categories:

- single mode use (typically rail or truck)—due either to container availability or a desire for the protective features offered

- intermodal land-based uses—to allow for transfers from rail to truck if a rail line does not exist at either the customer's or supplier's site
- intermodal marine uses—to allow shipment of products across oceans

This section focuses on the marine transport of intermodal containers, but discusses the need to combine rail and road frequency analyses as well if these modes are also utilized.

A typical intermodal operation involving marine transport is depicted in Figure 2-10. This figure presents only an overview—there can be a number of other steps along the way within each individual transport mode. Containers may also be taken off the truck or railcar before they are unloaded.

In addition to the hazards already listed in Tables 2-3, 2-6, and 2-9, there are several additional hazards unique to intermodal containers that should be considered in a TRA. These include:

- forklift punctures
- dropping a container from a forklift or top lifter or from a crane
- fires in temporary storage areas
- overheating if left sitting on a dock for an extended period
- loss of refrigeration enroute
- impact from other containers
- fastener failure
- crane failure
- mishandling of container

The specific handling operations for a given container's movement may suggest additional hazards.

2.6.1. Failure Modes

Intermodal container failures may occur due to damage caused by:

- dropping the container during loading onto or unloading from ship,
- dropping another container on the container of concern during loading and unloading operations while it is aboard the ship,
- ship accidents in the harbor,
- ship accidents at sea,
- staging operations accidents in the port facilities, and
- rail and truck accidents associated with moving a container to or from a port.

Loading or unloading the product from the container is not covered in this guideline.

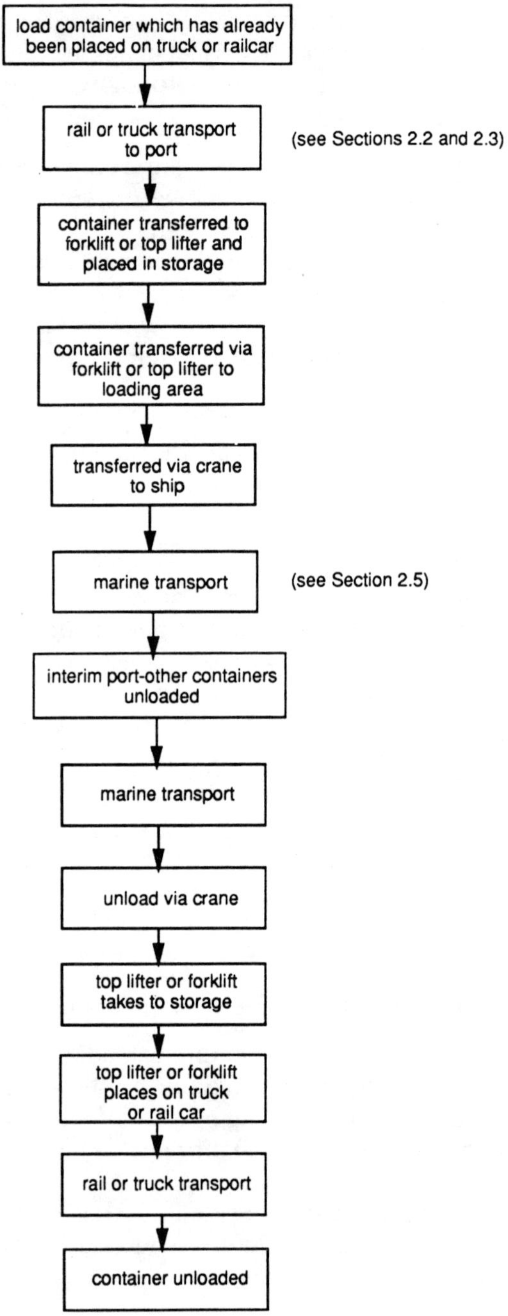

FIGURE 2-10. Typical Intermodal Transport Operation.

Other than the drop accidents and accidents associated with moving containers in storage areas, these events are similar to other truck, rail, or marine accidents. The drop accidents may be from cranes used to load and offload ships, or from top lifters used in storage areas at ports. In addition, other types of lifting devices may be used in such storage areas and may pose a risk of puncturing a container.

2.6.2. Parameters Influencing Accident Rates

For marine operations, the primary parameters of concern include the number of lifts and transfers, the number of interim offloadings, the number of harbor visits, the number of port visits, the type and condition of unloading crane being used, and the position of the container in the stack of shipboard cargo. Containers on top of stacks are exposed to overhead movements of other containers. When possible, HAZMAT cargoes are segregated to avoid such impacts both onboard a vessel and in port. By coordinating with the vessel operator, even more protection may be available.

Other parameters include pilot training and experience, shipboard navigation equipment, and crane operator training and experience. Rail and road operations are discussed in Sections 2.2 and 2.3.

2.6.3. Parameters Affecting Release Probabilities

The chance of release is influenced by the distance the container is dropped or the distance a second container is dropped before striking the intermodal container, as well as the basic design and construction of the container(s). The other parameters affecting releases due to accidents in ports and harbors are similar to those discussed in Section 2.5.3. Additionally, the position of the container in the stack is an important factor. Those closer to the edge of the ship are more susceptible to damage during impact events.

For operations at the port, the container design is key in determining whether or not there will be a release, as is the nature of the operations at the port. Whether a container is moved by forklift, top loader, cranes, etc., determines not only potential fall heights and orientations, but also the potential for puncture. Some port facilities are so large that there will be a number of lifts involved as the container is moved between different buffer and more permanent storage areas. For the more robust container designs, all of these interim handling activities may actually pose very little chance of a release.

2.6.4 Container Types

The primary intermodal container types are comprised of the following:

- ISO containers (liquid bulk)
- Single tank
- Double tank
- Insulated
- Refrigerated
- Box containers
  —20 foot length
  —40 foot length

These containers vary significantly in terms of their robustness. The frame structure around an ISO container not only increases its strength, but protects its fittings. Appendix A includes descriptions and sketches of these containers.

2.6.5. Accident Trends with Time

Very few data are available to establish accident rates, let alone trends, and generally must be gathered from individual ports. Crane design and operation has improved over time as well as vessel design, construction, and communications systems. However, there is insufficient record keeping to establish any patterns over time or even among facilities.

2.6.6. Confidence in Data

Hardly any data are available as few record keeping systems exist (although a few ports have offered anecdotal reports). However, the total number of lifts and transfers is generally well documented by the various authorities that manage and/or oversee port activities. Data for container accidents in transit are also very limited. Thus, this mode has unusually good denominator data, but limited numerator data. Anecdotal data calibrate well with the limited obtainable accident data and the results of engineering analyses.

2.6.7. Mode-Specific Issues

The primary issue for intermodal containers is the need to consider multiple modes of transport and to combine these in the overall risk analysis. The risks associated with truck and/or rail transport to and from a marine terminal must also be included with the marine risk. In addition, the nature of the risks in the temporary storage terminals is more controllable and amenable to mitigation strategies than other transportation risks. Whether a container lands on deck or in the water may influence material behavior upon release.

2.6.8. Non–Accident-Initiated Releases

Spontaneous failure of an intermodal container is very unlikely. Overheating and activating a relief device might be a problem for some materials, particularly if they are left on the dock for an extended period of time in the sun, which may be more likely than for other transport modes. Some containers used in European service will not have relief devices due to regulatory restrictions. The exact equipment design must therefore be reviewed carefully. Fittings may be subject to failure. Refrigerated containers may also have scenarios resulting from a failure of the refrigeration system due to either mechanical or human failures.

2.6.9. General Calculation Procedure

The overall release frequency is calculated by summing the following items over the entire route of concern.

Release Frequency
= Release probability during container movement × number of lifts
+ Release probability during loading/unloading × number of port calls
+ Release probability during harbor transit × number of port calls
+ Release probability during ocean transit (2-6)

where the release probability is the combination of the accident rate (probability) and conditional release probability. Any associated rail or truck accident releases must also be incorporated. The release during loading/unloading is for other containers moving overhead, which may be dropped and thus damage the intermodal container of concern.

For example, assume that we wish to examine the frequency of release associated with shipping a container from one port to another with two intermediate port visits and the following hypothetical probabilities for a total loss of the container contents:

Sample Data

2×10^{-6} per lift (for dropping the container)
1×10^{-7} per port call (for other containers dropped during loading/unloading)
6×10^{-7} per harbor transit (for accidents such as collisions)
6×10^{-7} per ocean voyage (for all accidents)

Then the total frequency of release is calculated by adding the following:

	Sample Results
Loading	2.0×10^{-6} per shipment
Ocean voyage	6.0×10^{-7} per shipment
Two intermediate port calls	2.0×10^{-7} per shipment
Four harbor transits	2.4×10^{-6} per shipment
Unloading	2.0×10^{-6} per shipment

This hypothetical set of data gives a total frequency or probability of 7.2×10^{-6} per shipment. Non–accident-initiated events (if any) could be added to this.

The chance of 1×10^{-7} per port call would be built up from the chance of dropping another container (say 2×10^{-6}), the number of other containers moved overhead per port call (say 10), and the chance of severely damaging the container (say one in 200) if a container were dropped on it.

2.7. Conducting a Frequency Analysis

As discussed in the various sections above, the basic calculation scheme given in Figure 2-1 actually has numerous permutations depending on the mode or modes of transportation of interest. Figure 2-11 gives just one example for rail, for one car.

If different segment accident rates are used, or if multiple cars are moved at a time, this sequence will have to be more detailed. If one car

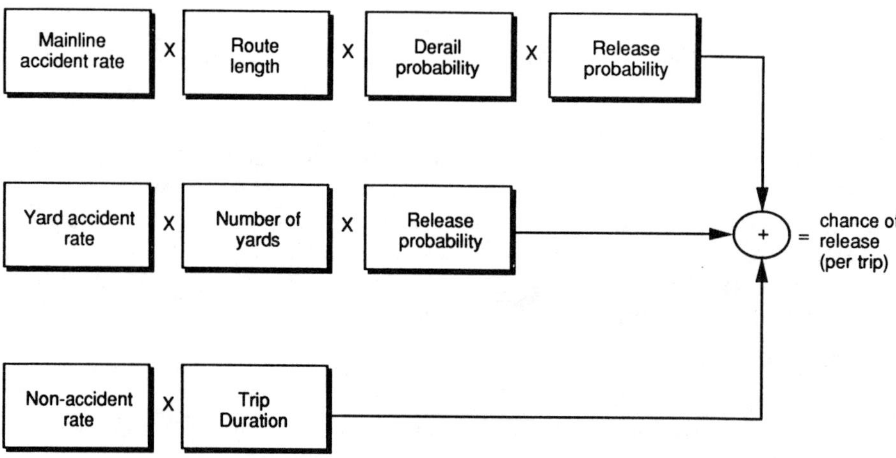

FIGURE 2-11. Example Rail Calculation Sequence.

is moved each week, the results should be scaled by a factor of 52 in order to determine annual risk.

The basic requirements for conducting a frequency analysis for a transportation movement are given in Table 2-14. The level of detail and scope of the study play a key role in determining just what data are needed. For instance, if the main goal is to determine the overall chance of a release, then an average accident rate may be adequate. If only one segment of a route has any population along it, then the accident rate for that segment may be most critical and historical data can be used to generate the accident rate. If various segments of a route are very different in terms of both roadway and population characteristics, then the route segments should be defined as whenever one of the parameters changes, and the accident rates determined accordingly.

Other general issues to be addressed prior to conducting a frequency analysis include whether to compute frequency per trip, per year, or for some specified number of HAZMAT trips. Frequencies per loaded trip are most useful when the total number of trips is not known, when route comparisons are being made, or if comparisons are being made to some standard that is available on a per trip basis. Frequencies per year allow more equitable comparisons among modes, since the total quantity moved per trip and therefore the total number of trips per year is usually

TABLE 2-14
Typical Data Requirements for TRA Frequency Analysis

Mandatory	*Optional*
Route length	Length by segment
Accident rate (average)	Accident rate by segment or cause
Release probability for specific container type	Further details on container: design-protective features, wall thickness, etc.
Number of trips per year	
Number of yard visits (rail)	Route characteristics such as type of road, track class, waterway type, etc.
Number of port calls (marine)	Nonaccident release data
Pipeline diameter (pipeline)	Route specific accident data
Type of release of concern—all vs. large vs. . . .	

different for different modes. The risks for a specified number of movements may be beneficial in analyzing a limited campaign—such as truck shipments for two years until a process is located at a site to eliminate the need for future transportation. Risks per year can also be most readily compared to background risk levels or risks at fixed facilities—or even added to them, if appropriate.

A second general issue relates to the consideration of strictly loaded containers or to unloaded containers as well. In general, the calculation sequence is exactly the same, and, if the return route is the same, the chance of a release can be doubled. However, the potential consequences of a "release" from an empty container are quite different. For most materials, the risks associated with releases from empty containers are negligible (or nonexistent if the container is purged) and are ignored. The specific concerns associated with your HAZMAT will guide you in determining whether or not to include the transport of empty containers. For problematic materials, it may really be non–accident-initiated releases that should be considered.

One of the most critical general issues is whether or not a quantitative analysis is even necessary. If the same mode of transportation is to be used and all one wishes to evaluate are various route alternatives, then a comparison of key features could be more appropriate.

If the route is clearly shorter, uses better roadways and has fewer people alongside it, no quantitative estimate of release frequency is going to add clarity to the ultimate decision. If there are major trade-offs among several routes, quantitative analyses may be helpful—but they may not need to be excessively detailed and further analyses of potential consequences and impacts may not be required.

Table 2-15 provides some guidance on when different levels of analysis may be required. Data, time, or resource restrictions may further modify the suggested level. The appropriate scope of the whole TRA was dis-

TABLE 2-15
Selecting the Appropriate Level of Frequency Analysis

Situation	*Level of Analysis*
One mode—simple route selection	Qualitative
One mode—complex route selection	Simple or detailed quantitative
Multiple modes—limited consequences	Simple quantitative
Multiple modes—extensive consequences	Detailed quantitative

cussed in Section 1.11 and clearly is of significance here as well. In general, however, cross-modal comparisons will need a consistent and greater level of detail in order to allow representative risk comparisons.

For instance, one mode may be far more likely than another to have a release, but far less likely to have a large release. Yet another mode may be just as likely to have a large release, but its routing may be largely through unpopulated areas. These sorts of trade-offs are not easily addressed qualitatively, and must draw on average accident rates, release probabilities, and distributions of release sizes at a minimum.

2.8 References

This section lists the references cited in this chapter. Additional references that may be of interest to the reader can be found in the topical bibliography and in the transportation data source sheets in Chapter 4.

Abkowitz, M., A. Eigher, S. Srinivasan. *Assessing the Release and Costs Associated with Truck Transport of Hazardous Wastes*. Prepared by Rensselaer Polytechnical Institute for the EPA, (1984).

ANSI/ASME B 31.8. *Gas Transmission and Distribution Piping Systems*. 1986 Edition.

Armstrong, J.H. *The Railroad—What It Is, What It Does: The Introduction to Railroading*. Simmons–Boardman Publishing; Omaha, NE, 1978.

Arthur D. Little, Inc. *Risk Assessment for Gas Liquids Transportation from Santa Barbara County*, prepared for Santa Barbara County Resource Management Department, July 1990.

British Standard 8010. *Code of Practice for Pipelines*, 1991.

CALTRANS; *Transportation of Hazardous Materials in California by Highway and Rail*, Report to the Legislature (1986).

Department of the Army, Corps of Engineers. "Waterborne Commerce of the United States, Calendar Year 1986." *Part 2—Waterways and Harbors—Gulf Coast, Mississippi River System and Antilles*. WRSC-WCUS-86-2.

ECO Engineering, Inc. *An Assessment of Tanker Transportation Systems in Cook Inlet and Prince William Sound*. Prepared for Alaska Oil Spill Commission; (1989).

Ervin, R.D., et al. *Future Configurations of Tank Vehicles Hauling Flammable Liquids in Michigan Highways*. Prepared by University of Michigan for Michigan Department of State Highway and Transportation, PB81-143281, (1980).

European Gas Pipeline Incident Data Group. "Gas Pipeline Incidents: A Report of the European Gas Pipeline Incident Data Group." *Pipes and Pipelines International*, 1970-1988.

Geffen, GA et al. *Assessment of the Risk of Transporting Propane by Truck and Train*. Battelle Memorial Institute/Pacific Northwest Labs. PNL-3308. Available from NTIS, Contract No: AC06-76RL01830, (1980).

Harwood, D., E. Russell. "Procedures for the Development of Truck Accident Rates and Release Probabilities for Use in Hazmat Routing Analysis." Institute for Risk Research;

Waterloo, Ontario. *The International Consensus Conference on the Risks of Transporting Dangerous Goods*, April 6-8, 1992; (1993).

IMCO. *Tanker Casualties Report*. E. London. No. 78.16, (1978).

Institution of Gas Engineers. *Steel Pipelines for High Pressure Gas Transmission*. IGE/TD/1, Edition 2, 1984.

Insurance Institute for Highway Safety. *Big Trucks and Highway Safety*, (1985).

J.J. Henry Co. *An Analysis of Oil Outflows Due to Tanker Accidents*. Prepared for U.S. Coast Guard, AD-780-315, (1973).

Jones, D. J.; et al. *An Analysis of Reportable Incidents for Natural Gas Transmission and Gathering Lines—1970 through June 1984*. American Gas Association; NG-18, Report No. 158, (1986).

Oceanographic Institute of Washington. *Risk Analysis of the Oil Transportation System*. Seattle, Washington. NTIS #OIW-OCW-7201, (1974).

Office of Technology Assessment. *Transportation of Hazardous Materials*, OTA-SET-304; (1986).

Porricelli, J., V. Keith. *Tankers and the US Energy Situation: An Economic and Environmental Analysis*. Marine Technology, October 1974.

Saccomano, F.F., A.Y.W. Chan. *Economic Evaluation of Routing Strategies for Hazardous Road Shipment*. Transportation Research Record No. 1020, (1985).

Sandwell. *A Risk Analysis of Tanker Traffic Movements within the Port of Vancouver*. Prepared for the Vancouver Port Corporation, (1991).

Title 49, Code of Federal Regulations, Part 193. *Liquefied Natural Gas Facilities: Federal Safety Standards*.

Townsend, N.A. and G.D. Fearnehough. *Control of Risk from UK Gas Transmission Pipelines*. Presented at the 7th Symposium on Line Pipe Research, AGA, Houston, October 1986.

U.S. Coast Guard. *Marine Casualty Information Reporting System* (CASMAIN), 1981-1991, (1992).

U.S. Coast Guard. Office of Merchant Marine Safety. *Draft Regulatory Analysis and Environmental Impact Statement for Design Standards for New Tank Barges and Regulatory Action for Existing Tank Barges to Reduce Oil Pollution Due to Accidental Hull Damage*. Documents #75-083 and #75-083a, May 1979.

U.S. Federal Railroad Administration. *Railroad Accident/Incident Database*, 1975-present, FRA Office of Safety, (1992).

3

Special Topics

There are a number of concerns that are either unique to or of special importance to TRA. Using the framework set out initially in Figure 1-1, Figure 3-1 identifies the special topics addressed in this chapter and their relationships to the steps of a TRA. These topics generally support quantitative TRAs but are also partially applicable to qualitative TRAs.

The selected topics covered in this chapter are:

- *Selection of Release Scenarios*—this step is important to most TRAs and is critical to quantitative TRAs as it is the fundamental process for selecting both incidents and incident outcomes.
- *Population Densities*—this topic addresses the treatment of population in the TRA and is therefore fundamental to all but the simplest qualitative TRAs.
- *Route Segmentation*—this process helps to set and control the number of unique incident outcome cases that are to be considered and also influences the extent of frequency and consequence analyses that must be conducted.
- *Ignition Algorithms*—for flammable releases, semiquantitative and quantitative TRAs need some approach for handling ignition of the releases, making this step vital.
- *Treatment of Marshaling Yards*—this topic is primarily of concern for rail movements, although equivalent facilities also exist in truck and barge operations. The need for consideration is dependent on the presence of such a facility and its usage level.
- *Transloading Operations*—this topic is also only pertinent for operations where transferring from one transport container to another takes place.
- *Unique Consequence Models*—these are an issue in many, but certainly not all studies. They address the need for modeling un-

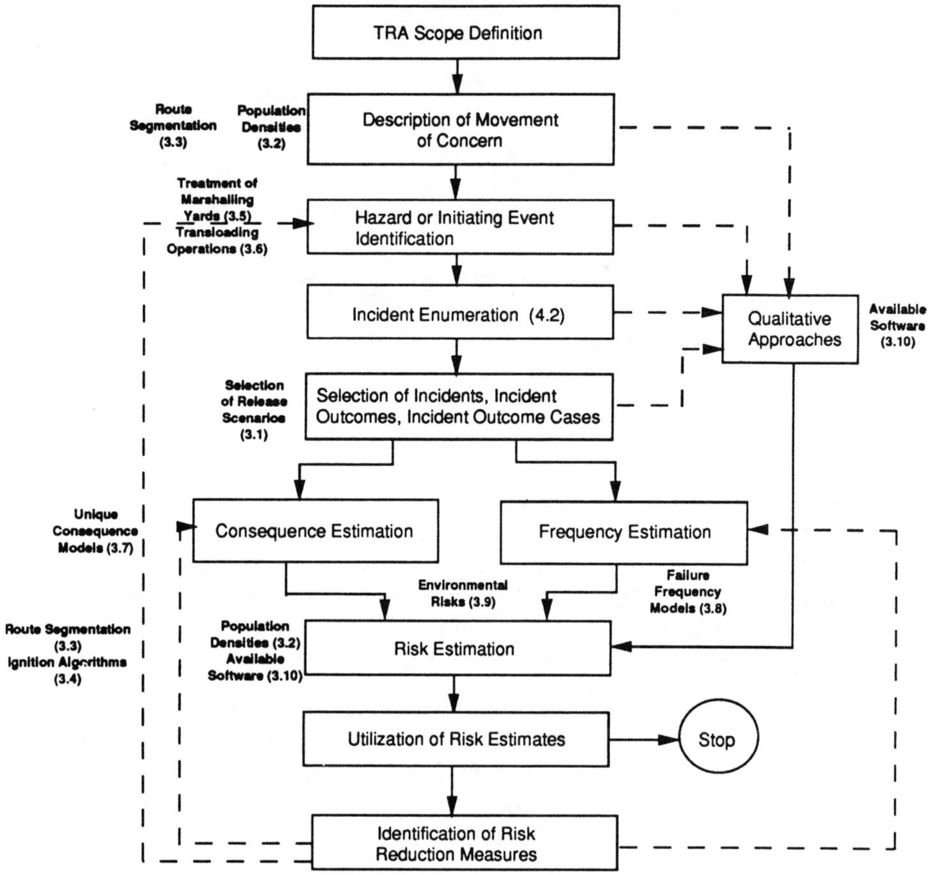

FIGURE 3-1. Roles of Special Topics.

confined spills, spills on water, BLEVEs of the same container which fuels the fire, pipeline cratering, etc.

- *Failure Frequency Models*—these are additional tools for calculating accident rates and release probabilities in situations where the strict use of historical data is considered inadequate or inappropriate.
- *Environmental Risks*—in addition to public safety risks, more and more TRAs are starting to address environmental risk. A brief overview of the issue is presented here, focussing on how pieces of a public safety TRA can be used in an environmental TRA.
- *Available Software*—there are a number of TRA specific tools available in addition to the consequence models and frequency assessment tools also used in CPQRA. Some available TRA software is described.

3.1. Selection of Release Scenarios

Many reports of transportation releases are inadequate for risk assessment purposes. While they may indicate the total volume lost, they rarely indicate the rate of release or the quantity in the container from which a release occurred. One does not know if all the material came from one compartment or if several compartments were involved. The size of a hole, tear, or puncture is also rarely specified—except in some pipeline databases. Thus, guidance is needed and is provided below in selecting representative release scenarios that are both reasonable and consistent with the limited data available. The actual frequency data are described in Chapter 4.

In addition, available spill data allow only the release quantity to be estimated. This means that the orifice size and/or release rate need to be back-calculated based on the ultimate spill size. If the release duration is assumed, then the flow rate and orifice size can be calculated. The degree of back-calculation necessary is determined by the input required for the particular model being used. If the flow rate and duration of the release are all that are needed for the model, then no further calculations are required. If the orifice size is important, and flow rate and duration have been specified, it may be necessary to estimate an orifice size, calculate the associated flow rate and then repeat this procedure until the desired flow rate or duration is found.

Pipeline data generally give leak and rupture failure rates, and may specify the size of the leak.The rupture is often assumed to equal the cross-sectional area, and significant leaks are commonly taken as 1- to 2-inch diameter holes, as these match the failure data collection scheme. Such holes may represent equivalent areas for extended leaks along a weld or may represent small punctures or the breakage of fittings. The other modes need more consideration and are discussed in the following sections.

3.1.1. Truck Data

The Hazardous Material Information System (HMIS) of the U.S. Department of Transportation has maintained a database on the frequency and size of reported releases of LPG. For the 11-year period of 1976–1986, 70 known releases of LPG were reported in highway accidents, or about 6.4 per year. Of these releases, 34 were 1000 gallons or less, while 36 were larger than 1000 gallons (Wilson Hill, 1987).

During the 11-year period 1976 through 1986, the HMIS has reported that 1154 releases of gasoline occurred on the highway due to accidents, or an average of 105 per year. The distribution of these releases by size indicate that small releases (1–1000 gallons) account for 35 percent of all spills; and large releases (>1000 gallons) account for 65 percent of all spills.The relatively low number of small releases is likely the result of underreporting of such incidents.

Another source of spill size data is an analysis of LPG releases which occurred between 1971 and early 1981 (Croce, 1982). These releases were reported to DOT and the Material Transportation Bureau and include vehicle accidents, equipment failures, human errors and loading/unloading incidents. For a subset of 23 vehicle accident-induced releases the breakdown by spill size indicated in Table 3-1 was obtained.

The bimodal nature of these results has been observed for a number of modes of transportation. It should also be noted that most of the large spills were in excess of 80 percent of the capacity—10 of the 13 largest spills. While limited, this data set indicates that large spills are more likely than small ones. This again may be the result of underreporting small spills.

Several studies, by the Office of Technology Assessment (OTA, 1986), by Quantalytics, Inc. (Swoveland, 1986), and by Midwest Research Institute (Harwood, 1987) have raised concerns about the underreporting biases of the HMIS database and of other databases as well. The database is assembled by HMIS from voluntary reports of truck incidents by the interstate motor carrier firms. Intrastate carriers are exempt from reporting.

The degree of underreporting is, of course, uncertain. The MRI report (Harwood, 1987) cited several comments on underreporting found in other sources:

- estimates by OTA suggest that the underreporting is substantial;
- a DOT source estimated that 20 percent of all accidents are reported; and
- a comparison of HMIS data and a hazardous spill database developed by the Bureau of Motor Carrier Safety (BMCS) of the Federal Highway Administration indicate that about one-half of the spill accidents in each database is missing from the other database.

TABLE 3-1

Distribution of LPG Spill Sizes in Vehicle Accidents (Croce, 1982)

	Number of Spills with Indicated Capacity Lost		
	0–10%	*11–60%*	*61–100%*
Trucks <7000 gal	3	2	7
Trucks ≥7000 gal	3	2	6

Quantalytics (Swoveland, 1986), in their report, assume that there is an 80 percent response rate in the HMIS database for more serious accidents involving external puncture or external pressure/heat, and a 40 percent response rate for all other serious accidents. If these corrections are applied to the reported HMIS data, the LPG spills would be distributed into 65 percent small and 35 percent large. (The 34 small incidents are assumed to be 40 percent of the actual incidents of this size—or there were 85 such events. The 36 large events are 80 percent of the actual incidents of this size—or there were 45 such events. Of the newly derived total of 130 events, 65 percent are thus small and 35 percent are large.)

It was also noted by MRI (Harwood, 1987) that the underreporting of small spills of some hazardous materials may be the result of a 1981 change in the incident reporting requirements. According to this change, small-quantity spills of electric battery acid and paint no longer needed to be reported. Because of this change, it is speculated that some trucking firms believed mistakenly that small-quantity spills of other materials were also exempt from reporting. To the degree that this is true, it is likely that small releases of LPG, either in vapor form or in liquid form which subsequently vaporizes quickly and without incident, would be less likely to be reported than small releases of gasoline which would vaporize slowly and which can cause pavement damage at a minimum. Such very small releases generally do not contribute significantly (if at all) to TRA risk levels. The absence of very small releases is of greatest concern when the available data are used for another material which poses significantly greater hazards such that even small releases are a risk.

3.1.2. Marine Data

The most extensive marine data are for oil spills, and the historical record has demonstrated that while most spills are very small, most of the oil spilled each year is from a handful of large spills. Despite the size of many oil tankers, from 1982 to 1989 97.8 percent of oil spills were less than 10,000 gallons. Table 3-2 presents some data from two recent years for tankships and tank barges (NRC, 1991). In this report *minor* is defined as less than 10,000 gallons spilled with no special environmental threat, *medium* as 10,000 to 100,000 gallons, and *major* as over 100,000 gallons. Clearly, for some materials, a significant hazard can be posed by a minor or medium spill. No further discussion of smaller spills was included in this report, however.

Impact-type marine accidents of sufficient severity to cause a spill generally result in damage to localized areas of the barge or tanker structure, such that the credible spill size would be limited to a maximum of the contents carried in one cargo tank. Some rare events, such as a

TABLE 3-2
Distribution of Oil Spill Size in the United States)(NRC, 1991)

	Number of Spills in Indicated Size Ranges		
	Minor spills	*Medium spills*	*Major spills*
Tank ships			
1988	260	4	1
1989	259	3	3
Tank barges			
1988	578	8	5
1989	629	8	2

high-energy collision in deep water, or a foundering in deep water, could result in a complete loss of the contents of the vessel.

Some impact-type casualties, such as collisions at or above the water line, may result in spills much smaller than the total contents of the breached tank. Other types of casualties, such as material or structural connection failures, or breaks in piping, valves, fittings, etc., may result in relatively small release quantities. Most of these small releases are only a few gallons in size but the sizes vary and can be up to as much as 10 percent of the total tank contents. It should be kept in mind that this total quantity might not seem to be minor or "small," but the rate of release may be very low while the duration may be sufficiently long to produce this outcome.

Many very small releases are experienced in barge and tanker operations; these are generally not related to any sort of accident or casualty. Such releases are often not addressed, particularly because they are too small to pose risks to the public.

3.1.3. Rail Data

Aalyses of rail car spills have broken spill sizes down into spills of:

- 1 to 100 gallons
- 101 to 1000 gallons
- 1001 to 10,000 gallons
- greater than 10,000 gallons

As with many modes of transportation most spills are small, but the distribution is somewhat bimodal. That is, spills are either very small or relatively large. A detailed review of the set of smaller spills found that they were clustered at the lowest end of the range, spilling less than 1 percent of the contents. Even for this subset of small spills, they were clustered again at the lowest end.

One recent report (Raj, 1993) evaluated tank car punctures to develop the likelihoods of various hole sizes by car type, based on data from 1965 to 1988. These data include accidents which occurred both before and after the shelf coupler requirements. Table 3-3 summarizes the distribution of hole sizes for various car types.

3.1.4. Summary

Given the bimodal split in spill sizes and the general size of most transportation containers, there are several suggested approaches for selecting spill sizes. The suggestions for pipelines were given in Section 3.1.

For screening purposes, assuming a release of the total contents of a single container allows differences between modes to be identified and has some credibility. Opportunities for isolation and containment or prevention of the release of contents are more limited in transportation accidents. To simplify the process, an instantaneous loss can be assumed. If this is not considered sufficiently realistic, a short duration (1 to 3 minutes) loss of the total contents can be used. However, if a simplified consequence model cannot properly handle such finite duration releases, the nature of the hazard should be considered to determine if a continuous or instantaneous release assumption is more appropriate.

TABLE 3-3
Distribution of Tank Car Puncture Sizes (Raj, 1993)

	DOT Car Type				
Hole Area (10^{-4} m^2)	*111 noninsulated*	*111 insulated*	*105*	*112/114*	*112/114-105 (S,J,T)*
3	2.3%	3.1%	10.0%	9.4%	9.1%
10	2.6%	2.6%	2.0%	4.1%	0%
18	2.8%	1.1%	6.0%	5.3%	4.5%
26	1.2%	2.4%	8.0%	4.7%	4.5%
81	23.5%	25.3%	24.7%	14.8%	30.3%
270	3.7%	2.6%	2.0%	4.7%	4.5%
780	9.9%	5.8%	8.0%	20.5%	9.1%
860	17.5%	16.3%	20.7%	8.1%	25.8%
2500	36.4%	40.8%	18.6%	28.4%	12.2%

For more comprehensive TRAs, there is an infinite range of possible spill scenarios. To keep the analysis manageable, one can start with two basic release scenarios. One is the instantaneous or virtually instantaneous total loss of a container's contents. The second is a much smaller release. Based on the data given in Sections 3.1.1 to 3.1.3, it is suggested that this scenario be considered equal to the loss of 10 percent of a single tank's contents. From the data given in the sections above, it can be recognized that this is a reasonably conservative representation of the smallest spill sizes while still being credible. It also allows container size variations to be considered, which a specified release size would not. Losing such a quantity is obviously not something that happens instantaneously. Since most small continuous releases will achieve steady-state dispersion distances within roughly 10 minutes, this spill can be modeled as a 1 percent loss per minute for 10 minutes. While this rate might actually continue beyond 10 minutes, the additional duration of a small spill is not likely to change the predicted hazard zones, except for toxic dose response relationships, where a longer duration will increase the area receiving a critical dose.

If it is necessary or desired to get a more detailed estimate of the risks, one can reduce the size of the 10 percent spill and add an intermediate spill size or add a smaller spill scenario to the two already considered. For example, in the case of the rail spill sizes discussed above, most rail cars are 20,000 to 30,000 gallons. A 10 percent spill is thus 2000–3000 gallons. This falls in the next to largest category as shown in Table 3-3. In this case, the 10 percent spill might be modeled as 300 gpm for 10 minutes and left in the analysis. The next size smaller spill (101-1000 gal) could then be added as well. A mid-range spill size of 30 gpm for 10 minutes might be assumed for the small spill. Another approach would use consequence modeling to determine the minimum spill size which poses public safety risks.This size would then need to be matched to the available release probability data.

Another means of increasing the level of detail and thoroughness is to add a situation involving a release from more than one compartment or tank. This might address spills from multiple rail cars—in which case the data from Table 3-3 could be quite helpful in generating the spill size combinations—or adjacent tanks or compartments on a barge or tanker. Tandem trailers also pose some potential for multiple tank spills. Generally such events are much less likely than a release from one container, plus a small release from a second container might be lost in terms of its contribution if the first release is large. Thus it is not uncommon to only consider integer values of multiple container losses. Whether releases are simultaneous or sequential also needs to be taken into account.

Exceptions to the above—in terms of relative likelihood—include thin-walled tankers where a collision might easily penetrate two or more compartments, and interconnected tanks. In the latter case, the release would continue for a longer period of time, but the orifice size and therefore rate of release would not be affected.

In the case where compartments contain different materials, it will be necessary to understand how the materials will interact. If there is no reactivity concern, the modeling might assume that all the released material is the same, and use the material with the greatest potential hazard. If there are many different rail cars or drums potentially involved, another approach is to use the material which represents the greatest proportion of the cargo.

Non–accident-initiated release scenarios such as valve leaks and fitting breaks can be selected as in *Guidelines for Chemical Process Quantitative Risk Analysis* (CPQRA—CCPS, 1989), and such data sources as those listed in *Guidelines for Process Equipment Reliability Data* (CCPS, 1989) can be used for quantification.

3.2. Population Densities

It is necessary to know the population exposure along the route in order to estimate the consequences and the risk resulting from a transportation incident. If an average individual risk estimate (see discussion in Section 5.2.1) is all that is desired, extensive population data are not required. However, it is still necessary to determine the approximate number of people whose individual risk is being estimated.

The exposed population is often defined using a population density. Population densities are an important part of a TRA for several reasons. The most notable is that the density is typically used to determine the number of people affected by a given incident with a specific hazard area. Population data are available in different forms from several sources. Some examples of sources of population data along a specified route are census reports, detailed maps, zoning data, aerial or satellite photographs, videotapes of the route, or actual inspection of the route by the analyst conducting the TRA (i.e., traversing the route oneself). Section 4.4.1 discusses obtaining these data in greater detail.

In the absence of specific population data, default categories can be used. Table 3-4 gives one set of possible values. These values are based on detailed examinations of census data. For example, the average population density outside of standard metropolitan statistical areas is approximately 20 people per square mile (U.S. Department of Commerce, 1984). The same reference gives 2900 people per square mile for those inside

TABLE 3-4
Representative Default Population Densities

Designation	*Density*	*Description*
Remote	20 people/sq mile	Nonmetropolitan area with scattered housing; farms
Rural	100 people/sq mile	Small village or town; recreation areas
Suburban	1000 people/sq mile	Typical suburbs; mixed use areas
Urban	3000 people/sq mile	Small city; densely populated suburbs; congested commercial areas
Extremely High	10,000 people/sq mile	Very dense city area

central cities. Other categories or representative densities can be assigned based on the actual data for specific routes. In cases where densities are being assigned based on visual inspections or with the aid of maps or aerial photographs, the categories should be defined in terms of commercial and industrial development as well. For example, "suburban" might be expanded to include mixed commercial/residential areas or moderately dense industrial areas. In general, it is a good idea to verify the validity of visual assignments to categories by spot checking against census data. (See Section 4.4.1 for a discussion of population data sources.)

The population density can be averaged over the whole route or the route can be subdivided into any number of segments with a separate population density for each individual segment. (See Section 3.3 for further discussion of route segmentation.) The route is defined as a corridor, and its width is often called a bandwidth. For this application, the bandwidth is twice the maximum downwind hazard distance for the material released (i.e., the maximum downwind distance on either side of the route). This defines the area and therefore the population over which the risks are distributed and is illustrated in Figure 3-2 (page 122). An example of this is the maximum downward dispersion distance for a toxic cloud under Stability Class F conditions. When looking at the population distribution along a route, the analyst must consider the full bandwidth of concern (the bandwidth associated with the worst case consequence) and how the population density varies over this distance (i.e., is the population clustered right up against the edge of the route or is it spread out for some distance from the route?). If there is a river or forest or other unpopulated area to one side of the route, the density should be obtained for the populated side only and an adjustment for the reduced number of wind directions of concern can be made in the risk estimation process.

However, this does not mean that areas along riverfronts or forests should be completely ignored. There may also be environmental concerns that need to be addressed. (See Section 3.9 for a further discussion on environmental risks for TRA.)

In regards to transportation of hazardous materials by any mode, but especially by barge, buffer distances may need to be considered. Buffer distances are areas where no population exists, such as in a pipeline right-of-way or a green belt along a freeway. For example, a buffer distance should be evaluated for the cases where a barge is involved in an incident and a hazardous material is released in the middle of a waterway, on the side of the waterway opposite the population of concern, and on the near side of the waterway. The result of an incident like this would be that the immediate area around the vessel is uninhabited and the released material would have to disperse some varying distance downwind before reaching any of the population of concern.

Generally one considers all three of these cases and then distributes the release frequency among them. Alternatively, one can just consider the worst case, which would be the shore nearest the population. Given that groundings generally do not occur in the middle of a body of water and that after a collision vessels may drift out of control, the two shore line cases might be judged to be more likely than the case for the middle of the body of water. For very short routes, or along pipelines, it may be practical to use actual population distributions as obtained from census data or other data collection activities, instead of using more general population densities.

Special attention should be paid to temporal variations in population (e.g., rush hour versus rest of the day for highway transportation) and day/night variations in population that result from different land use (e.g., industrial versus residential). In addition, it may be desirable to consider on-road and off-road populations separately as has been done by the U.K. Health and Safety Commission (Advisory Committee, 1991). Purdy (1993) has developed a scheme that separates on-road populations into two groups: motorists built-up behind an accident giving a very high population density, and other side motorists with a higher than normal density due to slow traffic resulting from curiousity about the accident. Assumptions were made about vehicle lengths and types to derive estimates for average population densities in each category. These estimates also consider the separation between the on-road and off-road populations, and sections of ribbon development (e.g., shopping centers) occurring in narrow strips alongside a particular road.

Another approach may be to assume that on-road populations mirror the land development along the road (e.g., road segments with dense areas of development alongside will also have high density on-road populations). Alternatively, the on-road population can be separated from the off-road population by assuming that the on-road population has no ability to escape. A good example of this would be the traffic on a freeway backed up behind an accident. The need for these splits into population density cases will be determined by the depth of study and the amount of variability posed between the cases—such as day and night populations. If a movement is conducted only at a specified time of day or on certain days of the week, it may be possible to consider this when gathering the population data.

Sensitive populations also need to be discussed. It cannot be assumed that all humans will have the exact same response to equal exposures of a hazardous material. Obviously, some portions of a given population will be more sensitive than others. Areas where this is likely to be a concern are segments of routes that are close to facilities such as hospitals, nursing homes and schools. These populations represent both sensitive receptors (individuals who may show a particular sensitivity to an exposure to a hazardous material as a result of their size, metabolism, physical condition, etc.) and individuals who cannot easily be evacuated. Information about the location of sensitive populations can be obtained from local Fire Departments or Local Emergency Planning Committees (LEPCs). Other populations where rapid evacuation may not be possible include prisons, sports stadiums, auditoriums, etc. Depending on their frequency of use, these sensitive populations may or may not warrant special attention.

Finally, when discussing population data and potential consequences to people resulting from a hazardous materials transportation incident, individuals such as drivers, pilots, train engineers and crews are usually excluded from the analysis. This results from both the nature of the risk borne by this group (being an occupational risk it is considered more voluntary and more subject to their control) and because these individuals could be affected by the event itself, not just the HAZMAT release. In other words, a member of a train crew could be fatally injured in a derailment whether or not a hazardous material was released. In order to keep this individual's death from skewing the results it would either be excluded from the public risk or presented separately to illustrate its contribution to the overall risk level, and to avoid mixing public and employee risks.

3.3. Route Segmentation

There are several variables that are considered when analyzing a route (population density, accident rate, meteorological conditions, type of roadway or track or waterway, etc.). Depending on the level of detail of the data available for the TRA, these variables are very rarely constant over the whole route. An example is population density. Most moderate to long routes travel through areas of very high density (urban and suburban areas) and areas of low density (rural and remote areas). In these cases, the changes in population density along the route generally need to be reflected in the TRA, in order to get a more accurate estimate of the risk. Segments represent a compromise between one set of average conditions and continuously changing conditions.

The basic assumption for a route segment is that all factors are constant. When one or more factors change, a new segment needs to be defined. Small or moderate differences between route segments can have a significant effect on the final risk estimate. Segments can be defined by changes in such variables as population density, accident rate, railroad track class, type of highway or road, meteorological conditions, buffer distance, etc. Should any variable change, a new segment would be specified. As the number of segments analyzed increases, the risk estimates can become more accurate and better reflect the actual risks present along the given route. However, the number of incident outcome cases requiring analysis also increases.

Generally, the accident scenarios do not change along the route except for special locations such as railyards and harbors. Likewise, the consequence modeling results will remain the same unless there is a major change in relative humidity, atmospheric temperature, terrain or wind speed along the route. Thus, the main parameters that will influence the definition of route segments are the population density and the accident/release frequency. This then accounts for the differences in the magnitude of the consequence associated with a release in the middle of a large urban area versus a release in a sparsely populated rural area, or a change in road type or rail track class that causes a portion of the route to have a different accident rate than the previous segment.

Figure 3-2 gives an example of a route segmented by population density. The different population densities are illustrated by different patterns. In addition, a bandwidth is superimposed over the route. If a particular population density had been limited to only one side of the route, the segment could have been artificially separated into two subsegments each with a length proportional to the time the wind blows to those directions. In areas where the bandwidth is large enough to encompass several population densities, there are several options. One option is to create a new average population density for the whole area. Another alternative is to determine the relative likelihood of the different potential hazard distances. If most of the time the hazard is significantly less than the distance associated with the worst case, then it may be appropriate to only use the density associated with the nearer zone and apply it throughout the bandwidth. This may be slightly conservative or slightly non-con-

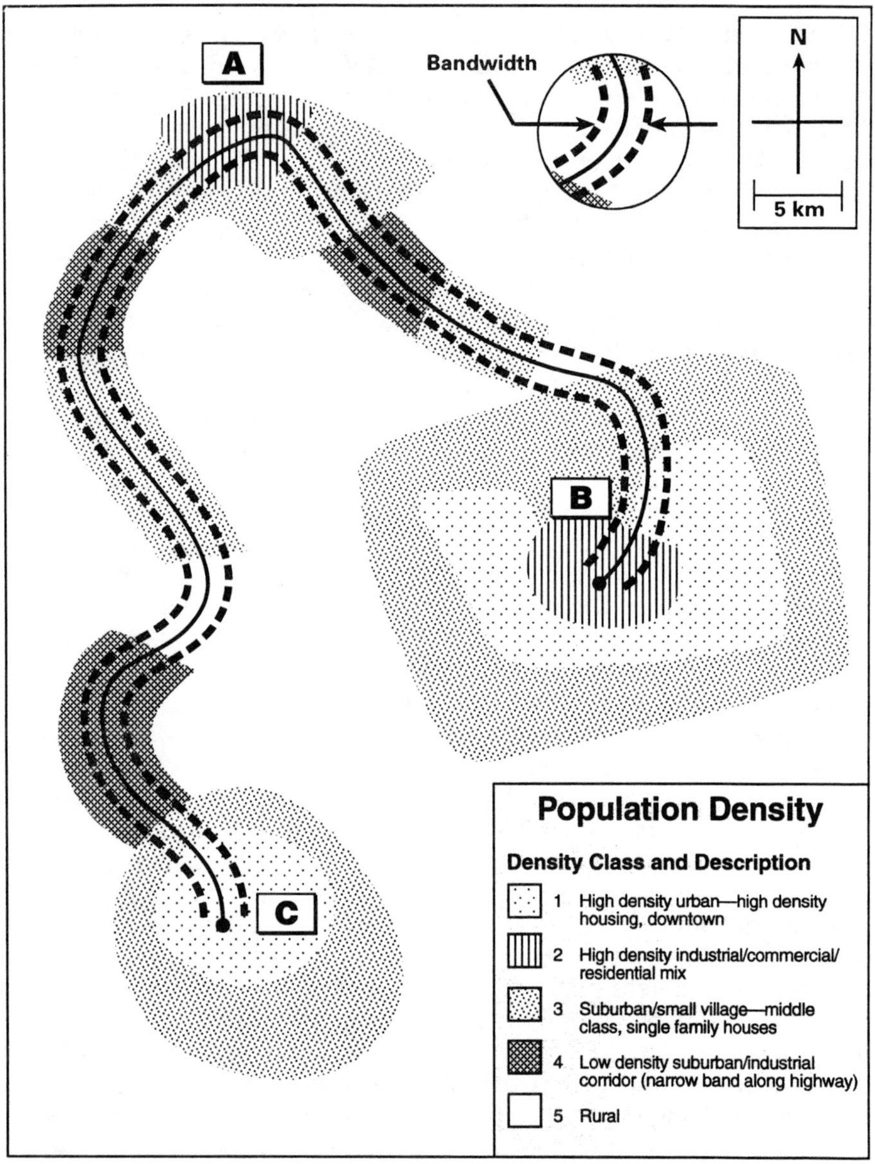

FIGURE 3-2. Sample Route Segmentation

servative depending on whether the density was higher along the route or away from the route. In the example shown in Figure 3-2, variations in other parameters such as accident rates could cause the segments depicted to be further subdivided.

3.4. Ignition Algorithms

The potential for ignition of a flammable material should be considered. The timing of ignition, immediately or after a flammable cloud has formed and started to disperse, must also be addressed. Immediate ignition probabilities are generally a function of material properties, the size of the release, and the nature of the incident, and are assigned considering these parameters, historical data, and engineering judgment. Immediate ignition may produce a pool fire, flame jet, fireball, etc. depending on the nature of the release.

Generalized approaches are needed for delayed ignition because the exact location of the release, and therefore the locations of the ignition sources which may be encountered, cannot be determined in advance and specified in the analysis. This is a fundamental difference between TRA and CPQRA in the area of ignition probabilities and methodologies.

Although the probability of immediate ignition is a function of the properties of the material involved in the incident, the size of the release, and the energy of the accident, ignition probability is usually determined based on both historical data and professional engineering judgment. This is a result of the lack of references and reliable data in this area. This judgment takes into account the nature of the event causing the release. For example, smaller spills from trucks that are typical of fitting breaks are apt to result from events without significant amounts of frictional energy or sparking, reducing the chance of immediate ignition. However, there may still be the potential for immediate ignition from the engines of the vehicle(s) involved as well as from other vehicles in the immediate vicinity. In the case of larger, instantaneous spills from trucks, a higher probability of immediate ignition would be expected. The justification is due in part to the increased size of the release and in part to the increased amounts of energy and sparking expected from an incident of such magnitude.

The overall likelihood of ignition for flammable liquids has historically been 24 percent for truck releases (Rhoads, 1978). This covers a broad spectrum of materials, release sizes, and release causes. It is likely to be dominated by gasoline and similar materials. A possible approach is to first split this number into a lower value for the more common small spills and a higher value for the less common large spills while still achieving the same weighted average. The relative energies of accidents for different modes of transportation also need to be recognized. For example, consider a one-vehicle truck accident versus a high speed rail derailment involving many rail cars.

Croce (1982) included a review of LPG vehicle accidents over a roughly 10-year period. Of 50 accidents reporting sufficient details for analysis, 37 had no ignition reported while 12 did (one had a report of "no outcome"). This implies 24 percent ignition, as found by Rhoads.

One might use a 15 percent chance of immediate ignition for small truck spill of gasoline and a 50 percent chance for a large spill. This gives the "correct" weighted average if there is a roughly 70 percent chance of a small spill. These values might be increased to 25 percent for small truck spills and 75 percent for large truck spills of LPG, based on the relative flammabilities of the materials. Other materials like ethylene oxide might support a further increase. For other modes, the spill sizes and energies involved both increase. Hence, for LPG the values might increase to 30 and 90 percent for the small and large spills respectively, for rail accidents.

These figures are also reasonably consistent with the *Second Canvey Island Report* (U.K. Health & Safety Executive, 1981) which has used onsite ignition probabilities of 0.1 if there were *no* ignition sources, 0.2 if there were *very few* ignition sources, 0.5 if there were *few* and 0.9 if there were *many*.

One approach to developing ignition probabilities considers the size of the release and nature of the accident in conjunction with material-specific properties such as:

- NFPA flammability rating,
- NFPA reactivity rating,
- flame velocity,
- minimum ignition energy,
- flammability limits, and
- ignition temperature.

These factors are then used to determine an internally consistent set of ignition probabilities for a variety of materials, by mode and spill size.

For pipeline ruptures where a large percentage of rupture events are due to third party damage a relatively high probability of immediate ignition is generally used. This is supported by the data given in Table 3-5. These figures represent the total chance of ignition, not just immediate ignition. However, these releases would not be expected to disperse to the same distances as would chemical releases. In addition, they also may be on pipelines which are further removed from industrial areas, than is expected for chemical pipelines.

The analyst should develop a rationale for selecting the immediate ignition probabilities and then strive to achieve consistency across modes and materials. The materials' NFPA ratings offer one way to group them into somewhat consistent sets. Including additional properties will make the sets more consistent. This will reduce the variability in relative risk assessments, but remain a contributor to the uncertainty found in an absolute risk assessment.

TABLE 3-5
Overall Probability of Ignition of Pipeline Failures

Data Source	*Probability of Ignition*	
U.K. Gas (Townsend, 1986)	Leaks:	0.1
	Ruptures:	0.5
U.S. Gas (Jones, 1986)	All sizes:	0.16
	Ruptures:	0.26
European Gas (European Gas Pipeline Incident Data Group, 1988)	Pinholes/cracks:	0.02
	Holes:	0.03
	Ruptures <16 in.:	0.05
	Ruptures ≥16 in.:	0.35
	All sizes:	0.03

Delayed ignition takes other factors into account. Delayed ignition probabilities can also be determined as a function of the cloud area or the location. In general, as the size of the cloud increases, the probability of delayed ignition decreases. This is due to the likelihood that the cloud has already encountered an ignition source and ignited before dispersing over a larger area (i.e., the cloud reaches an ignition source relatively close to the point of origin).

Simmons (1974) developed delayed ignition probabilities as a function of cloud area for LNG and LPG, as given in Table 3-6. This work includes the chance of immediate ignition. As this distribution does not consider the theoretical maximum cloud size for a specific release as an explicit factor, its application is limited. Moreover, the size of the release and the number of ignition sources encountered are not explicitly considered in his work. The values presented in Table 3-6 are based on historical incidents with estimations of cloud areas when these were not provided. A curve was then fitted to the data, and the values given in the table were calculated.

In a transportation incident, it is very likely that, if a cloud does not ignite immediately, the moving cloud would encounter several ignition sources along its path. In this case the probability of ignition would be given as a function of the cloud location and the ignition sources encompassed. Ignition sources can be classified as follows:

- Open flames
- Electrical sources
- Hot surfaces
- Impact effects

TABLE 3-6
Ignition Probability as a Function of Cloud Size[a]

Area covered by cloud (m^2)	*Cumulative Ignition Probability*
<30	0.5223
30–100	0.6396
100–300	0.7365
300–1000	0.8249
1000–3000	0.8864
3000–10,000	0.9343
10,000–30,000	0.9630
30,000–100,000	0.9813
100,000–300,000	0.9907
300,000–1,000,000	0.9959
1,000,000–3,000,000	0.9981
3,000,000–10,000,000	0.9992

[a]Simmons (1974).

Some examples of these sources include automobiles and other moving sources, furnaces, rooftop heating units, pilot lights on appliances, smoking materials, traffic lights, welding activities and others.

Even more so than for CPQRA, calculating ignition probabilities in a TRA is a difficult problem. While for CPQRA, an open flame is considered to be a certain ignition source, this degree of certainty may not be warranted in a TRA. For instance, the pilot light on an appliance may or may not be encountered by a dispersing vapor cloud. Other ignition probabilities are even less certain, as they are intermittent in nature.

A generalized approach is therefore typically needed due to the fact that the exact location of the release and the distribution of potential ignition sources along the route cannot be specified. Several approaches have been developed to take such uncertainties into account. In the *Second Canvey Island Report* (U.K. Health & Safety Executive, 1981), the conditional probabilities given in Table 3-7 above were used (conditional on the fact that prior ignition has not occurred).

While these probabilities were used for releases from fixed facilities, they are for the dispersing cloud after it has left the site and is starting to encounter populated areas. This would appear to be quite consistent with its use in a TRA. These values were selected based on judgment.

TABLE 3-7
Delayed Ignition Probabilities[a]

Type of Ignition	*Conditional Probability*
Edge/edge: Edge of cloud is at edge of population area when ignition occurs	0.7
Central: Cloud is right over population area at time of ignition	0.2
No ignition	0.1

[a]U.K. Health & Safety Executive (1981).

Concern over the high likelihood of ignition (and the potential understating of toxic risks in cases where ignition does not occur) could lead to a modified distribution—for example, where *edge/edge* ignition is reduced to 0.5 and *no ignition* is increased to 0.3.

The dependency on cloud size can be taken into account by screening whether the hazard distance is even expected to reach the population. This also allows a buffer zone to be taken into account if the transportation corridor is relatively isolated.

In a report on transportation of dangerous goods by rail in the U.K. (Advisory Committee, 1991), a mixture of historical data and engineering judgment was used to generate the ignition probabilities given in Table 3-8. In their opinion, the authors felt that the chance of immediate ignition was the same for LPG and gasoline, and that the overall chance of ignition for liquified flammable gases was the same as for large releases from facilities.

In addition to fire hazards, explosion hazards must also be considered. Explosion potential is a function of the chemical and physical properties of the material in combination with the available mass, degree of confine-

TABLE 3-8
Ignition Probabilities for Gasoline and LPG (by Rail)[a]

	Type of Ignition		
	Immediate	*Delayed*	*None*
Gasoline			
Small spill	0.1[b]	0	0.9[b]
Large spill	0.2	0.1	0.7
LPG			
Small spill	0.1	0	0.9
Large spill	0.2	0.5	0.3

[a]Advisory Committee (1991).
[b]Based on historical data.

ment, turbulence, and other variables. In transportation scenarios, most of these variables are not known in advance and assumptions must be made in selecting the parameters used in the consequence models. Explosion probabilities are sometimes based on reviews of past accidents, be they fixed facility or transportation events. As a result, the likelihood of a vapor cloud explosion following ignition used in TRA studies generally ranges from 0 to 50 percent. Based primarily on judgment and material properties, explosion probabilities given ignition for materials like LPG are often taken as 33 percent and for materials like ethylene oxide are taken as 50 percent. Sensitivity analyses can be run to test the effects of varying the explosion probabilities.

3.5. Treatment of Classification Yards

Classification yards are used by many companies as holding or storage areas for rail cars in the vicinity of a site prior to or after their being assembled or disassembled for a train. Such yards are generally offsite and less secure than the plant itself. The number of cars involved is likely to fluctuate depending on prices and purchasing habits of the parties involved and the train schedules. While less commonly used for long-term or intermediate length storage, truck and barge marshaling or staging areas also exist.

Given that very little switching or moving of rail cars (or other types of containers) occurs in such areas, they are usually treated as fixed sites and risk analyses can be conducted per the *CPQRA Guidelines* (CCPS, 1989). Fault trees can be constructed or historical accident data used for vessel leaks and failures, and the fraction of time the rail car (or barge or truck) is present onsite factored into the logic. While any one rail car might only spend several days in the yard, there may be hundreds of rail cars moved through the yard each year. This can yield several equivalent vessel-years of residence time.

The time that the rail car sits in a yard at one end or the other of its route may also be the most likely time for the hazardous outcomes of such events as overfilling, contamination, lack of inhibitor, and others to occur. These scenarios can therefore be quantified and added to the other yard risks. It may be appropriate in some circumstances to distribute such events to both the initial yard and the route itself, with the appropriate distribution suggested by the time it takes for the scenario to develop. For example, if an overfill takes a certain amount of temperature rise before causing a problem, it might be estimated that it would typically take X days for that temperature increase to occur. By comparing X to the average time spent in the yard, the appropriate split between possible event locations can be established.

The difficulty generally comes in making sure that these yards are addressed/covered, not in the analysis itself. Since they are often considered a distribution responsibility, yet act like a storage tank farm they tend to be forgotten when it comes to prioritizing activities for risk assessments.

3.6. Transloading Operations

Transloading hazardous materials from one container to another is not that common, but may need to be covered in some studies. These transfers may involve different sizes of vessels—such as a tanker to a barge—or different modes of transportation. (In some cases the containers themselves may be transferred.) These operations are generally similar to loading and unloading operations as covered by CPQRA, and can usually be addressed by CPQRA frequency, consequence, and risk techniques. The primary additional considerations are listed below.

- The potential for being impacted by the other vessel in marine transloading situations, or for rail cars to be bumped or pulled away prematurely.
- Weather conditions such as high winds or high sea states can also cause vessels to separate and rupture hoses.
- Incompatible or poor fitting match leading to leakage upon transfer.
- The potential for cranes or forklifts to drop containers or for forklift tines to puncture a container if this type of forklift is used.
- The potential for the wrong cargo to be transloaded.
- Increased buffer distances to populated areas, especially for marine trans-shipments.
- The availability of two cargos for event escalation—i.e., the ignition of a small leak from one container may impinge on the second container.

Many of these can be addressed by adding events to a fault tree and then using human reliability data, meteorological data, and/or some of the sources given in Chapter 4 to quantify the fault tree.

In general, the suggested approach is to use CPQRA techniques supplemented by TRA techniques. The TRA techniques of greatest potential benefit will be some of the consequence modeling guidance (such as spills on water) and the hazard identification and associated frequency estimation. One would generally conduct two small TRAs for the two different containers and/or transport modes and then address the transloading operations themselves. The overall risks would be additive for most risk measures, with location specific individual risk estimates needing to be handled differently.

3.7. Unique Consequence Analysis Models

Consequence analysis for risk assessment purposes was addressed in *CPQRA Guidelines* (CCPS, 1989), and the literature on consequence modeling is quite extensive (CCPS, *Characteristics of Vapor Cloud Explosions*, 1994; CCPS, *Consequence Modeling Proceedings*, 1991; CCPS, *Vapor Cloud Dispersion*, 1987). The purpose of this section is not to duplicate those efforts. However, there are some aspects of consequence analysis that may be particularly common to the transportation of hazardous materials. These will be discussed here. In addition, there may be other concerns which are material specific as opposed to scenario specific.

Table 3-9 lists some common concerns which may need to be incorporated into TRA applications. In many circumstances, these scenarios will need to be addressed through fault trees or other techniques and then

TABLE 3-9
Material Specific Concerns

Hazard Types	*Representative Materials*
Contamination • Heel (residual material) • Wrong product • Corrosion • Wrong container	Monomers; water reactive materials; moisture in cryogenic shipments
Extended Time • Insufficient Stability (inhibtor) — liquid phase — vapor phase • Loss of oxygen or improper padding • Decomposition • Vapor generation by solids (time and temperature related)	Ethylene oxide Methyl acrylate Ethyl acrylate Ethyleneamine Hydrogen peroxide
In-transit heating (solar and/or atmospheric)	Anhydrous HCl
Hydraulic overfill	Any
Delayed autocatalytic reaction	Water and acetic anhydride; ethylene oxide
Freezing and thawing	Acrylic and methacrylic acid; lack of glycol in styrene shipments during winter months
Inadvertent heating	LNG via ship; glacial and methacrylic acid; heat-sensitive materials
Loss of refrigeration	Cryogenic products on ships
Shock-sensitive materials	Organic peroxides

modeled and combined with accident-initiated releases. Contaminants, lack of inhibitors and other such problems would result from improper loading or unloading activities. Fault trees estimating the likelihood of such events would need to be prepared and then appropriate consequence models used to determine the timing and nature of a release. Material specific hazards such as temperature sensitivity will also need to be handled separately from accident-initiated causes. This might entail determining the susceptibility of a rail car to solar heating if left on a siding for an extended period of time, or the proximity to a small fire involving another rail car—a situation that might not yield a BLEVE for most materials, but which might trigger a temperature sensitive material to initiate a runaway reaction or overpressure.

Instances of concern to transportation incidents include scenarios such as unconfined liquid pool spreading, releases on water, cratering from underground pipeline ruptures, and subsea pipeline ruptures among others (see Table 3-10). The reader is also referred back to Section 3.1 for a discussion on the selection of release scenarios.

Unlike facility-based incidents discussed in *CPQRA Guidelines* (CCPS, 1989), diking or other forms of spill containment are not expected to be present at the location of a transportation related release. As a result, liquid pools are most likely to be unconfined, although terrain and other situation specific effects may limit pool spreading. When modeling this scenario, this potential lack of containment needs to be addressed. One option for the pool model is, therefore, to not have a containment area specified. Alternatively, if the model does not have this capability or if one prefers to specify an area, the specified area can be based on judgment, past experience, or the expected area of a piece of highway between drainage areas. Since the rate of evaporation from a pool is proportional to the pool's area, unconfined pool spreading will result in a greater rate of evaporation of the spilled material. Pool fires may also be much larger when unconfined. Unconfined spills, however, may be unduly conservative. Most roadways are designed to drain off and collect rainwater and many rail car spills occur in accidents where the ground is torn up and as a result, may limit the spill area. Therefore, the results of unconfined pool models may overestimate the actual pool area.

One possible approach is to start with an unconfined pool, or an area greater than the possible pool size in a model which requires an area to be specified. If the area actually calculated by the model seems incompatible with the selected release scenario—as a result of highway curbing, uneven terrain, etc.,—then the model can be rerun assuming either a specified area or a fixed percentage. For instance, one might restrict the area to 500 or 1000 m^2 regardless of material, or state that the area will be equal to 30 or 50 percent of the unconfined pool area. The first approach

TABLE 3-10
Transportation Specific Concerns

Concern	Applicability to Specific Transport Modes: Rail	Road	Marine	Pipeline
Spill on water	Over or near a body of water	Over or near a body of water	Virtually all spills	Undersea lines (underwater releases); across under rivers
Unconfined pools	In an undisturbed flat area	In an undisturbed flat area	On water	Above ground liquid lines in flat area
Cratering (soil overburden removal)	N/A	N/A	N/A	Ruptures or large leaks from underground pipelines under sufficient pressure
BLEVE-induced catastrophic vessel failure	Possible if more than one rail car is involved or sufficient quantity in single car with small leak to feed fire	Possible if sufficient quantity in car with small leak to feed fire or if double tank trailer or burning fuel leak	Possible if multiple discrete tanks	N/A
Toxic products of combustion or reaction	Dependent on material and whether ignition occurs	Dependent on material and whether ignition occurs	Dependent on material and whether ignition occurs	Dependent on material and whether ignition occurs

offers the advantage of simplicity, while the second approach allows differences in spill quantities and material characteristics to be incorporated.

Other approaches that people have taken include using the ditch area on the side of the road as the pool area. One rule of thumb uses an area of 1000 feet by the width of the ditch as a pool area. Another approach to estimating pool sizes is based on anecdotal information obtained from past train derailments. This method involves using a rail car length as a pool radius and estimates the pool area based on that value.

Pool, jet, and vapor cloud fires can pose both direct thermal radiation hazards and secondary toxic hazards from the products of combustion.

The nature of these hazards and the methodologies for modeling them are not unique to transportation (Mudan, 1988). However, they may be more common in transportation due to the increased likelihood of ignition as the result of the accident. Furthermore, the potential for the release to be immediately proximate to a populated area or even an elevated population in high rises or on hillsides may increase the hazard. In these instances, it is important to consider not only the ground level hazards, but also the height of the toxic plume's centerline relative to the elevation of the potential receptors.

Spills on water are much more likely in transportation incidents, especially in the marine (barge, tanker, etc.) mode of transportation, even through some spills could yield jet releases or be confined on deck. Therefore, material properties such as solubility, specific gravity, water reactivity and toxicity (if environmental or drinking water hazards are of concern) all need to be evaluated. In addition, wave action and its effect on material spreading has an influence on how the material will disperse in the water. Unfortunately, no one model takes all of this into account. There are several references that deal with the modeling of chemical and oil spills on water and the reader is referred to these for more details. (Thibodeaux, 1980; Fay, 1971; Southwest Research Institute, 1982, 1983). Environmental risks are also of concern for water spills, as mentioned in Section 3.9.

Pipeline transportation presents some unique applications for consequence analysis. Underground pipeline incidents pose the potential for cratering. In the event of a pipeline release where the line pressure may cause a crater to be formed, several things need to be considered. The first is whether the amount and rate of material released is enough to generate a crater, and if so, the size of the crater. There is no single acknowledged approach for determining the size of a crater. If one has estimated that the momentum of the initial discharge is sufficient to remove the overburden, then one approach is to select several crater sizes and test their effects on exit velocity and ultimate hazard distances. One can then use the crater diameter which produces the set of maximum hazard distances as a conservative estimate.

Once the crater is formed, material will continue to be released from the ruptured pipeline. Depending on the type and size of the line and the pressure, the material can disperse in any one of several ways ranging from shooting straight ahead at the end of the rupture to being sprayed upward after a line puncture from drilling equipment. In addition, the crater or some other obstruction may absorb the momentum of the jet release, causing it to entrain less air and increase its overall dispersion hazard. The properties most important in the estimation of downwind travel of materials are the source orientation, buoyancy, and momentum.

Melhem (1992) gives one approach for conservative estimation of hazard distances. Ooms (1972, 1974) developed some of the first jet dispersion models. Subsequent validation work has been performed by Havens (1988).

Another application unique to pipeline transportation has to do with underwater releases resulting from subsea pipeline ruptures. In order to model releases of this nature properly, the analyst must have an understanding of the flow dynamics involved. A release from an underwater pipeline is similar to that of a liquid discharge driven by the pressure difference between the source and the surroundings in that two phase flow will not occur with liquefied gases if the water depth is sufficient to suppress vaporization (Melhem, 1992). Pressure from the pipeline will drop as the release is isolated, allowing gravity intrusion of water into the lower level of the pipeline. This will cause liquid to be expelled from the upper section of the pipe until the segment inventory is depleted. Other factors that need to be addressed in order to estimate the release duration and pipe inventory are the liquid compressibility and elastic contraction of the pipe (Melhem, 1992). Other inputs to the modeling includes the temperature, pressure, and phase of the contents, pumping capacity, release geometry, release orientation and accident detection time (for pump shutdown or valve closure).

Above ground pipelines can be addressed in the same manner as piping systems in facility based assessments, with the recognition that the diking and grading commonly found in facilities is unlikely to be present. There will also frequently be a much longer length of line to empty even after pumps are shut down and valves are closed.

A BLEVE may be of concern in a transportation incident. A BLEVE is usually the result of an external flame impinging on the shell of a container, weakening the shell and resulting in a sudden rupture. The result is a sudden release of a large mass of pressurized liquid into the atmosphere. The reader is referred to previously discussed references (CCPS, *Characteristics of Vapor Cloud Explosions*, 1994; Melhem, 1992; CCPS, *CPQRA*, 1989) for methodologies by which to model this phenomenon. One possible TRA scenario involves multiple rail cars in a train accident where the material from a damaged car ignites and the flame from that fire is impinging on an adjacent car that itself is not leaking. This situation is similar to what is encountered in dealing with tank farms at the facility level.

The frequency of such an event can be estimated from the chance of a multiple car spill with judgmentally based adjustment factors for the likelihood of impingement on other cars. The consequences can be modeled as for any external source of heating on a vessel—remembering that relief valves may be inoperable if a rail car has been overturned by the accident.

Another scenario applicable to transportation incidents, results from the fire impinging on the vessel being produced by material leaking from that same vessel. Given the rate of discharge from the vessel versus the time it will take for the vessel to fail, the volume remaining in the vessel at the time of failure must be calculated. This remaining volume will be a deciding factor in whether a BLEVE is a legitimate scenario that needs to be considered in the TRA. There may not be enough material left in the vessel at the time of failure for a BLEVE to occur, and the pressure buildup versus time must be calculated—taking into account the concurrent release through the accident-induced hole.

In transportation incidents, the BLEVE potential for a leaking container will vary by incident and depends on the container size and type. In general, leaking tank trucks will have little potential to BLEVE unless there is an external (separate) fuel source such as a fuel tank, second trailer or the tank contains a compressed gas or pressurized liquid fuel. The volume of most tank trucks is usually small enough so that, for all but the slowest leaks, all of the material has been released from the container before the container BLEVEs. Rail cars, however, usually have a large enough volume so that there would be enough material left in the container at the time of failure to present a BLEVE hazard. For rail cars and certain materials in tank trucks, therefore, a more detailed calculation of BLEVE potential would be required.

When conducting the actual modeling for a BLEVE scenario, the reader should consider the consequences from thermal hazards (if the released material is flammable) as well as overpressure. For flammable materials, the release is likely to be ignited by the external fire, and the thermal radiation hazard may exceed that from overpressure. In addition, nonflammable materials can BLEVE and produce a resultant overpressure hazard.

Another scenario that can result is a runaway reaction inside the vessel. This can result from such incidents as a failure to put an inhibitor in the tank (see Table 3-9). These instances, however, are usually the result of actions that take place outside of the actual transportation of the material and are not discussed here. Moreover, their modeling is no different than at a fixed facility.

3.8. Failure Frequency Models

There are a number of specific analytical methods for determining failure frequencies that may be very beneficial in a transportation risk analysis. These can be grouped into three main types of analysis:

- testing methods
- mechanical predictive methods
- logical procedures

These are rigorous techniques and should only be used when quantitative analysis is necessary and historical data or extrapolations are inadequate. This might include a new type of container, additional shielding, etc. The testing methods are generally based on past experiments to determine structural integrity, from which various relationships have been developed. These relationships are then applied or extrapolated for a particular type of container.

The mechanical predictive methods are techniques which draw upon past engineering analyses to develop models or simulations for estimating the likelihood of container failure. The Minorsky method for ships is one such method (Minorsky, 1959).

Logical procedures include fault tree analysis and other hazard evaluation procedures (CCPS, HEP, 1992). They can also include logic-based adjustments to accident rates or conditional release probabilities to account for the implementation of risk reduction measures.

Each of these types of analyses is described further below. In addition, pertinent appendices discuss some of the techniques in greater detail.

3.8.1. Predictive and Testing Methods

Sandia Laboratories (Clarke, 1976) undertook a major study in the early 1970s to determine the severities of transportation accidents. This study entailed mechanical models, statistical analyses and judgmental factors. The containers analyzed were small packages, but the approach and relationships may be somewhat transferable to other (small) containers. Later studies dealt with larger containers (Dennis, 1987).

The purpose of the study was to provide a description of the fire, crush or puncture environments that containers would be exposed to in accidents and not container responses (i.e., likelihoods of failure) to those environments. Hence, the results must be interpreted carefully. The study addresses accident severity and the frequency of the accident concurrently. The results for the different environments are expressed similarly; they use joint probability distribution curves to illustrate how often (on a per mile basis) one might experience an event that exceeds a specified crush or puncture force, impact velocity, or fire duration.

The authors of the study believed that their evaluations were most accurate for air, truck, and trains, in that order and discouraged comparisons between the modes. For qualitative purposes, they grouped their severities into five categories. Table 3-11 illustrates some of their basic

TABLE 3-11
Probabilities (per mile) of Accident of Specified Severity[a]

Category	Truck
Minor: crush force <20,000 lb	2.4×10^{-6}
Moderate: crush force 20,000–50,000 lb	4.5×10^{-8}
Severe: crush force 50,000–100,000 lb	7.2×10^{-9}
Extra severe: crush force 100,000–150,000 lb	3.5×10^{-9}
Extreme: crush force >150,000 lb	1.2×10^{-9}

[a]Clarke (1976).

results. The severities take into account: crush force (in pounds); puncture potential as a ratio of relative velocity to the radius of the potential penetrator; and fire duration (in minutes) for a representative hydrocarbon fuel fire temperature of 1850°F. Impact velocities and immersion depths were also considered.The crush force ranges are given in Table 3-11 for a sense of increasing severity.

Another approach that involves the physical testing of a container is known as structural integrity testing. This involves tests of the mechanical properties of the particular material of construction or the structural design, and can provide indirect indication of failure potential by verifying the strength of the material or structure.

This type of testing is quite common for pipelines, and is increasingly being used on ISO-containers and rail cars.

3.8.2 Mechanical Predictive Methods

A number of well-known techniques of this type include the Minorsky method, a method developed for ships by Sandia, and a method developed for trucks and trains by Rhyne (1994). They often involve game theory and simulation principles along with engineering analyses, as the systems being analyzed involve two "randomly" moving targets in the case of marine collisions. This is in contrast to the much more fixed systems involved in rail, highway, and pipeline modes of transport.

The Minorsky method uses data from actual collisions to define relationships between the energy of the collision and some measures of the degree of structural damage. It was first developed in 1959 by V. U. Minorsky, a naval architect, and he concluded that a correlation could be established between a "structural resistance" factor and the total change in kinetic energy of the colliding vessels before and after the collision (Minorsky, 1959).

This total change in kinetic energy or the amount of energy absorbed in the collision can be related, using energy and momentum conservation principles, to the displacements and velocities of the ships. One of the steps is the development of critical impact velocities (for various sizes of striking ships) which are sufficient to cause structural damage to a given ship construction. These values are based on a worst-case assumption of a perpendicular collision (beam-on) where the striking ship hits the barge or tanker of concern in the tankage area. Such collisions are the most severe, are typical of major collisions, and simplify the analytical process.

Minorsky attempted to relate the calculated energy for specific collisions with various structural parameters, and found that a very good correlation was obtained when the structural resistance factor was defined as the amount (volume) of structural material involved in the collision which was likely to deform and absorb energy as opposed to those which would just bend or twist. This could include (as appropriate): decks and double bottoms in both vessels, the component of the hull plate of the striking vessel in the direction of the collision, and generally all main structural members in the longitudinal direction of the striking ship and in the transverse direction of the struck ship.

His work established a linear relationship between the structural resistance factor and the absorbed energy for the major collisions studied, while in many of the low-energy collisions there was considerable scattering of the data points. This scatter has been attributed to the underestimation of actual speed at impact, and determined not to be of particular importance.

Studies with scale models have generally shown Minorsky's results to be reasonable and somewhat conservative in those instances in which the ship's sidewall structure is more complex than those representative of oil tanker designs. Generally, this correlation with model studies has added to the level of confidence and validity of the application of the Minorsky method in collision risk studies.

3.8.3. Logical Procedures

3.8.3.1. Other Techniques

The logical procedures used to identify hazards for fixed facilities, such as fault tree analysis or failure modes and effects analysis can also be used in transportation situations. They provide modeling and estimation techniques for failures which are sufficiently complex to require analysis of combined effects of events, or detailed evaluations of individual equipment items or subsystems. The reader is referred to the *HEP Guidelines* (CCPS, 1992) for details on these methods.

Typical applications of these techniques in a TRA are for situations involving contamination, overheating or overpressure of a container. One example would be a highway accident that resulted from contamination, where the contamination was a result of improper loading of the truck. An event tree or a fault tree (see Chapter 3 of the *CPQRA Guidelines*) could be used to determine the overall event likelihood, based on the probabilities of each contributing factor.

Event trees are another logical procedure that can be very helpful in TRA studies. The various branches on the event tree can represent different ignition timing/locations, different meteorological conditions, day versus night, whether or not an explosion occurs, and even release sizes. This then determines the array of potential outcomes of a given accident or release.

3.8.3.2. Data Adjustments

Risk reduction measures are often quantified in terms of adjustments to either accident rates or conditional release probabilities. The determination of the logical adjustment may occasionally be through the use of specific historical or statistical data, but most often there are limited or no factual data available. In such instances, it may be appropriate to:

- *conduct a sensitivity study to see if a variable is significant*—this entails changing the value of the variable to other possible values. For instance the probability of immediate ignition could be changed from 0.5 to 0.1 and 0.9 in two separate sensitivity runs.
- *bracket the adjustment with available data*—this might involve taking data for a more robust and a less robust container as the brackets for a modified container design. An intermediate release probability could then be chosen or the risks could be expressed as falling in the band created by the two sets of results for the known containers.
- *extrapolate from available data*—this generally applies when one has data up to a point and the design or material properties go beyond that point. One example would be having information on ignition likelihood for a flammable material such as gasoline, and needing to derive an appropriate value for the even more flammable ethylene oxide. Another example would be having data on container or pipeline thickness up to a point and having the ability to use even thicker construction.
- *determine the maximum possible effect by reviewing the contributions of specific accident causes*—for example, if the data showed

that external interference caused 50 percent of pipeline accidents, a "perfect" means of protecting a pipeline from such events would only halve the overall failure rate. Given that a perfect system is unlikely to exist, the overall reduction is expected to be less than a factor of two.

- *use engineering judgment*
- *use some combination of the above*

3.9. Environmental Risks

Environmental risks are an area of increasing research and development. Traditionally, the potential consequences that have been addressed for transportation related incidents are acute human health effects from toxicity, thermal radiation and explosion overpressures. However, with the potential for contamination of soils, drinking water sources, surface water, and groundwater, additional consequences and risks may also need to be addressed in a risk analysis. Increasing regulatory attention may ultimately force more environmental risks to be evaluated.

The assessment of environmental risks is often an analysis that is conducted after the fact and one that is used to monitor the extent of contamination that is already present to help guide clean-up efforts. While the assessment of consequences has a methodology which is fundamentally different from that for a TRA, there are many portions of a TRA which are still applicable.

The focus of an ecological risk assessment is on the effects that chemicals pose in different environments. As a result, much has been learned about how chemicals will behave in the environment as well as the parameters necessary to estimate such behavior (U.S. EPA, March 1989; Paustenbach, 1989). Physical properties such as a material's solubility, specific gravity, adsorption coefficient and octanol water partition coefficient help determine a material's interaction with water and soil.

The environmental consequences and resultant risks will therefore vary along different portions of the route. The variance in the risks will be driven by the changes in the environment surrounding the route (e.g., changing soil type, crossing a river, etc.). The same principles of route segmentation (see Section 3.3) can be applied here with new segments defined when a route changes soil type or stops running along a body of water. In addition, environmental consequences can be predicted based on what is known about how the materials behave in the environment. Once the potential environmental consequences are chosen and the route is segmented based on relevant environmental factors, the process of a TRA is similar to that described in other parts of this book.

The chance of an accident, the chance of a spill given an accident, the distribution of spill sizes, and the shipment characteristics are unchanged. Population data may not be needed at all. Climatic data are less likely to include wind direction, and more likely to include ground temperature, potential rainfall, river current data, and geographic features such as lake diameter or river width. Risk measures may include FN curves and expected values where the parameter of concern is area contaminated, clean-up cost, or some other similar measure.

3.10. Available Software

The following is a list of some current commercial software products and contacts and their known features. Table 3-12 identifies which programs are applicable for the different modes.

Program: ADLTRS™

Source: Arthur D. Little, Inc., Acorn Park, Cambridge, MA 02140

System: PC version, IBM compatible, 286/386/486, 640 KB RAM, 2 MB Hard Disk Space, Microsoft Windows™ 3.1, DOS 3 or greater, Mouse

Description: The program offers PC based tools to evaluate and categorize the risks associated with differences between chemicals, transportation modes and routes. It offers rankings that can be used for a large number of movements. Easy-to-use techniques consider the chemical, transport mode, container type, distance, route characteristics and annual volume. Risks may be expressed in terms of human hazards from toxic, flammable and/or explosive chemicals, or environmental hazards to rivers, lakes, soil or groundwater. Final results are placed in risk categories to establish relative ranking.

Program:HazTrans®

Source: Abkowitz and Associates, Inc., 2100 West End Avenue, Suite 640, Nashville, TN 37203

System: PC Version, IBM Compatible, Microsoft Windows™ 3.0, DOS, Mouse

Description: The program is a multimodal, hazardous materials management tool that is shipment, material, and container specific. It is compatible with and uses spatial data available from the U.S. Census Bureau, the U.S. Geological Survey and the National Weather Service. It enables the user to address risk assessment, routing, and scheduling, emergency preparedness, evacuation planning and incident management.

Program: PC* HazRoute®

Source: ALK Associates, Inc. 1000 Herrontown Road, Princeton, NJ 08540

System: PC Version IBM Compatible, 386 or higher, 35 MB Hard Disk Space, 8 MB RAM, VGA Monitor, Microsoft Windows™ 3.1.

Description: ALK's PC* HazRoute® is a Microsoft Windows™ application that helps shippers and carriers evaluate and select rail and highway routings for their hazardous materials. It analyzes the cost and risk factors involved in hazardous materials movement over a current or proposed route, and generates alternative routes based on the user's selection of risk criteria. ALK's PC* HazRoute® is based on ALK's U.S. highway and rail proprietary databases, population statistics, and safety data.

Program: SAFETI

Source: Technica, Inc., Software Products Division, 1400 N. Harbor Blvd., Suite 440, Fullerton, CA 92635

System: IBM PC 386, 3 MB Extended Memory, Hard Drive, Math Coprocessor, EGA or VGA Monitor, DOS 3.3 or higher

Description:SAFETI is an integrated set of computer programs designed to automate the risk assessment of chemical and petrochemical facilities involving the manufacture, storage, and transport of toxic and flammable materials. Calculates the consequences of possible accidental releases and their imact based on event frequency to produce measures of risk such as Risk Contour Plots and FN curves.and many county roads.

A PC-based model has also been developed by the Inter Industry Rail Safety Task Force to address TRA by rail. Information on the model can be provided to members by contacting their representatives at Chemical Manufacturers Association, Association of American Railroads or Railway

TABLE 3-12

Program Applicability by Mode

Mode	*ADLTRS™*	*HazTrans®*	*SAFETI*	*PC* HazRoute®*
Rail	✔	✔	✔	✔
Highway	✔	✔	✔	✔
Vessel	✔	✔	✔	Future
Pipeline	✔		✔	

Note: HazTrans® and ADLTRS™ files may be exchanged via the Microsoft Windows (Version 3.1) Operating System™.

Progress Institute. The U.S. Coast Guard is also in the process of developing a computer model to do risk analysis for vessel transport through ports and harbors.

3.11. References

This section lists the references cited in this chapter. Additional references that may be of interest to the reader can be found in the topical bibliography.

Advisory Committee on Dangerous Substances. "Health and Safety Commission," *Major Hazard Aspects of the Transport of Dangerous Goods*, Report and Appendices, HMSO. (1991).

Center for Chemical Process Safety. *Guidelines for Evaluating the Characteristics of Vapor Cloud Explosions, Flash Fires, and BLEVEs*. American Institute of Chemical Engineers. New York (1994).

Center for Chemical Process Safety. *Guidelines for Process Equipment Reliability Data, with Data Tables*. American Institute of Chemical Engineers. New York (1989).

Center for Chemical Process Safety. *Guidelines for Use of Vapor Cloud Dispersion Models.* American Institute of Chemical Engineers. New York (1987).

Center for Chemical Process Safety. *Proceedings of the International Conference/ Workshop on Modeling and Mitigating the Consequences of Accidental Releases of Hazardous Materials.* American Institute of Chemical Engineers. New York (1991).

Center for Chemical Process Safety. *Guidelines for Chemical Process Quantitative Risk Analysis*. American Institute of Chemical Engineers. New York (1989).

Center for Chemical Process Safety. *Guidelines for Hazard Evaluation Procedures, Second Edition with Worked Examples.* American Institute of Chemical Engineers. New York (1992).

Clarke, R. K. et al. *Severities of Transportation Accidents*. Sandia Laboratories. NTIS SLA-74-0001 (1976).

Croce, P. A. et al. *A Feasibility Study of a Sealed Safety Monitor for Trucks Carrying LNG and Other Hazardous Materials.* Prepared by Arthur D. Little, Inc. for the U.S. Department of Energy, Office of Operational Safety Programs DOE/EV/10502-1. (1982).

Dennis, A. W. et al. *Severities of Transportation Accidents Involving Large Packages;* Sandia Laboratories. NTIS SAND-77-0001. (1987).

European Gas Pipeline Incident Data Group. *Gas Pipeline Incidents: A Report of the European Gas Pipeline Incident Data Group*. Pipes and Pipelines International, 1970-1988.

Fay, J. "Physical Processes on the Spread of Oil on a Water Surface." *Proceedings of the Joint Conference on Prevention and Control of Oil Spills.* American Petroleum Institute. (1971).

Harwood, D. *Present Practices of Highway Transportation of Hazardous Materials—Task C, Interim Report, Analysis of Existing Databases.* Midwest Research Institute. Project Number 8669-S (1987).

Havens, J. *A Dispersion Model for Elevated Dense Gas Jet Chemical Releases.* NTIS No. PB 88-202379. National Technical Information Service, Springfield, VA (1988).

Jones, D. J. et al. *An Analysis of Reportable Incidents for Natural Gas Transmission and Gathering Lines—1970 Through June 1984.* American Gas Association. NG-18, Report No. 158 (1986).

Melhem, G., P. Croce. *Advanced Consequence Modeling: Emission, Dispersion, Fires and Explosions.* Van Nostrand Reinhold, New York (in preparation).

Minorsky, V. U. "An Analysis of Ship Collisions with Reference to Protection of Nuclear Power Plants." *Journal of Ship Research* (1959).

Mudan, K., P. Croce. "Fire Hazard Calculations for Large Open Hydrocarbon Fires." *SFPE Handbook of Fire Protection Engineering.* ISBN 0-87765-353-4, National Fire Protection Association, Quincy, MA (1988).

National Research Council, R&D Panel. *Analysis of Statistics Concerning Oil Pollution Incidents in the U.S.* National Academy of Sciences (1991).

Office of Technology Assessment. *Transportation of Hazardous Materials.* OTA-SET-304 (1986).

Ooms, G. "A New Method for the Calculation of the Plume Path of Gases Emitted by a Stack." *Atmosphere Environment,* Vol. 6 (1972).

Ooms, G. et al. "The Plume Path of Vent Gases Heavier Than Air." *The First International Symposium on Loss Prevention and Safety Promotion in the Process Industries*, ISBN No. 0-444-41230-1 (1974).

Paustenbach, D. *The Risk Assessment of Environmental and Human Health Hazards: A Textbook of Case Studies*. John Wiley & Sons, New York (1989).

Purdy, G. *The Measurement of Risk from Transporting Dangerous Goods by Road and Rail; in Transportation of Dangerous Goods: Assessing the Risks.* Institute for Risk Research. Waterloo, Ontario (1993).

Raj. P. and C. Turner. *Hazardous Material Transportation in Tank Cars: Analysis of Risks.* NTIS: FRA/ORD-92/34. May 15, 1993.

Rhoads, R. E. *An Assessment of the Risk of Transporting Gasoline by Truck. U.S. Department of Energy.* Contract No. EY-76-C-06-1830 (1978).

Rhyne, W. R. *Hazardous Materials Transportation Risk Analysis: Quantitative Approaches for Truck and Train.* Van Nostrand Reinhold, New York (1994).

Simmons, J. A. *Risk Assessment of Storage and Transport of Liquiefied Natural Gas and LP-Gas*. Final Report for Contract 68.01.2695 for U.S. Environmental Protection Agency (1974).

Southwest Research Institute. *Analytical and Experimental Study to Improve Computer Models for Mixing and Dilution of Soluble Hazardous Chemicals.* U.S. Coast Guard, Report No. CG-D-31-83 (1982).

Southwest Research Institute. *Revision and Experimental Verification of the Hazard Assessment Computer System Models Spreading, Movement, Dissolution and Dissipation of Insoluble Chemicals Spilled onto Water.* U.S. Coast Guard, Report No. CG-D-35-83 (1983).

Swoveland, C., J. Cawdery. *LPG Transport R&D Risk-Benefit Analysis, Volume I: Analysis and Results.* Quantalytics, Inc., Transport Canada Report No. 74583 (1986).

Thibodeaux, L. *Spill of Soluble, High Density, Immiscible Chemicals on Water.* US Department of Transportation. Report No. DOT-OS-70040 (1980).

Townsend, N., G. Fearnehough. *Control of Risk from U.K. Gas Transmission Pipelines.* British Gas plc, Research and Development Division, Engineering Research Station ERS E531 (1986).

U.K. Health & Safety Executive. *Canvey: A Second Report—A Review of Potential Hazards from Operations in the Canvey Island/Thurrock Area Three Years After Publication of the Canvey Report.* HMSO, ISBN No. 0-11-883459-2 (1981).

U.S. Department of Commerce. *Metropolitan Statistical Areas—Supplementary Report—1980 Census of Population.* PC80-SI-18 (1984).

U.S. Environmental Protection Agency. *Ecological Assessment of Hazardous Waste Sites: A Field and Laboratory Reference.* Washington, DC (1989).

Wilson Hill Associates. Data sheets forwarded by Wilson Hill Associates, Inc. to L. Perkins, Chevron USA, dated May 13, 1987. (Wilson Hill Associates maintained the HMIS under contract to DOT.)

4

TRA Data

The data sources that can be used in a TRA are presented in this chapter, along with guidance on their use. Some data sources are specific to one mode of transportation, while others may address a number of different modes. TRA data sources may provide information on the rate of accidents, the number of accidents, the likelihood of releases, the number of releases, the total miles traveled by a given mode (called *exposure* data), or a combination of these.

The data provided may be specific to a particular type of vehicle, rail car, barge, etc. or to a specified type of roadway, track, or river, or may be very general in nature. This chapter therefore discusses how to adjust and enhance data for modified circumstances, what options may be available in the absence of data, and what ongoing data collection efforts are known to exist.

In addition to TRA-specific data on accident rates, release probabilities, etc., the other basic data needs for conducting a TRA study are also addressed. These include information on: population densities and distributions; meteorological conditions such as temperature, relative humidity, wind speed and atmospheric stability class; route information such as length, road or rail type, current speeds for marine movements, buffer distances to populated area, actual roads taken, actual yards where switching activities occur, etc.; and shipment characteristics such as volume, pressure, temperature, composition, number per year, etc. This chapter presents guidance on obtaining information for many of these concerns.

This chapter is focused on how to obtain data and use them, as opposed to presenting a set of recommended data. Chapter 2 provides basic data that can be used in a screening level study and the *Guidelines for Process Equipment Reliability Data* (CCPS, 1989) provide additional guidance on equipment failures that may be associated with non–accident-initiated releases. The reader is also referred to the topical bibliography.

4.1 Taxonomy for Accident Data

The taxonomy for accident data provides a logical framework by which to identify and to organize data values, and provides a convenient structure for locating specific information of interest. As presented here, it is more detailed than the available data resources. The extra detail is provided, however, to allow for a consistent organization and to permit the orderly introduction of new data sources that may become available. With respect to data sources, the data for each transportation mode may be from one or more major categories representing the rate of occurrence, release probability and potential exposure.

The data sources are described in terms of the modes they address, the basis of their data, and the type of data they provide—accident rates, release probabilities, exposure data, etc. Accident rates may be broken down to represent the type of route (e.g., divided versus undivided highway, Track Class 2 versus 4, river versus open ocean). Release probabilities are classified by variables such as container type, shipping mode (e.g., pressurized versus unpressurized) and possibly characteristics such as speed and route conditions. Release probabilities for non–accident-initiated events such as fitting or valve failures are also considered as data classes within the taxonomy structure. Additionally, some required data may be found in the *Guidelines for Process Equipment Reliability Data* (CCPS, 1989) or other sources of data on process equipment failure rates.

Exposure data may be classified based on characteristics such as HAZMAT versus non-HAZMAT (to ensure compatibility with incident data); quantity of material being shipped (i.e., ton-miles); route and/or shipping characteristics for each mode of failure indicated. Table 4-1 summarizes the taxonomy for TRA accident data.

It is important to recognize that, in some instances, multiple sources may be required for the derivation of a single data value. For example, failure rate data is represented by ratios relating the number of failures to some measure of the number of operations or distance traveled. It is quite possible that the numerator and denominator data will be derived from different sources. Section 4.6 addresses how data from various sources are pulled together.

4.2 Historical Data Sources

This chapter contains a profile of forty-one different sources of data on accident rates, release probabilities, numbers of accidents or releases, and exposures (i.e., miles traveled). These data sources include articles, reports, databases, published risk assessments, government databases and government reports. They cover a time period of roughly 20 years.

Each data source is described in terms of the following characteristics:

- Title
- Reference Number (4-1 to 4-41)
- Sponsor/Author
- Industry
- Time Period
- Frequency of Update
- Data Boundary
- Taxonomy Reference (cross-reference to Table 4-1)
- Nature of Data
- Data Type and Access
- Definition of Spill/Selection Criteria for Incidents
- Description
- Caveats on Use

This allows the reader to obtain pertinent references, to briefly assess their applicability to the situation at hand or to be forewarned of potential limitations. Sample data suitable for use in screening level studies have been provided in Chapter 2, for each mode.

A summary of the data sources profiled is given in Table 4-2. This table first lists the sources that address multiple modes of transportation and then addresses mode specific sources in the following order—pipeline, rail, road, and marine. These sources cover many types of data, but focus on incidents and releases more than exposure information. Table 4-1 breaks out the different topics into the scheme by which the individual data sources are classified. In some cases further subdivisions are shown in parentheses. The items in parentheses are common subdivisions that could apply to every entry in that particular section. For example in Section 2.1 on Rail Accident Data, all of the subsections (2.1.1–2.1.4) can be further broken out by Train Speed and Track Class, and all of the Road Accident Data subsections (Sections 3.1.1–3.1.2) can be further subdivided by vehicle speed.

Table 4-3 takes the highest level categories from the taxonomy and indicates which sources address each category. In general, these sources do not provide enough information about mitigation and control effectiveness to quantify their benefits.

In the past, data sources and publications have followed their own individual taxonomy schemes, which vary from one another and from the scheme being proposed here. Table 4-3 shows the relationship between the Data Taxonomy discussed in Section 4.1 (and illustrated in Table 4-1) and the individual transportation data sources applicable to each topic. The applicable transportation data sources correspond to numbered sheets presented at the end of this chapter which describe each source in detail. Each number represents a different transportation data source.

TABLE 4-1
TRA Data Taxonomy

1.0 PIPELINES
1.1 Accident Data
    1.1.1 Pipe (Subdivided by Diameter & Wall Thickness)
        1.1.1.1 Seamless Steel Pipe
        1.1.1.2 Electric Resistance Welded Steel Pipe
        1.1.1.3 Submerged Arc Welded Steel Pipe
        1.1.1.4 Steel Pipe with Partial Containment
        1.1.1.5 Steel Pipe with Total Containment
        1.1.1.6 Other
1.2 Release Data
    1.2.1 HAZMAT
        1.2.1.1. Pressurized Liquid
        1.2.1.2 Non-pressurized Liquid
        1.2.1.3 Gas
    1.2.2 Non-HAZMAT
        1.2.2.1 Pressurized Liquid
        1.2.2.2 Non-pressurized Liquid
        1.2.2.3 Gas

2.0 RAIL
2.1 Accident Data (Subdivided by Train Speed & Track Class)
    2.1.1 Class I Railroad
        2.1.1.1 Collision
        2.1.1.2 Derailment
    2.1.2 Class II Railroad
        2.1.2.1 Collision
        2.1.2.2 Derailment
    2.1.3 Class III Railroad
        2.1.3.1 Collision
        2.1.3.2 Derailment
    2.1.4 Yards
2.2 Accident-Initiated Release Data
    2.2.1 Derailment Probability (Subdivided by Speed and Position)
    2.2.2 Derailment Release Probability-Mainline
    (Subdivided by Speed and Protective Measure)
        2.2.2.1 Car Specification 1
        2.2.2.N Car Specification N
    2.2.3 Yard Derailment Release Probability
        2.2.3.1 Car Specification 1
        2.2.3.N Car Specification N
2.3 Non–Accident-Initiated Release Data
    2.3.1 Fitting Release
    2.3.2 Valve Release
    2.3.3 Rupture Disk Release
    2.3.4 Piping Release
    2.3.5 Vessel Failure Release

TABLE 4-1
TRA Data Taxonomy (cont.)

3.0 ROAD
3.1 Accident Data (Subdivided by Vehicle Speed)
- 3.1.1 Urban Routes
  - 3.1.1.1 Multilane Divided Highway
  - 3.1.1.2 Multilane Undivided Highway
  - 3.1.1.3 Local Street (One way)
  - 3.1.1.4 Local Street (Two way)
  - 3.1.1.5 Freeway/Limited Access
- 3.1.2 Rural Routes
  - 3.1.2.1 Divided Highway
  - 3.1.2.2 Undivided Highway
  - 3.1.2.3 Freeway/Limited Access
  - 3.1.2.4 Local Street (Two way)

3.2 Accident-Initiated Release Data (Subdivided by container type)
- 3.2.1 Single Vehicle Noncollision
- 3.2.2 Single Vehicle Collision
- 3.2.3 Multivehicle Collision

3.3 Non–Accident-Initiated Release Data
- 3.3.1 Fitting Release
- 3.3.2 Valve Release
- 3.3.3 Rupture Disk Release
- 3.3.4 Piping Release
- 3.3.5 Vessel Failure Release

4.0 BARGE
4.1 Accident Data (Subdivided by Traffic Density & Waterway Type)
- 4.1.1 Inland-waterway
  - 4.1.1.1 Grounding
  - 4.1.1.2 Ramming
  - 4.1.1.3 Allision
  - 4.1.1.4 Fires/Explosions
  - 4.1.1.5 Collisions
  - 4.1.1.6 Other
- 4.1.2 Ocean-going
  - 4.1.2.1 Grounding
  - 4.1.2.2 Ramming
  - 4.1.2.3 Allision
  - 4.1.2.4 Fires/Explosions
  - 4.1.2.5 Collisions
  - 4.1.2.6 Other

4.2 Release Data (Subdivided by Traffic Density & Waterway Type & BargeType)
- 4.2.1 Inland-waterway
  - 4.2.1.1 Grounding
  - 4.2.1.2 Ramming
  - 4.2.1.3 Allision
  - 4.2.1.4 Fires/Explosions
  - 4.2.1.5 Collisions
  - 4.2.1.6 Other

TABLE 4-1
TRA Data Taxonomy (cont.)

4.0 BARGE (cont)
4.2 Release Data (Subdivided by Traffic Density & Waterway Type & BargeType) (cont.)
- 4.2.2 Ocean-going
  - 4.2.2.1 Grounding
  - 4.2.2.2 Ramming
  - 4.2.2.3 Allision
  - 4.2.2.4 Fires/Explosions
  - 4.2.2.5 Collisions
  - 4.2.2.6 Other

4.3 Non–Accident-Initiated Release Data
- 4.3.1 Fitting Release
- 4.3.2 Valve Release
- 4.3.3 Rupture Disk Release
- 4.3.4 Piping Release
- 4.3.5 Vessel Failure Release

5.0 OCEAN-GOING VESSEL
5.1 Accident Data (Subdivided by Traffic Density and Harbor Type)
- 5.1.1 Harbor Operation
  - 5.1.1.1 Grounding
  - 5.1.1.2 Ramming
  - 5.1.1.3 Allision
  - 5.1.1.4 Fires/Explosions
  - 5.1.1.5 Collisions
  - 5.1.1.6 Other
- 5.1.2 Ocean-going Operation
  - 5.1.2.1 Grounding
  - 5.1.2.2 Ramming
  - 5.1.2.3 Allision
  - 5.1.2.4 Fires/Explosions
  - 5.1.2.5 Collisions
  - 5.1.2.6 Other

5.2 Release Data (Subdivided by accident type: Grounding, Ramming, Allisions, Capsize, etc.; Include Nonaccident Release)
- 5.2.1 Single Hulled Tankers
- 5.2.2 Double Hulled Tankers, Double Bottomed Tankers
- 5.2.3 LPG/Chemical Tankers
- 5.2.4 LNG Tankers
- 5.2.5 OBO (Oil/Bulk/Ore) Ships
- 5.2.6 Container Ships
- 5.2.7 Other Ships

5.3 Non–Accident-Initiated Release Data
- 5.3.1 Fitting Release
- 5.3.2 Valve Release
- 5.3.3 Rupture Disk Release
- 5.3.4 Piping Release
- 5.3.5 Vessel Failure Release

TABLE 4-1
TRA Data Taxonomy (cont.)

6.0 INTERMODAL CONTAINERS
6.1 Accident Data (Subdivided by category: dropped; container impact; accident damage; in harbor/at sea; staging; storage; etc.)
- 6.1.1 ISO Containers
- 6.1.2 Single Tank
- 6.1.3 Double Tank
- 6.1.4 Insulated
- 6.1.5 Refrigerated
- 6.1.6 Box Containers

6.2 Release Data
- 6.2.1 ISO Containers
- 6.2.2 Single Tank
- 6.2.3 Double Tank
- 6.2.4 Insulated
- 6.2.5 Refrigerated
- 6.2.6 Box Containers

6.3 Non–Accident-Initiated Release Data
- 6.3.1 Fitting Release
- 6.3.2 Valve Release
- 6.3.3 Rupture Valve
- 6.3.4 Piping Release
- 6.3.5 Vessel Failure Release

7.0 OTHER TOPICS
7.1 Population Data
7.2 Climatic Data
7.3 Route Data

TABLE 4-2
Summary of Data Sources for Transportation Risk Analysis

Transportation Data Source	Title	Mode	Nature of Data
4-1	Severities of Transportation Accidents	Air, Highway, Rail	Accident rates, release probabilities
4-2	An Assessment of the Risk of Transporting Propane by Truck and Train	Highway, Rail	Accident rates, release probabilities
4-3	Annual Report of Hazardous Materials Transportation	Multiple	Number of incidents
4-4	FACTS: A Databank for Accidents with Hazardous Materials	Multiple	Number of incidents
4-5	Worldwide Oil Spill Incident Database	Pipeline, Marine	Number of incidents
4-6	Hazardous Material Information Syatem	Multiple	Number of incidents
4-7	Major Hazard Aspects of the Transport of Dangerous Substances	Highway, Rail, Marine	Methodology, accident rates, release probabilities
4-8	Performance of Pipework in the British Sector of the North Sea	Pipeline	Failure rates, release probabilities
4-9	An Analysis of Reportable Incidents for Natural Gas Transmission and Gathering Lines: 1970 through June 1984	Pipeline	Failure rates, release probabilities, number of incidents, exposure data
4-10	An Analysis of Reportable Incidents for Natural Gas Transmission and Gathering Lines: June 1984 through 1990	Pipeline	Failure rates, release probabilities, number of incidents, exposure data
4-11	Pipeline Reliability	Pipeline	Failure rates, release probabilities
4-12	Pipeline Reliability: An Investigation of Pipeline Failure Characteristics and an Analysis of Pipeline Failure Rates	Pipeline	Failure rates, release probabilities
4-13	Safety of Interstate Natural Gas Pipelines	Pipeline	Number of incidents, number of releases

TABLE 4-2
Summary of Data Sources for Transportation Risk Analysis (cont.)

Transportation Data Source	*Title*	*Mode*	*Nature of Data*
4-14	Gas Pipeline Incidents: A Report of the European Gas Pipeline Incident Data Group	Pipeline	Failure rates, release probabilities
4-15	Performance of Oil Industry Cross-Country Pipelines in Western Europe: Statistical Summary of Reported Spillages—1991	Pipeline	Failure rates, release probabilities
4-16	Analysis of Tank Cars Damaged in Accidents 1965–1986; Analysis of Aluminum Tank Cars Damaged in Accidents 1965–1986;	Rail	Release probabilities
4-17	Accident/Incident Bulletin	Rail	Number of incidents, number of releases
4-18	Railroad Accident/Incident Database	Rail	Number of incidents, number of releases
4-19	Statistical Trends in Railroad Hazardous Materials Transportation Safety, 1978 to 1986	Rail	Accident rates, release probabilities
4-20	Annual Report on Hazardous Materials Transportation by Rail	Rail	Accident rates, number of incidents, exposure data
4-21	Truck Accident and Fatality Rates Calculated from California Highway Accident Statistics for 1980 and 1981	Highway	Accident rates
4-22	Fleet Accident Rates	Highway	Accident rates, number of incidents, exposure data
4-23	An Assessment of the Risk of Transporting Gasoline by Truck	Highway	Accident rates, release probabilities, number of incidents, number of releases, exposure data
4-24	Summary of Motor Vehicle Accidents in the Petroleum Industry for 1988	Highway	Accident rates, number of incidents, exposure data
4-25	Present Practices of Highway Transportation of Hazardous Materials—Task C Analysis of Existing Databases	Highway	Accident rates, release probabilities, number of incidents, number of releases, exposure data

TABLE 4-2
Summary of Data Sources for Transportation Risk Analysis (cont.)

Transportation Data Source	Title	Mode	Nature of Data
4-26	Various State Traffic Accident Analyses	Highway	Number of incidents, number of releases, exposure data
4-27	Procedures for the Development of Truck Accident Rates and Release Probabilities for Use in Hazmat Routing Analysis	Highway	Accident rates, release probabilities
4-28	Waterborne Commerce of the United States	Marine	Exposure data
4-29	Polluting Incidents in and Around U.S. Waters	Marine	Number of incidents, number of releases
4-30	Commercial Vessel Safety, Risk Assessment Study. Volume I, Survey of Data for Marine Risk Assessment	Marine	Accident rates, release probabilities, number of incidents, number of releases, exposure data
4-31	Lloyds Maritime Information Services, MARDATA	Marine	Number of incidents, number of releases
4-32	Marine Casualty Information Reporting System (CASMAIN)	Marine	Number of incidents,
4-33	Marine Investigations (MIN) Module of MSIS and CASMAIN	Marine	Number of incidents, number of releases
4-34	Fire/Explosion Probabilities of Liquid Gas Ships	Marine	Accident rates
4-35	Tanker Casualties	Marine	Accident rates
4-36	Worldwide Tanker Spill Database	Marine	Number of releases
4-37	TIGER/Line Census Files, 1990		Population data, route data
4-38	Summary Tapes (STS)		Population data
4-39	1992 County and City Extra; Annual Metro, City, and County Data Book		Population data
4-40	NOAA Environmental Data Service		Climatic data
4-41	Statistical Abstract of the United States		Climatic data, population data

TABLE 4-3
Relationship of Taxonomy to Transportation Data Sources

Taxonomy	*Applicable Transportation Data Sources*
1.0 PIPELINES	
1.1 Accident Data	4-4, 8, 9, 10, 11, 12, 13, 14, 15
1.2 Release Data	4-4, 5, 6, 8, 9, 10, 13, 14, 15
2.0 RAIL	
2.1 Accident Data	4-1, 2, 4, 7, 17, 18, 19, 20
2.2 Accident-Initiated Release Data	4-2, 3, 4, 6, 7, 16, 17, 18, 19, 20, 29
2.3 Non–Accident-Initiated Release Data	4-4, 6, 7, 19
3.0 ROAD	
3.1 Accident Data	4-1, 2, 4, 7, 21, 22, 23, 24, 25, 26, 27
3.2 Accident-Initiated Release Data	4-2, 3, 4, 6, 7, 23, 27, 29
3.3 Non–Accident-Initiated Release Data	4-4, 6, 7
4.0 BARGE	
4.1 Accident Data	4-4, 28, 30, 31, 32, 33
4.2 Release Data	4-3, 4, 5, 6, 29, 30, 31, 32, 33
4.3 Non–Accident-Initiated Release Data	4-4, 6, 31, 33
5.0 OCEAN-GOING VESSEL	
5.1 Accident Data	4-4, 7, 28, 30, 31, 32, 33, 34, 35
5.2 Release Data	4-3, 4, 5, 6, 7, 29, 30, 31, 32, 33
5.3 Non–Accident-Initiated Release Data	4-4, 6, 7, 31, 33
6.0 INTERMODAL CONTAINERS	
6.1 Accident Data	4-4
6.2 Release Data	4-4, 6
6.3 Non–Accident-Initiated Release Data	4-4, 6
7.0 OTHER TOPICS	
7.1 Population Data	4-37, 38, 39, 41
7.2 Climatic Data	4-40, 41
7.3 Route Data	4-37

4.3. Adjustments and Enhancements to Data

Typically, accident, release, and exposure data represent averages of data over a wide variety of circumstances. Most often they therefore do not fully represent circumstances specific to the situation being evaluated and adjustments may need to be applied to reflect such conditions. For example, effects of extreme weather, local route conditions, severity of incident of concern may all be specific to the problem being analyzed but are not likely to be accommodated in overall average statistics regarding accidents rates or consequences. In such situations, the application of factors to adjust a generic data set to the particular set of circumstances is appropriate.

In some cases where sufficient data are available, the adjustment factors may be estimated from historical experience by comparing the statistics with and without the influencing circumstances being considered. In other cases, engineering analysis may support the development of appropriate adjustment factors by considering the physics of the failure mechanism and making adjustments based on the relative influence of the factor on the key physical parameters resulting in failure. For example, the effect of wall thickness on probability of puncture might be estimated from the alteration in the key strength parameters affecting puncture. Section 3.8.2 described methodologies that address variabilities in mechanical strength. Alternatively, if one has data on normal gasoline trucks (MC 306/406) but is actually using a far more robust pressure vessel, it may be sufficient to state qualitatively that the container in use has a lower likelihood of puncture.

In general, the precision need in such adjustments depends on the purpose of the study and the overall uncertainty in the underlying data. This will help determine whether to estimate the adjustments qualitatively or quantitatively. In some instances, there may even be some data available which indicate a trend for the situation at hand, but which are not sufficient to determine a specific accident rate or release probability. In such instances, using more generic data may be a good starting point and then the trends and engineering judgment can be combined to select the appropriate level of adjustment.

An important caveat in applying adjustment factors to data is that the effect of multiple factors may not be independent. Specifically, the presence of one factor may either enhance or reduce the effectiveness of another. For example, in the case of pipeline transportation, the frequency of pipeline failures decreases as the pipe diameter increases and also as the pipewall thickness increases. However, the diameter and wall thickness are not independent. In order to maintain the same stress level or pressure, the required pipewall thickness increases proportionally with the pipeline diameter. Thus, the effects cannot be estimated individually. Similarly, the benefits of head shields and shelf-couplers are very related. Past studies on the benefits of head shields showed very large improvements (i.e., reductions) in the likelihood of release given an accident. However, these studies were done prior to the widespread incorporation of shelf-couplers. Reducing the number and changing the nature of head punctures via the incorporation of shelf-couplers changes the potential benefits available from head shields significantly. Thus, the improvements of each are not simply additive.

When adequate data are available, the effects of combined risk control options may be estimated directly. More often, available resources will not be adequate and judgment must be applied to ensure reasonable results in the presence of multiple adjustments. The application of limits to total adjustments to data is a useful tool in this regard. For example, a large overall adjustment to accident rates in particular is probably unrealistic and it is therefore reasonable to limit total adjustments to no more than a factor of two or three except under unusual circumstances where large adjustments are justifiable or some evidence is available.

Although the above has emphasized the use of adjustment factors to reflect risk control option effectiveness, another application is to use a limited data set to adjust a larger, more generic data set. For example, a large database describing general accident rates might be adjusted by using the experience with a particular operator's fleet to give the benefit of data specific to the carrier being analyzed while retaining the greater detail provided in the larger database.

4.4. Sources of Additional Data

The amount and quality of the data used in the conduct of a TRA are critical to the quality and accuracy of the final product. Oftentimes clear, concise data specific to the needs of the user are not readily available. The analyst may also not be acquainted with some data sources. Therefore, a number of additional sources of data are discussed in the following sections. These sources are generally more movement specific than are accident rates and conditional release probabilities, and as a result, may not be extractable from landmark studies. Situation-specific concerns and differences often are addressed using such parameters as population density, climatic conditions, and route characteristics.

4.4.1. Population Data

Population data are often incomplete and are not always found locally at a level of detail adequate to accurately estimate the risks in that area. If this is the case, there are several sources of data that may provide valuable information. Data can be obtained from the U.S. Census Bureau in varying degrees of detail. Population data can be found broken down by state, county, city, income, gender, age, marital status, race, education, and many other factors. Census data are also broken down beyond the city and town level into tracts and blocks. A census tract is a small, relatively permanent statistical subdivision of a county. Census tracts are delineated for all metropolitan areas (MAs) and other densely populated counties by local census statistical areas committees following Census Bureau guidelines. Census tracts usually have between 2500 and 8000 persons, and when first delineated, are designed to be homogeneous with respect to population characteristics, economic status and living conditions. (This can be helpful if considering the vulnerability of various types of construction to different levels of over pressure, or whether or not windows are likely to be open and allow toxic vapors to enter.) Blocks are small areas bounded on all sides by visible features such as streets, roads, streams, and railroad tracks, and by invisible boundaries such as city, town, and county limits (U.S. Department of Commerce, 1991). This information is available on CD-ROM disks which are available from the Census Bureau.

The census data, while seemingly very detailed, may still not give information to the level of detail that would allow an analyst to get information on the population density within a certain corridor, and also are usually several years old by the time they get widely published. Data of the sort desired may be obtained by using Geographic Information Systems (GIS) and overlaying the route of concern on the GIS map. The GIS programs can then use the census data to calculate the desired information. One source of such data is *Map 'n' Go© 1994* from DeLorme Mapping. Information may also be available from aerial photos showing the location and proximity of populated areas to the route of concern. Photographs may also be used to estimate industrial and commercial populations. One way to estimate populations is by looking at the number of cars or parking spaces in certain areas and to assume one or two persons per car. This can be particularly helpful in rural areas or foreign countries where detailed statistics are not available. Another useful source of information is annual city and state fact books which may be published at a city, state, or national level. One example is the *1992 County and City Extra; Annual Metro, City and County Data Book* (Slater, 1992).

Two other areas of concern for population estimates are on-road population estimates and day/night shifts in industrial areas. This type of information cannot be obtained from census data and must be obtained from other sources such as state highway commissions and local business agencies or alliances. Onroad population estimates can be derived from traffic density data, vehicle counts, or even highway design capacities. This type of information, however, is especially subject to seasonal and day/night shifts in traffic patterns. In addition, accidents may cause traffic to slow or people to congregate thereby increasing the population in the immediate area. Other areas where people congregate such as sporting events, fairs, etc., may also be of concern. Zoning maps can be another source of population information. If an area is zoned industrial, it is more likely to be subject to daily shifts in population due to working day schedules. Business associations, fire departments, state emergency planning commissions (SEPC), or local emergency planning committees (LEPC) may have information on industrial populations.

4.4.2. Climatic Data

The environment in which an incident occurs can be a deciding factor on whether the incident actually occurs as well as significantly influencing the ultimate consequences. Environmental factors that may contribute to these effects include terrain and weather.

Incidents may be more likely to occur during adverse or extreme weather conditions (e.g., dense fog, storm, etc.), and, as with CPQRA, the weather conditions also have a major effect on the way a release spreads. Meteorological data can be obtained from the National Oceanic and Atmospheric Administration (NOAA), the National Climatic Center (NCC) and from airports or even some plant sites in the areas of concern. On-line data are available from the TTN bulletin board (U.S. EPA Source Receptor Branch, Research Triangle Park, NC). Stability Array (STAR) Program data are also available on-line and from NOAA (see Transportation Data Source 4-40). These data include wind directions, wind speeds and atmospheric stability categories. Figure 4-1 is a typical 16 point wind rose. This shows the percentage of the time the wind blows from each direction and the frequency of calms. The wind data may also be received in tabular form. The radial bars in Figure 4-1 show the percentage of time the wind blows **from** various directions and the frequency of calm conditions (no wind) is entered in the center. We know from this windrose, for example, that 10 percent of the time the wind was observed there were calm conditions and that 5 percent of the time the wind prevailed from the north.

Depending on the length of the route, weather conditions can be expected to vary from region to region. These variations may be minor, however, allowing constant meteorological conditions to be applied to the whole route. In addition, transportation may only take place during daylight hours, allowing the range of possible meteorological conditions to be narrowed. It is usually impossible to isolate a single average condition that adequately represents all weather conditions, and many TRAs employ at least two weather conditions: one stable (e.g., 4 mph, Stability F) and the other characteristic of average daytime conditions (e.g., 10 mph, Stability D). However, a single conservative weather condition could be used for screening calculations. The atmospheric temperature and relative humidity can also significantly influence the consequences of a release. It may be necessary to use two different atmospheric temperatures - one for winter and one for summer. Many annual fact books such as the *Statistical Abstract of the United States* published by the U.S. Department of Commerce include summary weather data for key cities. To simplify the TRA, small differences in average conditions are not usually worth taking into account, thus if there are differences along a route as it moves from the

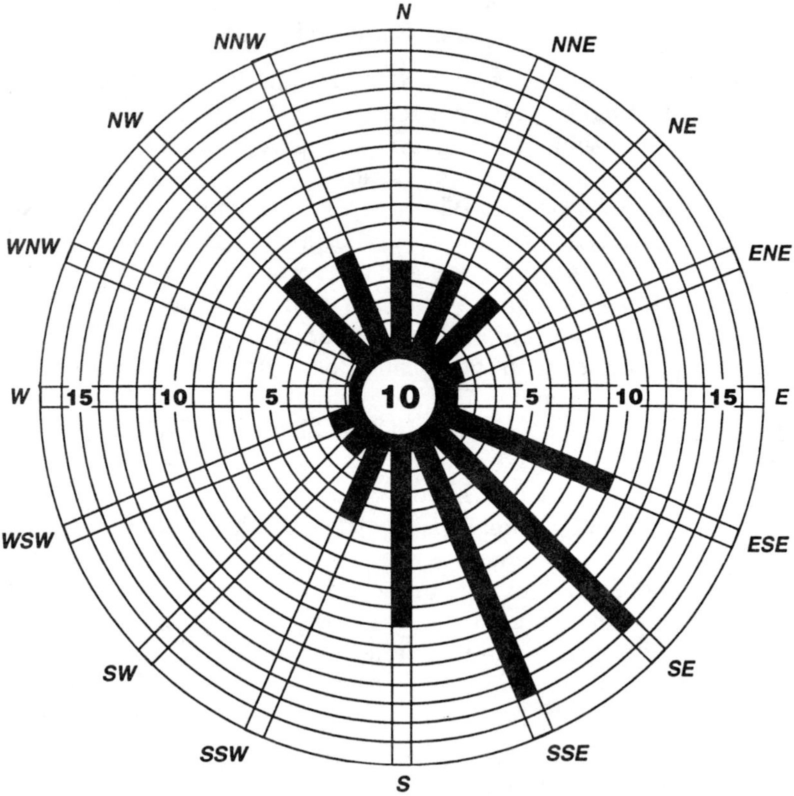

FIGURE 4-1. Typical Wind Rose.

Gulf Coast to the northeast, for example, it may be possible to partition the route into two parts with similar conditions.

With regard to terrain influences, the topography of the area in which the incident occurs can affect the dispersion of a cloud, channel the flow of a liquid release, hinder means of egress for exposed persons trying to escape and even be a contributing factor to the incident itself. In general, large obstacles (ranging from tall buildings to mountains to levees) need to be taken into account when doing the consequence analysis. This may be only a qualitative consideration given the state of consequence modeling and the varying number and type of obstacles encountered along a particular route. Data sources for this material include United States Geological Survey (USGS, Reston, VA) topographical maps, almanacs, aerial photographs or satellite imaging (LANDSAT, Lanham, MD; SPOT, Reston, VA), and possibly a visual inspection of the route (i.e., driving the route).

4.4.3. Route Data

As discussed in Chapter 3, the type of data available on the route of concern is extremely important. The quality and completeness of this data will have a large effect on the outcome of the risk assessment. Just knowing the origin/destination and length of a route and its general population characteristics are usually not enough in a detailed TRA, although it may be sufficient for simpler studies.

For road transportation, the type of roads must be addressed. Is the route all freeway, two lane roads, city streets, or a combination of two or more of these? This type of data should be available from the shipper or carrier of the material of concern. Sometimes alternate routes are given to the driver and it is left up to his or her discretion as to which route to take. In these cases, several routes may need to be analyzed. Data on the type of road used should be available from a standard road atlas and population data can also be derived from this type of atlas in combination with census data or other sources described previously. Visual inspection of the route can also identify any unusual hazards along the particular route (e.g., hazardous intersections, rail crossings, tight curves, narrow or no shoulder, etc.). Geographical information systems can be very helpful in identifying and characterizing routes. In addition, population characteristics may be different for opposite sides of a roadway and this should be addressed in the characterization of the route. Seasonal variations in traffic density or even rush hour/non-rush hour variations may also be important to ascertain.

For rail transportation, acquisition of necessary route data becomes a little more complex. Again, specific routing information should be available from the carriers and from car tracking reports. Details usually include the railroad(s) traveled on, the number and location of yard visits, etc. Care should be taken to discriminate between yards where activities are limited to crew changes and those where the cars of concern are reconfigured. Once the route is known it can be traced out on a standard rail or commercial atlas or via a GIS database to allow the population distribution to be determined. Visual inspection of the route and/or discussions with the railroad can identify any unique hazards along the route such as unprotected railroad crossings.

For pipeline transportation, more so than other modes, information is best obtained from the company that operates the pipeline. They can provide the necessary information about the pipeline right-of-way (ROW), such as where the pipeline runs, the width of the ROW, the location of valve stations, the location and size of compressor stations, the location and number of utility crossings, the presence of other pipelines in the ROW, etc.

For vessel transportation, information should be available from the carrier concerning the specifics of the route such as intermediate ports of call, mooring or anchorage areas and other routine stops along the route. Seasonal variations in current flows or limitations such as low water levels and/or freezing are usually also known on the basis of operating experience. The U.S. Army Corps of Engineers also maintains some of this information about navigable waterways.

4.4.4. Shipment Data

In addition to the shippers themselves, several other sources exist that have information on the shipments of hazardous materials. Trade organizations and associations may be a good source of information. Some of the more common sources for this information are:

American Petroleum Institute (API)
2101 L Street NW
Washington, DC 20037

American Trucking Association (ATA)
2200 Mill Road
Alexandria, VA 22314

Association of American Railroads (AAR)
50 F Street NW
Washington, DC 20001

Chemical Manufacturers Association (CMA)
2501 M Street NW
Washington, DC 20037

Compressed Gas Association (CGA)
Crystal City Gateway #1, Suite 501
1235 Jefferson Davis Highway
Arlington, VA 22202

International Maritime Organization
4 Albert Embankment
London SE1 7SR, United Kingdom

Pharmaceutical Manufacturers Association
1100 15th Street, NW
Washington, DC 20005

Railway Progress Institute (RPI)
700 N. Fairfax Street
Alexandria, VA 22314

Synthetic Organic Chemical Manufacturers Association (SOCMA)
1330 Connecticut Ave., NW, Suite 300
Washington, DC 20036

In addition there are material-specific organizations that can also supply information for certain materials of concern such as:

Chlorine Institute
2001 L Street NW
Washington, DC 20037

Ethylene Oxide Industry Council
2501 M Street NW, Suite 330
Washington, DC 20037

Fertilizer Institute
1015 18th Street NW
Washington, DC 20036

Vinyl Institute
Wayne Interchange Plaza II
155 Route 46 West
Wayne, NJ 07470

Finally, U.S. federal, state, and local governments need shipment data to help them set regulations, plan for emergencies, and draft regulations. Data and information systems pertaining to hazardous materials transportation are kept by federal agencies, regional federal offices, different departments of state governments and some local government offices. To determine what data sources exist at the federal level, the Office of Technology Assessment (OTA, 1986) examined then current databases. Table 4-4 presents a summary of the data resources on commodity flow. OTA found that the U.S. federal data collection activities are numerous and diverse with each of them providing modal transportation information of varying completeness. These sources provide useful information on regional flows of hazardous materials transportation in the United States, and a sound basis for state or local commodity flow data collection.

The Standard Transportation Commodity Code (STCC) referenced in Table 4-4 is a numerical designation used to identify a specific material in transportation. These codes are used to track commodity flows of specific materials throughout the country. Table 4-5 gives some representative STCC codes for some commonly transported hazardous materials. The seven-digit STCC is the full number, the five-digit STCC drops the leading two digits.

TABLE 4-4
Commodity Flow Database

Databases	*Kept by*	*Years*	*Modes*	*Commodity codes*	*Conversion table*	*Strengths*	*Weakness/ Drawbacks*
Commodity Transportation Survey (CTS)	Bureau of the Census	1977	All	5-digit STCC	Yes	• Multimodal • Consistent selection procedure for all sample data points for all modes • Cross-checked against the census of manufacturers	• Only 5-digit level of commodities • No hazardous materials flags • Only shipments from manufacturing sites to first destinations • Only "principal" mode is reported
Truck Inventory and Use Survey	Bureau of the Census	1977, 1982	Highway	Simple classes	No	• Covers all trucks used in the U.S. • Contains hazardous materials-related data items • Sample biased toward heavy trucks	• No flow data • Only rudimentary commodity information • Tractor database, not a trailer database—reflects tractor use, not trailer use
Motor Carrier Census	Bureau of Motor Carrier Safety, FHWA	Most recent 5 years	Highway	Hazard classes	No	• Comprehensive listing of carriers and truck fleet operators	• No flow data • Mileage and fleet size data are sparse
TRANSEARCH, FREIGHTSCAN, etc.	Consulting firms	Varies	All	Varies, up to 7-digit for rail	Yes	• Cross-checked against other production/consumption data • Melding of the best available for each mode	• Truck flows predominantly based on the CTS data (see above) • Not in the public domain
National Motor Truck Database	Consulting firms	1977 to present	Highway	Varies, up to 7-digit STCC	Yes, where commodity code is provided	• Focuses on long-distance highway flows • True flow data • Describes vehicle used to carry the commodity	• Purposely excludes short-haul truck movements especially in the Northeast • Not in the public domain
Waybill File	Interstate Commerce Commission	At least past 12 years	Rail, TOFC/ COFC	7-digit STCC	Yes	• Well-organized sample (1%) of all rail flows • Database is consistent enough for trend analyses • Contains some routing information	• Not all hazardous material flows use the special Hazardous Materials STCC

TABLE 4-4
Commodity Flow Database (cont.)

Databases	*Kept by*	*Years*	*Modes*	*Commodity codes*	*Conversion table*	*Strengths*	*Weakness/ Drawbacks*
Waterborne Commodity Statistics	Army Corps of Engineers	At least 12 years	Water, domestic and international	4-digit WCSC code	Only to a limited extent	• "100%" sample of all vessel movements • Complete routing information	• Only 163 commodity codes in all, so level of detail is weak • Conversion table has some incorrect cross-references
TRAIN II	Association of American Railroads	Current	Rail, TOFC/ COFC	7-digit STCC	Yes	• "100%" data on all movements for participating railroads • Routing information	• Not specifically designed to record car movement histories • Not in the public domain
Hazardous Waste Shipment Data	States, for EPA	Varies	Primarily highway	Either EPA codes or OHMT	No	• "100%" sample of all hazardous waste shipments • Actual flow data	• Many states do not computerize the data • No consistency to commodity code usage • No routing information

ACRONYMS: EPA = U.S. Environmental Protection Agency; FHWA = Federal Highway Administration; OHMT = Office of Hazardous Materials Transportation, Research and Special Programs Administration; STCC = Standard Transportation Commodity Code; TOFC/COFC = trailer on flatcar (piggyback)/container on flatcar; WCSC = Waterborne Commerce Statistics Center (U.S. Army Corps of Engineers).

SOURCE: Office of Technology Assessment.

TABLE 4-5
Standard Transportation Commodity Codes (STCC) Designations for Commonly Transported Hazardous Materials

Material	*STCC*	*Material*	*STCC*
LPG	4905752	Chlorine	4904120
Anhydrous Ammonia	4904210	Vinyl Chloride	4905792
Sulfuric Acid	4930040	Styrene	4907265
Sodium Hydroxide	4935243	Methanol	4909230
Phosphoric Acid	4930247		

(AAR)

It should also be noted that if one is conducting a study on one particular hazardous material, there may be other materials shipped in the same types of containers and for which more data are available. Information on typical containers and their specifications is included in Appendix A—Standard Container Illustrations and Specifications. These data will still be relevant to the situation of interest.

Commodities in the STCC are classified according to the industry which normally produces them, not by end use. The first five digits of the STCC coincide with the Commodity Classification for Transportation Statistics, a commodity adaptation of the Standard Industrial Classification (SIC) codes published by the U.S. Office of Management and Budget (OMB), which was developed for use in the Census of Transportation and adopted by the Interstate Commerce Commission (ICC) as the mandatory reporting form for ICC regulated carriers. The sixth and seventh digits of the STCC provide specific commodity identification. Table 4-6 lists some of the agencies from which commodity flow information described in Table 4-4 may be obtained.

4.4.5 International Data

The preceding sections have focused on obtaining data for transportation movements in the United States. Most governments will have agencies or departments that are very similar to their United States counterparts. In addition, there may be information available from private organizations, consulting firms, and/or various software products. References to such sources and products can often be found in publications, articles, and proceedings which cover international conferences.

TABLE 4-6
Sources for Commodity Flow Databases

Bureau of the Census Customer Services Washington, DC 20233	U.S. Army Corps. of Engineers District New Orleans Attn: CELMN-ED-S PO Box 60267 New Orleans, LA 70160-0267
Bureau of Motor Carrier Safety Federal Highway Administration 400 Seventh Street Washington, DC 20590	Association of American Railroads 50 F Street, NW Washington, DC 20001
Interstate Commerce Commission 12th and Constitution Ave., NW Washington, DC 20423	

Many of the accident rate and release probability data sources that appear at the end of this chapter address transportation on an international basis. This is particularly true for pipelines and marine movements, which have common environments and/or more international standards. Thus, the data for these two modes can often be used directly in international studies. For rail and highway movements, the accident rates are influenced by differences in the supporting infrastructures. The release probabilities are significantly affected by differences in both typical speeds and container design specifications. For these two modes, more specific local data should be sought from appropriate data sources (such as the carriers or local governments) and/or publications.

These guidelines have not focused on international sources of information although several have been obtained through the course of the data collection efforts.

4.5. Completeness and Accuracy of Data Sources

As with any type of data source, the completeness, accuracy, and applicability to the task at hand needs to be addressed.

For highway transportation of hazardous materials, the larger databases on hazardous materials accidents/incidents include the Research and Special Programs (RSPA) Hazardous Materials Incident Reports (HMIR), the Bureau of Motor Carrier Safety's (BMCS) truck accident database, and the State Traffic Accident Records System (STARS). In order to understand the data available from these sources, the reporting requirements must be analyzed and understood. The RSPA and BMCS databases are federally funded, rely on voluntary reporting of incidents/accidents and have different reporting criteria (Hobeika, 1991; Harwood et al., 1990). As a result, accidents/incidents are not uniformly reported in either database. In addition, states also have differing reporting requirements. The resulting regional differences cast some doubt on the use of national default values for truck accident rates. It has also been concluded by some that state derived values for accidents/incidents should be used in a TRA if available (Hobeika, 1991). Another approach is to use carrier specific data which will also be collected in a consistent fashion, although the carrier may cover a region rather than just one state.

The OTA report (1986) included an examination of the completeness and accuracy of the HMIS database. Their conclusion was that secondary sources needed to be analyzed to ensure that there was an accurate representation of the number of accidents. This stems in large part from a significant underestimation of the number of hazardous material releases. Problems were found with the databases that were used as secondary or comparative sources as well as with HMIS.

For rail transportation, the secondary sources of data are the official databases of railroad accidents and HAZMAT releases maintained by the Federal Railroad Administration (FRA) and RSPA. Both FRA and RSPA are agencies of the U.S. Department of Transportation. For reported accidents, the FRA database is complete with respect to HAZMAT cars damaged and derailed in accidents, but provides no information regarding the type of HAZMAT car involved or the nature of the damage. The RSPA database of HAZMAT releases identifies the chemical involved but also provides no information about the tank car type (beyond what can be inferred from the chemical) or nature of the damage. A match between the two databases was obtained in 80 percent of the reported accidents (using date and location as the matching parameters).

These findings are more or less significant depending on the nature of a particular TRA. If one is focusing on catastrophic spills from marine vessels, for example, there are probably very few incidents that would not be reported. However, if one were trying to obtain information about accidents and small releases as well, then the completeness of available data sources may be both more questionable and more important.

The completeness and accuracy of data sources must be considered in the context of the unique needs of the analysis being undertaken. All data sources, including those discussed here and included in the data source profiles must therefore be evaluated considering the manner in which they will be applied in the analysis. In particular, it will be noted that the data representing different transportation modes will be in different formats and often in different levels of completeness from each other. For example, reporting thresholds vary for different modes of transportation and will influence the completeness of data and their applicability for specific analyses. When data from multiple agencies or geographic areas are compared, different reporting criteria may again result in difficulty in combining the data to produce meaningful estimates. Table 4-7 summarizes some of the key concerns for each transport mode under consideration. In comparing or combining data resources, it is important that issues such as these be addressed either by ensuring that the sources are compatible or that appropriate adjustments have been made using, for example, the techniques described for adjusting data sources or for obtaining estimates in the absence of data.

Some of the variables listed in Table 4-7 are broken out in the data sources themselves; for example, pipeline failure rates may be given in terms of both overall rates and by diameter in some sources. A second source may only give overall rates, but knowledge of the industry could suggest that they would fall within a limited range of diameters. Then,

TABLE 4-7
Key Areas of Concern in Comparing or Combining Data Sources

Transportation Mode	*Key Areas of Concern/Potential Differences*
1.0 Pipelines	Reporting Threshold Comparable Pipe Characteristics (Age, Diameter, . . .) Study Date
2.0 Rail	Train Length Reporting Threshold (Accident and Release) Comparable Protective Equipment (Shelf Couplers) Comparable Maintenance and Inspection Practices Study Date
3.0 Road	Reporting Threshold (Accident and Release) Comparable Type of Route/Environmental Conditions Comparable Level of Driver Training Study Date
4.0 Barge	Reporting Threshold (Accident and Release) Traffic Levels and River Conditions (Currents) Barge Design/Placement Study Date
5.0 Ocean-Going Vessel	Reporting Threshold (Accident and Release) Vessel Design Navigation System Local Traffic Control Systems Study Date
6.0 Intermodal Containers	Reporting Threshold (Accident and Release) Container Design Container versus General Cargo Ships Placement on Ship Crane Design Study Date

rather than comparing overall rates from the two sources, one can compare the total over the diameters of concern from the first source with the overall rate from the second.

Other specific concerns are set out in the Transportation Data Source descriptions under "Caveats on Use," and are discussed in many of the studies listed in the Topical Bibliography. In general, one should keep in mind the time period of the data. Were there major changes in design characteristics (shelf couplers, double hulls, etc.), reporting thresholds (major changes such as from $5000 to $50,000), operating practices (no driver training versus skilled drivers), or control systems (on-board tachometers, local navigational systems in ports) between the time periods of the two studies of concern? If so, this information will both help reconcile the differences between the data sources and indicate which one is more applicable and should be used.

Similarly, being aware that data are for vessels on inland waterways versus ocean crossings can indicate the most applicable source of data, the need to adjust data to make them more suitable, and/or that the available data can be considered to bound or bracket the situation at hand rather than accurately represent it.

In many cases, even for the same mode, data may need to be pieced together from different sources in order to obtain the information desired for the transportation study.

For example, if one's objective were to determine accident frequencies for trucks in the state of California, one approach might include:

1. Identification of specific route segments
2. Obtain accident counts from the State Traffic Accident Surveillance and Analysis System (TASAS), for those segments. Accident counts should be for all vehicles as well as a variety of truck types, in order to provide a potential match with the total vehicle or truck miles.
3. Obtain Annual Average Daily Traffic (AADT's) counts for trucks and for all vehicles for those same segments from the Caltrans Annual Average Daily Truck Traffic on the California Highway System Report.
4. Divide the accident counts by the number of years for which data was collected to get an annual figure. Multiply the AADT's by the segment length (for each segment of concern) and by 365 days per year to get the total miles traveled per segment per year, and divide the annual accident count by the total miles traveled to get the accident rates for each segment.

This method, however, is not without complications. There are inconsistencies in the definitions of trucks between the TASAS and AADT databases and the two cannot be completely reconciled. This may be addressed by using a broader range of truck types than is of concern to the specific study in order to obtain consistency between the two databases. Thus there would be a match between the numerator and denominator. In addition, TASAS requires segment definitions to be specified to the thousandth of a mile and will not accept a segment if it crosses Caltrans district boundaries.

4.6 Data Normalization

Any one analysis may need to use:

- accident rate data from one source and release probabilities from another
- numerator data from one source and denominator data from another
- a composite accident rate from a number of sources
- a mix of data sources and engineering judgments
- company-specific accident rates and release probabilities from public data sources
- company-specific accident rates and engineering judgment regarding release probabilities

Each of these situations has its own special concerns and issues to be aware of before using the data.

In the first instance, an accident rate might be derived from a broad population, such as all trucks on divided highways. However, a second source might provide the likelihood of release given an accident for a number of specific truck types. This case allows the analyst to couple two data sources without many concerns other than assuring a comparable definition of an accident.

In the situation where the numerator and denominator data come from two different sources, there are a number of issues. The first is basic data compatibility. If the numerator is HAZMAT releases and the denominator is all miles traveled on a given mode—for both HAZMAT and non-HAZMAT materials—there are numerous potential problems. For one, the denominator is overstated for the selected numerator, artificially lowering the releasing accident rate. For another, it may be more difficult to determine the appropriate differences for various container types when the chance of an accident is embedded in the releasing accident rate. Furthermore, the uncertainties in the numerator and denominator require fairly unique treatments as each represents a random variable (Haugen, 1968).

The proper derivation of a composite accident rate requires that either all the accident rates be defined on a common basis or that the differences in definitions be understood so that they can be taken into account.

The primary issue in all of the remaining situations, as in most of those already described above, is a common definition of accident. If the basic data on accident rates are for all accidents with any damage over $2000, but the chance of a release is for accidents involving at least $10,000 of damage, there is a fundamental mismatch. A much larger fraction of such severe accidents would be expected to yield releases than would be found in a broader class of accidents.

For instance, one data source might indicate that 10 percent of all accidents with \$2000 or more of damage might yield a release, while 30 percent of accidents with \$10,000 or more damage might yield a release. If the chance of the lower severity accident is 5×10^{-7}/mile, the associated chance of a release is 5×10^{-8}/mile. However, if the available data on accident rates did not include information on accident severity (i.e., the minimum amount of damage required for incorporation into the accident rate estimate), the 30 percent release figure which was meant for use with accidents exceeding \$10,000 in damage, might be applied to the base accident rate where accidents with \$2000 in damage are recorded. This would yield a chance of release of 1.5×10^{-7}/mile. This is a factor of three too high. In circumstances where there is such a mismatch of data sources, an adjustment factor must be incorporated.

Nothing explicit has been said about the distribution of damage sizes or the chance of a release for an accident where damages are between \$2000 and \$10,000. Given the previous information and making an assumption that no accident in the \$2000 to \$10,000 range could cause a release, the number of accidents between \$2000 and \$10,000 can be determined:

a = fraction of accidents between \$2000 and \$10,000
$1 - a$ = fraction of accidents with damage ≥\$10,000
b = fraction of accidents between \$2000 and \$10,000 where a release occurs (a subset of a)

and

Damage Level (\$)	*Chance of Release*
Greater than or equal to \$2000	overall average = 0.1
—between \$2000 and \$10,000	—b (unknown)
—greater than or equal to \$10,000	—0.3

By using a weighted average, the accident and release percentages can be calculated using the following equation:

$$a \times b + (1 - a) \times (0.3) = 0.1 \tag{4-1}$$

By assuming that $b = 0$ (i.e., no accident in the \$2000 to \$10,000 range will cause a release) and solving for a, it leaves a value for a of 2/3. This would give a distribution of two-thirds of the accidents having damage between \$2,000 and \$10,000 and one-third having damage in excess of \$10,000. The chance of release would then be calculated as shown in Table 4-8.

Alternatively, if one were to judge that 90 percent of the accidents were between \$2000 and \$10,000 (i.e., $a = 0.9$), using Equation (4-1) as above,

TABLE 4-8

Calculation of Single Contribution Releasing Accident Rate

5×10^{-7}/mile	(overall accident rate)
×	
1/3	(chance accident is over $10,000)
×	
0.3	(chance of a release given accident over $10,000)
=	
5×10^{-8}/mile	(releasing accident rate per mile)

one could solve for b and determine the percent of accidents between $2000 and $10,000 that would result in a release:

$$0.9 \cdot b + (0.1) \cdot (0.3) = 0.1 \tag{4-2}$$

Solving for b gives a value of approximately 7 percent. This means that 3 percent of the overall total of 10 percent releases would be resulting from accidents over $10,000, and the remaining 7 percent would be due to accidents between $2000 and $10,000.

The total chance of a release would remain at 5×10^{-8}/mile, but it would be broken down by the contributors shown in Table 4-9.

TABLE 4-9

Calculation of Multicontribution Releasing Accident Rate

Lower Severity Contribution	
5×10^{-7}/mile	(overall accident rate)
×	
0.9	(chance accident is between $2000 and $10,000)
×	
0.078	(chance of a release given accident is between $2000 and $10,000)
=	
3.5×10^{-8}/mile	(releasing accident rate from lower severity accidents)
Higher Severity Contribution	
5×10^{-7}/mile	(overall accident rate)
×	
0.1	(chance accident is over $10,000)
×	
0.3	(chance of a release given accident over $10,000)
=	
1.5×10^{-8}/mile	(releasing accident rate from higher severity accidents)

4.7 Options in the Absence of Complete Data

It is often the case that data pertaining directly to items under consideration are not available. There are four basic approaches to dealing with this problem. These are:

- estimating based on available data, engineering judgment, and/or analyses
- extrapolating from limited data
- bounding the situation using existing information
- collecting new data

In the first case, approximations from other sources representing similar items may be used to provide the estimates necessary to complete the analysis. An example of similar data might be the substitution of the failure probability of similar transport containers or the use of accident rates from one country to estimate rates in another country.

Engineering analysis and judgment may sometimes be applied to the adjustment of related data as was described in Section 4.5. In this case, however, the analysis must address the base failure rate that should be used rather than the relative effect of a measure on a known historical failure rate. Typically, this results in more extensive analysis and often is not practical. An example of such an approach might be a comparison of the distribution of strength parameters of a structural member to the distribution of stresses estimated by a technique such as finite element analysis in order to estimate the likelihood that the stress will exceed the strength. Application of such techniques generally requires considerable analysis and probably data collection on key material properties. It would typically be practical only for a very limited number of key items for which no other method is available. In general, the more practical approach would be to use such analysis to adapt or interpret limited data resources on similar items.

It is also possible in some cases to apply statistical techniques to analyze data or extrapolate from available data resources. For example, a regression analysis might be used to model the effects of several parameters on the item of interest and a new rate estimated by applying the parameter values representing the actual circumstances to the resulting regression model.

Sometimes, it is possible to bracket or bound the desired case by using best and worst case assumptions to over and underestimate the values, be they final results, accident rates, release probabilities, etc. For example, if failure rate data for a component were available in environments which are both more and less severe than that envisioned for the actual application, the actual rate would be expected to lie between the two estimates.

If the resulting spread is sufficiently narrow, it may suffice for application to the analysis; otherwise the answer might be selected judgmentally within the indicated range. For transportation these bounds might be old single-hulled oil tankers at one extreme and LNG carriers at the other. Failure rates for single-walled pipe and those for two entirely redundant pipes might be used to bracket the benefits potentially associated with a double-walled pipeline. In actuality, there would be some failure modes (such as excavation with very heavy equipment or seismic events) which could result in the failure of a double-walled pipe while other sources (such as corrosion) would probably not be significant as monitoring of the annular space would detect such failures before they could lead to a failure of the second line and cause a release.

The last category of activities is new data collection efforts. These might be starting from scratch on a new design or going back to old incident reports to see how many include or fall into a certain type of failure not previously uniquely identified. In terms of exposure data or population data, new efforts might include traffic surveys, driving the route, etc.

4.8. Ongoing Data Collection Efforts

Data collection efforts are always ongoing as there is always someone who needs better data. Such efforts are often very localized or limited in scope. However, there are a number of recent industry-wide efforts worth noting. This section emphasizes U.S. data collection. Among these is the recent project by the Inter Industry Rail Safety Task Force. As a part of an effort to build two separate models, one for risk screening and one for risk assessment, there has been a major task to develop:

- track class specific accident rates
- rail car specific release probabilities
- quantified risk reduction benefits
- speed dependent derailment fractions

These values are being developed through a combination of historical data, engineering judgment, and engineering analyses. Together they will represent the most comprehensive database for rail TRAs.

The Marine Transportation Task Group of the Chemical Manufacturers Association and an Ad-hoc Oil Pollution Act of 1990 (OPA '90) Work Group are just beginning a data acquisition project to obtain data on past incidents, as well as traffic volumes. This project will help put marine traffic on a similar footing with the rail industry, except for the efforts to quantify the benefits of risk reduction. The Marine Transportation Task Group is also beginning a barge data collection study.

The International Consensus Conference held in Toronto in April 1992 (Solway, 1992) attempted to gain consensus for rail and truck accident rates for use in TRAs, but rapidly concluded that such goals were unattainable without a much more specific study.

In addition to these industry-wide efforts, many carriers have begun to collect their own accident rate data. This trend may or may not involve the use of DOT or another agency's reporting thresholds. The larger companies not only use DOT (or other) reportable accident rates, often for communicating with suppliers, but many have their own definition of accident and associated rate. These rates may be inconsistent between carriers, but generally provide consistency for a given carrier across a region, even when state reporting thresholds vary. Such accident rate data can also be very helpful in trying to assess the effectiveness or benefits of different operating or training programs, as each carrier tends to be more uniform in such programs than the industry as a whole.

4.9 References

This section lists the references cited in this chapter. Additional references that may be of interest to the reader can be found in the topical bibliography. For this chapter, the individual transportation data source sheets should also be consulted if further information is needed on a particular topic.

Center for Chemical Process Safety. *Guidelines for Process Equipment Reliability Data, with Data Tables*. New York: American Institute of Chemical Engineers, 1989.

DeLorme Mapping. *Map 'n' Go© 1994.* Lower Main Street, P.O. Box 298, Freeport, Maine 04032 (207) 865-1234.

Harwood, D., J.G. Viner, and E. Russell. "Truck Accident and Release Rates for Hazmat Routing, Assessing Community Safety for Hazardous Materials Transport," *National Conference on Hazardous Materials Transportation*. St. Louis, Missouri, May 14–16, 1990.

Haugen, E. *Probabilistic Approaches to Design.* New York: John Wiley & Sons, 1968.

Hobeika, A., and S. Kim. "Databases and Needs for Risk Assessment of Hazardous Materials Shipments by Truck." *HAZMAT Transport 91, A National Conference on the Transportation of Hazardous Materials and Wastes.* The Transportation Center. Northwestern University, Evanston, IL, June 17-19, 1991.

Office of Technology Assessment. *Transportation of Hazardous Materials.* OTA-SET-304. (1986).

Slater, C. and G. Hall. *1992 County and City Extra; Annual Metro, City and County Data Book.* 1990 U.S. Census based population data.

Solway, J. "What Is Risk?" *Consensus Report, International Consensus Conference on the Risks of Transporting Dangerous Goods.* Toronto, Canada. April 6-8, 1992. ISBN

U.S. Department of Commerce. *Summary Tape File 1 on CD-Rom Technical Documentation, 1990 Census of Population and Housing.* 1991.

U.S. Environmental Protection Agency. *EPA. TTN and SCRAM Computer Bulletin Board.* Support Center for Regulatory Models, Source Receptor Branch.

TRANSPORTATION DATA SOURCE 4-1

TITLE: Severities of Transportation Accidents

SPONSOR/AUTHOR: R.K. Clarke, J.T. Foley, W.F. Hartman, D.W. Larson, Sandia National Laboratories.	INDUSTRY: Aircraft, Truck, Railroad
TIME PERIOD: Late 1960's to early 1970's	FREQUENCY OF UPDATE: Not Applicable

NUMBER AND TYPE OF RECORDS: Graphs and tables giving probability of accidents having certain severities.

DATA BOUNDARY: U.S. Transportation Industry

TAXONOMY REFERENCE: 2.1, 3.1	NATURE OF DATA: Accident Rates, Release Probabilities

DATA TYPE AND ACCESS:

Report No. SLA 740-001VS14

Cost = $44.50

National Technical Information
5285 Port Royal Road
Springfield, VA 22151

DEFINITION OF SPILL/SELECTION CRITERIA FOR INCIDENTS:

DESCRIPTION: This report documents the development of data on the severity as well as the frequency of accidents involving truck, rail and air transport. Volume 1 includes a summary giving the probability of occurrence of accidents as a function of accident severity. Volume 2 gives supporting data, calculations, and analysis. Severity is based on the orientation and force of the accident and the resultant potential to rupture the container.

CAVEATS ON USE:

TRANSPORTATION DATA SOURCE 4-2

TITLE: An Assessment of the Risk of Transporting Propane by Truck and Train.

SPONSOR/AUTHOR: Battelle Pacific Northwest Laboratory	INDUSTRY: Energy
TIME PERIOD: 1980	FREQUENCY OF UPDATE: Not Applicable

NUMBER AND TYPE OF RECORDS:

DATA BOUNDARY: Truck and Rail Transportation of Propane.

TAXONOMY REFERENCE: 2.1, 2.2, 3.1, 3.2,	NATURE OF DATA: Accident Rates, Release Probabilities

DATA TYPE AND ACCESS:

National Technical Information Service (NTIS)
5285 Port Royal Road
Springfield, VA 22151

Report No. PNL-3308/XPS
Cost $36.50

DEFINITION OF SPILL/SELECTION CRITERIA FOR INCIDENTS:

DESCRIPTION: The report presents the risk of shipping propane by truck and train. Accident rates, release probabilities and consequences were derived to generate overall risks produced by the transportation of propane. The risk analysis uses a predictive technique instead of the historical analysis method.

CAVEATS ON USE:

TRANSPORTATION DATA SOURCE 4-3

TITLE: Annual Report on Hazardous Materials Transportation

SPONSOR/AUTHOR: U.S. Department of Transportation	INDUSTRY: Transportation (multiple modes)
TIME PERIOD: Annual	FREQUENCY OF UPDATE: Annual / Usually June every year

NUMBER AND TYPE OF RECORDS:
Several thousand reports on incidents for each year by all modes.

DATA BOUNDARY: Hazardous Materials Transportation in U.S.

TAXONOMY REFERENCE: 2.2, 3.2, 4.2, 5.2	NATURE OF DATA: Number of Incidents

DATA TYPE AND ACCESS:

Report No. (None)
Cost $35.00

US Department of Transportation
Research and Special Programs Administration
Washington, DC 20590

DEFINITION OF SPILL/SELECTION CRITERIA FOR INCIDENTS:
Varies by mode

DESCRIPTION:

Provides DOT's analysis of reported incidents involving hazardous materials. All modes of transportation are included.

CAVEATS ON USE:

Not all incidents are reported.

TRANSPORTATION DATA SOURCE 4-4

TITLE: FACTS: A Databank for Accidents with Hazardous Materials

SPONSOR/AUTHOR: TNO Industrial Safety Databank	INDUSTRY: various
TIME PERIOD:	FREQUENCY OF UPDATE: ongoing

NUMBER AND TYPE OF RECORDS: Approximately 16,000 records on worldwide incidents for production, storage, transshipment, transport, use and disposal of chemicals

DATA BOUNDARY: see above

TAXONOMY REFERENCE: All	NATURE OF DATA: Number of Incidents

DATA TYPE AND ACCESS:
Off-line Database

TNO Environmental and Energy Research
Department of Industrial Safety
P.O. Box 342
7300 AH Apeldoorn, The Netherlands

Cost- Dependent on size of search

DEFINITION OF SPILL/SELECTION CRITERIA FOR INCIDENTS: Presence of danger and/or damage to the nearby residential population and the environment.

DESCRIPTION:
A databank containing technical data about incidents with hazardous materials that happened during such activities as production, storage, transshipment, transport, use and disposal of chemicals. Near-misses are also included. Data collection covers world-wide information. This is an off-line service and search requests must be submitted to TNO. Output can be obtained in hard copy or on computer files. The computer files are accessed using a program called PC-FACTS, also developed by TNO.

CAVEATS ON USE: Data is not comprehensive. The focus is on large scale accidents. Many smaller accidents that were not reported are not covered in this data set.

TRANSPORTATION DATA SOURCE 4-5

TITLE: Worldwide Oil Spill Incident Database

SPONSOR/AUTHOR: Oil Spill Intelligence Report

INDUSTRY: Oil

TIME PERIOD: 1978-Present

FREQUENCY OF UPDATE: Ongoing

NUMBER AND TYPE OF RECORDS: Information on oil spills gathered from government agencies, oil companies, and trade associations from around the world.

DATA BOUNDARY: Oil and petroleum based products

TAXONOMY REFERENCE: 1.2, 4.2, 5.2

NATURE OF DATA: Number of Incidents

DATA TYPE AND ACCESS:
Hardcopy of report or Database files in .dbf or ASCII format

Oil Spill Intelligence Report
37 Broadway
Arlington, MA 02174

Cost depends on number of records obtained

DEFINITION OF SPILL/SELECTION CRITERIA FOR INCIDENTS: Spills of oil or petroleum based products greater than 10,000 gallons

DESCRIPTION: Database containing the date, source type (tanker, barge, pipeline, or storage tank), amount and type of oil spilled, causes, damages and more from oil spills worldwide that involved at least 10,000 gallons of oil or petroleum based products. The user can either obtain the whole database or have searches conducted for them and obtain the hard copies of the output.

CAVEATS ON USE: Spills are not limited to transportation only. Production and storage incidents are also included. Transportation modes covered are barge, tanker and pipeline.

TRANSPORTATION DATA SOURCE 4-6

TITLE: Hazardous Materials Information System

SPONSOR/AUTHOR: U.S. Department of Transportation	INDUSTRY:
TIME PERIOD: 1971 - Present	FREQUENCY OF UPDATE: Ongoing

NUMBER AND TYPE OF RECORDS: Over 200,000 Incident Reports

DATA BOUNDARY: Hazardous Materials Transportation Incidents

TAXONOMY REFERENCE: 1.2, 2.2, 3.2, 4.2, 5.2, 6.2, 2.3, 3.3, 4.3, 5.3, 6.3	NATURE OF DATA: Number of Incidents

DATA TYPE AND ACCESS:
9 Track Tape - Mainframe
ASCII Format
Cost = $85 for whole database

Manager of Information Systems
U.S. Department of Transportation
Washington, D.C. 20950

DEFINITION OF SPILL/SELECTION CRITERIA FOR INCIDENTS: Hazardous material incident involving death, injury requiring hospitalization, property damage in excess of $50,000.00, evacuation, closure of a major transportation artery or facility, alteration of aircraft flight patterns, the release of a radioactive material or etiologic agent or a situation judged to merit notification.

DESCRIPTION: This is a database maintained by DOT that compiles incident reports on incidents which meet the above criteria.

CAVEATS ON USE:

TRANSPORTATION DATA SOURCE 4-7

TITLE: Major Hazard Aspects of the Transport of Dangerous Substances

SPONSOR/AUTHOR: UK Health and Safety Executive	INDUSTRY: Road and Rail Transportation with some Marine
TIME PERIOD: 1991	FREQUENCY OF UPDATE: Not Applicable

NUMBER AND TYPE OF RECORDS:

DATA BOUNDARY: Road and Rail Transportation. Marine Transportation was addressed where it could lead to major accidents which could affect non-employees or members of the public.

TAXONOMY REFERENCE: 2.0, 3.0, 5.0	NATURE OF DATA: Methodology, Accident Rates, Release Probabilities

DATA TYPE AND ACCESS:

	Health & Safety Executive		Her Majesty's Stationary Office
	Library of Information Services	or	Bernan-Unipub
	Baynards House		4611-F Assembly Drive
Report No. ISBN 011 885676 6	One Chepstow Place /Westbourne Grove		Lanham, MD 20706-4391
Cost = $18 + postage	London W2 4TF\United Kingdom		

DEFINITION OF SPILL/SELECTION CRITERIA FOR INCIDENTS:

DESCRIPTION: The report covers the consideration of the major hazard aspects of the transport of dangerous substances, identification of appropriate control measures, and advice on any additional action that might be necessary. There is also a comparison of relative risks between road and rail transport. Some accident rates and release probabilities can also be extracted.

CAVEATS ON USE:

TRANSPORTATION DATA SOURCE 4-8

TITLE: Performance of Pipework in the British Sector of the North Sea

SPONSOR/AUTHOR: UKAEA	INDUSTRY: Offshore Oil and Natural Gas
TIME PERIOD: January 1977-January 1983	FREQUENCY OF UPDATE: Not Applicable

NUMBER AND TYPE OF RECORDS: Failure rates based on 27 actual incidents for UK DOE reports.

DATA BOUNDARY: Offshore oil, gas and process fluid submarine pipelines within the UK Continental Shelf.

TAXONOMY REFERENCE: 1.1, 1.2.1	NATURE OF DATA: Failure Rates, Release Probabilities

DATA TYPE AND ACCESS:

Report No.: NCSR/GR/71; 8/87

The Editor
United Kingdom Atomic Energy Agency
Safety and Reliability Directorate
Culcheth, Warrington WA3 4NE
United Kingdom

DEFINITION OF SPILL/SELECTION CRITERIA FOR INCIDENTS:

The data analysis relates to pipeline incidents only for the three main categories of failure:
1) All reported incidents. Includes all reports of anchor, cable or trawl contact and also severe concrete coating damage.
2) Incidents causing shutdown. This group covers: (a) incidents where shutdown has occurred at some stage, either from damage repair or in anticipation of damage occurring, and (b) incidents where serious damage has occurred.
3) Incidents causing immediate shutdown. This group covers: (a) all incidents where any discharge from a pipe occurred and (b) faults in pipe laying which were not immediately found and repaired.

DESCRIPTION:

The aim of this project was to collect information on offshore oil, gas, and other related pipelines for the purpose of deriving reliability data. The data was gathered from various files, reports, and reference charts held at the Pipeline Inspectorate at the UK Department of Energy. Such files include construction files on new pipe, maintenance files concerning pipe work, and incident files logged by operators of any occurrence of pipeline leak or damage per the requirements of the UK DOE. The latter files provided the majority of the raw data used to compile this paper. This raw data includes incidents involving risers, valves, pig traps and main tie-in couplings (i.e., riser tie-ins) in addition to pipelines and summarizes more detail which is contained in the original report. Fault rates were found based on service, type, length, outside diameter and maximum allowed operating pressure.

CAVEATS ON USE:

TRANSPORTATION DATA SOURCE 4-9

TITLE: An Analysis of Reportable Incidents for Natural Gas Transmission and Gathering Lines; 1970 through June 1984

SPONSOR/AUTHOR: For American Gas Association By Battelle	INDUSTRY: Natural Gas
TIME PERIOD: 1970 through 1984	FREQUENCY OF UPDATE: Periodic

NUMBER AND TYPE OF RECORDS: Several thousand incidents of service and test failures.

DATA BOUNDARY: Gas Pipelines

TAXONOMY REFERENCE: 1.1, 1.2.1.3	NATURE OF DATA: Failure Rates, Release Probabilities, Number of Incidents, Exposure Data

DATA TYPE AND ACCESS:

Report
AGA Catalog No.: L51499

American Gas Association
Order Processing Department
1515 Wilson Blvd.
Arlington, VA 22209

DEFINITION OF SPILL/SELECTION CRITERIA FOR INCIDENTS:
Death or injury requiring hospitalization; Removal from service of any segment of transmission pipeline; Gas ignition; Test failure that occured while testing with gas or other medium; Property damage exceeding $5,000; Leak requiring immediate repair; Significant in judgment of operator.

DESCRIPTION:

This report is by Battelle Columbus Division to the Line Pipe Research Supervisory Committee of the American Gas Association. It presents an analysis of statistical data obtained from reports of leak or rupture (service) accidents and test failures in natural gas transmission and gathering lines during the period from 1970 through June 1984. All gas transmission companies were required to notify the Office of Pipeline Safety Operations in the event of a "reportable accident" as defined by the Code of Federal Regulations. The purpose of the study is to organize the reportable pipeline incident data into a meaningful format from which the safety record of the industry can be assessed.

CAVEATS ON USE: Reporting criteria changed in June 1984 to be significantly less comprehensive. As a result, there is inconsistency between this report and the one covering the period from June 1984 through 1989. In addition, while small leaks of natural gas may pose limited risks, the same size failure on a toxic chemcial pipeline may pose sufficient risk to merit its inclusion in the failure rate.

TRANSPORTATION DATA SOURCE 4-10

TITLE: An Analysis of Reportable Incidents for Natural Gas Transmission and Gathering Lines; June 1984 through 1990

SPONSOR/AUTHOR: For American Gas Association By Battelle	INDUSTRY: Natural Gas
TIME PERIOD: June 1984 through 1990	FREQUENCY OF UPDATE: Periodic

NUMBER AND TYPE OF RECORDS: 536 onshore pipeline incidents
85 offshore pipeline incidents

DATA BOUNDARY: Gas Pipelines

TAXONOMY REFERENCE: 1.1, 1.2.1.3	NATURE OF DATA: Failure Rates, Release Probabilities, Number of Incidents, Exposure Data

DATA TYPE AND ACCESS:	American Gas Association
Cost = $200.00	Order Processing Department
Report and/or	1515 Wilson Blvd.
computer files (.dbf format)	Arlington, VA 22209
AGA Catalog No.: L51664	

DEFINITION OF SPILL/SELECTION CRITERIA FOR INCIDENTS:
Fatality or injury requiring in-patient hospitalization; Property damage exceeding $50,000; Event resulting in emergency shutdown of LNG facility; Significant in judgment of operator.

DESCRIPTION:

This report presents an analysis of statistical data obtained from DOT Incident Reports for Gas Transmission and Gathering Systems for the period from June 1984 through 1990. The purpose of the study is to organize the reportable pipeline incident data into a meaningful format from which the safety records of the industry can be assessed. In addition, this data (the DOT Incident Report) has been computerized and the data files are also available from the AGA, through 1990.

CAVEATS ON USE:

Reporting criteria changed in June 1984. As a result, there is inconsistency between this report and the one for the period from 1970 through June 1984. Less stringent reporting criteria may fail to include certain failures that would be critical for materials other than natural gas. The more comprehensive data presented in the previous report (1970-1984) may be more applicable.

TRANSPORTATION DATA SOURCE 4-11

TITLE: Pipeline Reliability

SPONSOR/AUTHOR: de la Mare, R.; Anderson, O.

INDUSTRY: Petroleum and Natural Gas

TIME PERIOD: 1966-1977

FREQUENCY OF UPDATE: Not Applicable

NUMBER AND TYPE OF RECORDS: Data on frequency and cause of pipeline failures from a range of sources.

DATA BOUNDARY: Data for transmission lines carrying crude oil, natural gas, or refined petroleum products.

TAXONOMY REFERENCE: 1.2.1, 1.1

NATURE OF DATA: Failure Rates, Release Probabilities

DATA TYPE AND ACCESS:

Veritas Report No.: 80-0572

DNV Technica
P.O. Box 300
N-1322 Hovik
Norway

DEFINITION OF SPILL/SELECTION CRITERIA FOR INCIDENTS: Criteria varies due to comparison with several other data sources.

DESCRIPTION: This report is a comparison of North Sea pipeline failures to other major oil and gas pipeline systems throughout the world.

CAVEATS ON USE: Data used is not as comprehensive and detailed as its authors would have liked, and there are certain reservations regarding the scientific validity of the results.

TRANSPORTATION DATA SOURCE 4-12

TITLE: Pipeline Reliability: An Investigation of Pipeline Failure Characteristics and Analysis of Pipeline Failure Rates

SPONSOR/AUTHOR: Andersen, T. and Misund, A.	INDUSTRY: Petroleum and Natural Gas
TIME PERIOD: 1966-1981	FREQUENCY OF UPDATE: Not Applicable

NUMBER AND TYPE OF RECORDS: Data on frequency and cause of pipeline failures.

DATA BOUNDARY: Data specific to submarine and cross-country oil and natural gas pipelines worldwide.

TAXONOMY REFERENCE: 1.1	NATURE OF DATA: Failure Rates, Release Probabilities

DATA TYPE AND ACCESS: Journal Article	Journal of Petroleum Technology Vol. 35, No. 4, April 1983 pp. 709-717

DEFINITION OF SPILL/SELECTION CRITERIA FOR INCIDENTS:

DESCRIPTION: This article presents an overview of the causes and frequency of failure for submarine and cross-country pipelines handling oil and natural gas. It gives several tables and charts which include information on the type of pipeline, the cause of the failure and the number of failures. Data from failures in the US and North Sea are included. Failure rates based on the total length of piping are calculated.

CAVEATS ON USE:

TRANSPORTATION DATA SOURCE 4-13

TITLE: Safety of Interstate Natural Gas Pipelines

SPONSOR/AUTHOR: Federal Power Commission	INDUSTRY: Power
TIME PERIOD: 1950-1965	FREQUENCY OF UPDATE: Not Applicable

NUMBER AND TYPE OF RECORDS: 87 casualties/approx. 3000 pipe failures.

DATA BOUNDARY: US Interstate natural gas, large diameter, high pressure transmission pipelines.

TAXONOMY REFERENCE: 1.1, 1.2.1.3	NATURE OF DATA: Number of Incidents, Number of Releases

DATA TYPE AND ACCESS:

Report

DEFINITION OF SPILL/SELECTION CRITERIA FOR INCIDENTS:

DESCRIPTION: Military/Government type publication. It lists accidents with fatality scenarios that occurred during operation and maintenance of US interstate gas pipelines for 1950-1965. Also listed are individual pipework failures during that time, about 3000 entries, that have been compiled from various sources by the study committee.

Listing of accidents with fatalities and individual pipeline failures from 1950 to 1965.

CAVEATS ON USE: Pipework failure incident data is not all inclusive. Interstate pipelines were examined but not other types. Data may be out-of-date. This report is out of print, but can be obtained through inter library loans from several libraries throughout the United States.

TRANSPORTATION DATA SOURCE 4-14

TITLE: Gas Pipeline Incidents: A Report of the European Gas Pipeline Incident Data Group

SPONSOR/AUTHOR: European Gas Incident Data Group	INDUSTRY: Natural Gas
TIME PERIOD: 1970 - 1988	FREQUENCY OF UPDATE: Not Applicable

NUMBER AND TYPE OF RECORDS: Data on frequency and cause of pipeline failures

DATA BOUNDARY: Data on gas transmission pipelines only

TAXONOMY REFERENCE: 1.1, 1.2.1.3	NATURE OF DATA: Failure Rates, Release Probabilities

DATA TYPE AND ACCESS: Pipes & Pipelines International
July-August 1988
Journal Article pp. 11-14

DEFINITION OF SPILL/SELECTION CRITERIA FOR INCIDENTS:

DESCRIPTION: This report presents data collected from six European gas transmission companies on pipeline incidents. In particular, it presents the relationship between the frequency of incidents and the wall thickness of the pipeline for various incident types.

CAVEATS ON USE:

TRANSPORTATION DATA SOURCE 4-15

TITLE: Performance of Oil Industry Cross-Country Pipelines in Western Europe: Statistical Summary of Reported Spillages - 1991

SPONSOR/AUTHOR: CONCAWE The Hague	INDUSTRY: Oil
TIME PERIOD: 1991	FREQUENCY OF UPDATE: Annual

NUMBER AND TYPE OF RECORDS: Data on reported incidents from oil pipelines in Western Europe.

DATA BOUNDARY:

TAXONOMY REFERENCE: 1.1, 1.2	NATURE OF DATA: Failure Rates, Release Probabilities

DATA TYPE AND ACCESS:

National Technical Information Service
5285 Port Royal Road
Springfield, VA 22161

Report No.: PB93 187169

Cost = $17.50

DEFINITION OF SPILL/SELECTION CRITERIA FOR INCIDENTS: Spills greater than 1.0 cubic meters are counted. Spills less than this amount are addressed only when their environmental impact is noteworthy.

DESCRIPTION: This report is an annual report reviewing the performance of oil industry cross-country pipelines in Western Europe. It includes the details of the pipeline network and analyses of reported spill incidents. Causes are categorized as: mechanical failure; operational failure; corrosion; natural hazard; and third party hazard.

CAVEATS ON USE: Data is dependent on the number of companies participating in the survey.

TRANSPORTATION DATA SOURCE 4-16

TITLES: Analysis of Tank Cars Damaged in Accidents 1965 through 1986 (R-709)
Analysis of Aluminum Tank Cars Damaged in Accidents 1965 through 1986 (R-749)

SPONSOR/AUTHOR: Phillips, E.A. RPI/AAR Tank Car Safety Program	INDUSTRY: Railroads
TIME PERIOD: 1965-1986	FREQUENCY OF UPDATE: Not Applicable

NUMBER AND TYPE OF RECORDS:
Technical Reports presenting data

DATA BOUNDARY: Railroad tank car sustaining damage to the tank or its fittings in railroad accidents in the United States

TAXONOMY REFERENCE: 2.2	NATURE OF DATA: Release Probabilities

DATA TYPE AND ACCESS:

Reports R-709 and R-749

Association of American Railroads
Research and Test Dept.
3140 South Federal Street
Chicago IL 60616

Cost = $100 per report

DEFINITION OF SPILL/SELECTION CRITERIA FOR INCIDENTS:
Any tank car sustaining damage to the tank or the tank fitting should be reported to the RPI/AAR Tank Car database.

DESCRIPTION:

Using the RPI/AAR database of over 29,000 tank car damage reports, data are provided on the incidence of damage and hazardous material release by tank car type. These are the most recent published analyses of this database. (April 1993)

CAVEATS ON USE:
This is a voluntary industry effort, therefore completeness of reporting of damaged tank cars is unknown. Only periodic reports are publicly available - the database itself is proprietary. Note that the tank car damage reporting criteria are narrower than hazmat car damaged/derailed used for the FRA Railroad Accident Data.

TRANSPORTATION DATA SOURCE 4-17

TITLE: Accident / Incident Bulletin.

SPONSOR/AUTHOR: Federal Railroad Administration

INDUSTRY: U.S. Railroads

TIME PERIOD: 1975 - Present

FREQUENCY OF UPDATE: Annual / July each year

NUMBER AND TYPE OF RECORDS:
Graphs and tables summarizing the safety performance of US railroads for the year.

DATA BOUNDARY: US Common-Carrier railroads

TAXONOMY REFERENCE: 2.1, 2.2

NATURE OF DATA: Number of Incidents, Number of Releases

DATA TYPE AND ACCESS:
Report No. (None)

FRA Office of Safety
400 7th St. S.W.
Washington, DC 20590

Free

DEFINITION OF SPILL/SELECTION CRITERIA FOR INCIDENTS:

All accidents causing damage exceeding a specified dollar value of damage to railroad vehicles and facilities, currently (1992) approx. $6,200. Both hazmat and non-hazmat accidents are included.

DESCRIPTION:

This annual report provides numerous graphs and tables summarizing the incidence of railroad accidents, and personal injury incidents with respect to cause, location, severity, operating circumstances, etc. Several tables provide details of accidents involving hazmat shipments, however no information is provided on the type of hazmat involved.

CAVEATS ON USE:

TRANSPORTATION DATA SOURCE 4-18

TITLE: Railroad Accident/Incident Database.

SPONSOR/AUTHOR: US Federal Railroad Administration	INDUSTRY: Railroad
TIME PERIOD: 1975 - Present	FREQUENCY OF UPDATE: Annual / Usually around August each year

NUMBER AND TYPE OF RECORDS: Comprehensive data on each reportable railroad accident (currently approx. 3000/year)

DATA BOUNDARY: US Common-Carrier railroads - all accidents causing damage exceeding a specified dollar value to railroad vehicles and facilities.

TAXONOMY REFERENCE: 2.1, 2.2	NATURE OF DATA: Number of Incidents, Number of Releases

DATA TYPE AND ACCESS:

ASCII Computer files on Mainframe tapes - 9 Track

$35.00/tape

FRA Office of Safety
400 7th St. S.W.
Washington, DC 20590

DEFINITION OF SPILL/SELECTION CRITERIA FOR INCIDENTS:
All accidents causing damage exceeding a specified dollar value of damage to railroad vehicles and facilities, currently (1992) approx. $6,200. Both hazmat and non-hazmat accidents are included.

DESCRIPTION:

Provides details of train(s) involved in accidents, operating conditions (e.g. speed), accident cause, and accident consequences (number of cars in a train derailed). For hazardous materials shipments data includes a number of hazmat cars in the train, number derailed in the accident, number releasing hazmat and number of people evacuated as a result of a release. No information is given on identity of hazmat.

The requestor sends in a search request with a mainframe tape. The ouput is loaded on the tape in ASCII format.

CAVEATS ON USE: Damage thresholds change with time, plus underlying composition of railcar protective features also changes with time (i..e., data for 1970's and early 1980's represents a time period when few shelf couplers were in use).

TRANSPORTATION DATA SOURCE 4-19

TITLE: Statistical Trends in Railroad Hazardous Materials Transportation Safety 1978 to 1986.

SPONSOR/AUTHOR: Harvey, A.E., Conlon, P.C. and Glickman, T.S. Association of American Railroads	INDUSTRY: Railroads
TIME PERIOD: 1978-1986	FREQUENCY OF UPDATE: Not Applicable

NUMBER AND TYPE OF RECORDS:
Graphs, tables and descriptive text.

DATA BOUNDARY: US Common-Carrier railroads

TAXONOMY REFERENCE: 2.1, 2.2, 2.3	NATURE OF DATA: Accident Rates, Release Probabilities

DATA TYPE AND ACCESS:
Report R-640

Association of American Railroads
50 F Street N.W.
Washington, DC 20001

Cost = $30.00

DEFINITION OF SPILL/SELECTION CRITERIA FOR INCIDENTS:
a. All train accidents involving a hazardous materials car where railroad property damage exceeded a specified dollar value.

b. All releases of hazardous material from a tank car, however caused.

DESCRIPTION:
The report provides a comprehensive discussion of the causes of railroad accidents involving hazardous materials and releases of hazardous materials in railroad operations, however caused. Summary data are included (wholly derived from publicly available databases) and trends are discussed.

CAVEATS ON USE:
The report is now getting out of date for current use, especially since there were significant changes in the mix of train accidents and in tank car protective features between 1978 and 1985.

TRANSPORTATION DATA SOURCE 4-20

TITLE: Annual Report on Hazardous Materials Transportation by Rail.

SPONSOR/AUTHOR: Association of American Railroads, Bureau of Explosives	INDUSTRY: Railroads
TIME PERIOD: 1975-Present	FREQUENCY OF UPDATE: Annual/June every year

NUMBER AND TYPE OF RECORDS:
Summary data on both shipments of hazardous material by commodity, and number of releases or leaks

DATA BOUNDARY: US Common-Carrier railroads

TAXONOMY REFERENCE: 2.1, 2.2	NATURE OF DATA: Accident Rates, Number of Incidents, Exposure Data

DATA TYPE AND ACCESS:
Report No. (None) Association of American Railroads
50 F Street N.W.
Washington, DC 20001

Free

DEFINITION OF SPILL/SELECTION CRITERIA FOR INCIDENTS:
a. Hazardous material shipments in train accidents reportable to the Federal Railroad Administration.

b. Hazardous Material leaks or releases reportable to the Office of Hazardous Material Transportation of the US DOT.

DESCRIPTION:

Tabulation of top 125 hazardous material shipments by carloads, derived from railroad freight shipment tabulations of leakage frequency by material class and other accident and release data.

Narrative commentary on the year's data and trends.

CAVEATS ON USE:

TRANSPORTATION DATA SOURCE 4-21

TITLE: Truck Accident and Fatality Rates Calculated from California Highway Accident Statistics for 1980 & 1981.

SPONSOR/AUTHOR:
Smith, R., Wilmot, E.

INDUSTRY: Various

TIME PERIOD: 1980-1981

FREQUENCY OF UPDATE:

NUMBER AND TYPE OF RECORDS: Several thousand accidents from the California highway-accident database for 1980 & 1981.

DATA BOUNDARY: Truck transportation fatalities and accidents

TAXONOMY REFERENCE: 3.1

NATURE OF DATA: Accident Rates

DATA TYPE AND ACCESS: National Technical Information Service
5285 Port Royal Road
Springfield, VA 22161

Report No. DE 83004998

Cost = $19.50

DEFINITION OF SPILL/SELECTION CRITERIA FOR INCIDENTS:
Fatalities, injuries and property damage

DESCRIPTION: California state highway accident rates for three types of truck vehicles (pickup, truck without trailer, truck with trailer) were analyzed for 1980 and 1981 and for various road types in each of eleven state highway districts.

CAVEATS ON USE:

TRANSPORTATION DATA SOURCE 4-22

TITLE: Fleet Accident Rates

SPONSOR/AUTHOR:
National Safety Council

INDUSTRY: Many

TIME PERIOD: Annual

FREQUENCY OF UPDATE: Annual

NUMBER AND TYPE OF RECORDS:
Analysis of accidents reported for particular year with comparisons to previous years.

DATA BOUNDARY: Fleets who reported data, e.g., 1210 fleets in 1987.

TAXONOMY REFERENCE: 3.1

NATURE OF DATA: Accident Rates, Number of Incidents, Exposure Data

DATA TYPE AND ACCESS:

Product No. 229.27

Cost = $8.50 & Shipping & Handling

National Safety Council
444 N. Michigan Avenue
Chicago, IL 60611

DEFINITION OF SPILL/SELECTION CRITERIA FOR INCIDENTS:

Death, injury, or property damage.

DESCRIPTION:

Data is collected from participating fleets every month including number of miles driven, the cumulative number of accidents, and fleet size. Analysis is done and data is broken out by vehicle type and operation (e.g., trucks, buses & cars).

CAVEATS ON USE:

It is not known to what extent participating organizations reflect the experience of all fleets.

TRANSPORTATION DATA SOURCE 4-23

TITLE: An Assessment of the Risk of Transporting Gasoline by Truck.

SPONSOR/AUTHOR:
Battelle Pacific Northwest Laboratory

INDUSTRY: Energy

TIME PERIOD: 1978

FREQUENCY OF UPDATE: Not Applicable

NUMBER AND TYPE OF RECORDS:

DATA BOUNDARY: Truck transportation of Gasoline.

TAXONOMY REFERENCE: 3.1, 3.2

NATURE OF DATA: Accident Rates, Release Probabilities, Number of Incidents, Number of Releases, Exposure Data

DATA TYPE AND ACCESS: National Technical Information Service (NTIS)
5285 Port Royal Road
Springfield, VA 22151

Report No. PNL-2133

Cost =$27.00

DEFINITION OF SPILL/SELECTION CRITERIA FOR INCIDENTS:

DESCRIPTION: The report presents the risk of shipping gasoline by truck. Based on data available at the time, accident rates, release probabilities and consequences were derived to generate overall risks produced by the transportation of gasoline.

CAVEATS ON USE: Accident data used is from Bureau of Motor Carrier Safety (BMCS) database for years 1969-1972.

TRANSPORTATION DATA SOURCE 4-24

TITLE: Summary of Motor Vehicle Accidents in the Petroleum Industry for 1988

SPONSOR/AUTHOR: American Petroleum Institute	INDUSTRY: Petroleum
TIME PERIOD: 1988	FREQUENCY OF UPDATE: Not Applicable

NUMBER AND TYPE OF RECORDS: Accidents reported by 67 oil and gas companies.

DATA BOUNDARY: Motor Vehicle Accidents

TAXONOMY REFERENCE: 3.1	NATURE OF DATA: Accident Rates, Number of Incidents, Exposure Data

DATA TYPE AND ACCESS:

Report No. 855-19990

Cost = $15.00

American Petroleum Institute
Publications & Distribution Section
1220 L. Street N.W.
Washington, DC 20005

DEFINITION OF SPILL/SELECTION CRITERIA FOR INCIDENTS:

DESCRIPTION:

The report documents the number of accidents by vehicle type, mileage and company surveyed. Data is reported in motor vehicle accidents per million miles. Vehicle types are broken into passenger vehicles, service vehicles and product delivery trucks.

CAVEATS ON USE: Company participation in the survey is voluntary and fluctuates year by year.

TRANSPORTATION DATA SOURCE 4-25

TITLE: Present Practices of Highway Transportation of Hazardous Materials - Task C Analysis of Existing Databases

SPONSOR/AUTHOR: Harwood, D.	INDUSTRY: Highway Transportation
TIME PERIOD: 1987	FREQUENCY OF UPDATE: Not Applicable

NUMBER AND TYPE OF RECORDS: Several thousand incidents reported in four databases.

DATA BOUNDARY: Analysis of 4 highway transportation incident databases.

TAXONOMY REFERENCE: 3.1	NATURE OF DATA: Accident Rates, Release Probabilities, Number of Incidents, Number of Releases, Exposure Data

DATA TYPE AND ACCESS:

Report No. PB 91124016

Cost = $36.50

National Technical Information Service
5285 Port Royal Road
Springfield, VA 22151

DEFINITION OF SPILL/SELECTION CRITERIA FOR INCIDENTS:

DESCRIPTION:

The report presents the results of the analysis of existing accident, incident and exposure databases. The analysis focused on four databases directly relevant to highway transportation of Hazardous Materials.

- Research and Special Programs Administration (RSPA) Hazardous Materials Incident Databases
- Bureau of Motor Carrier Safety (BMCS) Accident Database
- Missouri Statewide Traffic Accident Reporting System (STARS)
- Bureau of the Census 1982 Truck Inventory and Use Survey (TIUS) Database

CAVEATS ON USE:

TRANSPORTATION DATA SOURCE 4-26

TITLE: Various - State Traffic Accident Analyses

SPONSOR/AUTHOR: State Department of Transportation

INDUSTRY: Transportation

TIME PERIOD: Generally Annual

FREQUENCY OF UPDATE: Ongoing

NUMBER AND TYPE OF RECORDS: Varies

DATA BOUNDARY: Varies, may be all vehicles or may be subdivided by vehicle classes. May include accident rates, accident counts, or vehicle counts.

TAXONOMY REFERENCE: 3.0

NATURE OF DATA: Number of Incidents, Exposure Data, Number of Releases

DATA TYPE AND ACCESS: Varies from hard copy reports to database output from the state to raw data available to user.

Individual State Departments of Transportation or other Authorities

DEFINITION OF SPILL/SELECTION CRITERIA FOR INCIDENTS: Varies

DESCRIPTION: Generally obtained via the motor vehicle safety offices within individual state Departments of Transportation.

CAVEATS ON USE: Accident Data may not be readily obtainable from one source. Several sources may need to be consulted and relevant accident data pieced together from these. For example, accident data for the state of California can be pieced together using information obtained from the California DOT (Caltrans) and the California Highway Patrol. (See Section 4.5)

TRANSPORTATION DATA SOURCE 4-27

TITLE: Procedures for the Development of Truck Accident Rates and Release Probabilities for Use in Hazmat Routing Analysis

SPONSOR/AUTHOR: Harwood, D., Russell, E.	INDUSTRY: Truck Transportation
TIME PERIOD: 1992	FREQUENCY OF UPDATE: Not Applicable

NUMBER AND TYPE OF RECORDS: Over 200,000 Incident Reports

DATA BOUNDARY: Hazardous Materials Transportation Incidents

TAXONOMY REFERENCE: 3.1, 3.2	NATURE OF DATA: Accident Rates, Release Probabilities

DATA TYPE AND ACCESS:
Paper Contained in Proceedings:
Transportation of Dangerous Goods:
Assessing the Risks
Cost=$60.00

Institute for Risk Research
University of Waterloo
Waterloo, Ontario
N2L 3G1

DEFINITION OF SPILL/SELECTION CRITERIA FOR INCIDENTS:

DESCRIPTION: The paper presents a procedure that can be used to develop estimated average truck accident rates and release probabilities for different highway and area types. The procedure is demonstrated using data from three states (California, Illinois and Michigan). These estimates can be used as default values in the absence of more specific data.

CAVEATS ON USE: The three states on which the default numbers are based all have different accident reporting criteria and as a result, have different accident rates.

TRANSPORTATION DATA SOURCE 4-28

TITLE: Waterborne Commerce of the United States

SPONSOR/AUTHOR: U.S. Army Corps of Engineers	INDUSTRY: Commercial Transportation
TIME PERIOD: 1989 is most recent compilation	FREQUENCY OF UPDATE: ongoing

NUMBER AND TYPE OF RECORDS: Several thousand records available by region of the country. Available in database form or hard copy.

DATA BOUNDARY: Marine transportation levels (i.e. denominator data)

TAXONOMY REFERENCE: 4.1, 5.1	NATURE OF DATA: Exposure Data

DATA TYPE AND ACCESS:
Manuscript Records, Computer Tape or Diskette

Cost: $15-20 depending on information

Commander
U.S. Army Engineer, District, New Orleans Computer Tape
Attn: CELMN-ED-S
P.O. Box 60267
New Orleans, LA 70160-0267

DEFINITION OF SPILL/SELECTION CRITERIA FOR INCIDENTS: N/A

DESCRIPTION: Vessel and cargo movement information reported to the Corps of Engineers by carriers engaged in the commercial transportation of goods on the navigable waterways of the United States. Prepackaged data and specialized search capabilities are available from the Corps.

CAVEATS ON USE:

TRANSPORTATION DATA SOURCE 4-29

TITLE: Polluting Incidents In and Around US Waters.

SPONSOR/AUTHOR: U.S. Coast Guard	INDUSTRY: Various
TIME PERIOD: 1982 and 1983	FREQUENCY OF UPDATE:

NUMBER AND TYPE OF RECORDS: Analysis of several thousand reports on oil and hazardous substance pollution in and around US waters.

DATA BOUNDARY: Oil and Hazardous Substance Pollution reports taken from the Coast Guard's Pollution Incident Reporting System (PIRS) Database.

TAXONOMY REFERENCE: 2.2, 3.2, 4.2, 5.2	NATURE OF DATA: Number of Incidents, Number of Releases

DATA TYPE AND ACCESS:

Report

Commandant (G-WP)
U.S. Coast Guard
2100 Second Street S.W.
Washington, D.C. 20593

DEFINITION OF SPILL/SELECTION CRITERIA FOR INCIDENTS: Criteria for oil spills is the creation of a sheen on the water. Criteria for hazardsous substances are the individual reportable quantities (RQ's).

DESCRIPTION: This publication provides general information concerning pollution discharges into US navigable waters for CY 1982 and 1983

CAVEATS ON USE: 1983 statistics in this report were preliminary at the time of publication and not all data for that year had been collected. Data for spills covers only those which entered the water.

TRANSPORTATION DATA SOURCE 4-30

TITLE: Commercial Vessel Safety. Risk Assessment Study. Volume I. Survey of Data for Marine Risk Assessment

SPONSOR/AUTHOR: Reiter, A. D. , Faragher, W.E.	INDUSTRY: Marine Transportation
TIME PERIOD: September 1979	FREQUENCY OF UPDATE: Not Applicable

NUMBER AND TYPE OF RECORDS:
Twenty three data sources are reviewed.

DATA BOUNDARY: Marine Transportation

TAXONOMY REFERENCE: 4.1, 4.2, 5.1, 5.2	NATURE OF DATA: Accident Rates, Release Probabilities, Number of Incidents, Exposure Data, Number of Releases

DATA TYPE AND ACCESS:

Microfiche Report
No. AD-A077 719/3/XPS
Cost = $9.00

National Technical Information Service
5285 Port Royal Road
Springfield, VA 22161

DEFINITION OF SPILL/SELECTION CRITERIA FOR INCIDENTS:

Varies by source

DESCRIPTION:

A survey and evaluation of data applicable to the evaluation of the risks involved in marine operations. Twenty three data systems in the following categories were reviewed: marine safety activities, marine pollution, marine traffic, vessel repair costs, vessel accidents, personnel injuries, vessel population and vessel violation injuries. Discussions of each system includes sources of data, procedures for collecting and recording the data, and potential error sources.

CAVEATS ON USE:

TRANSPORTATION DATA SOURCE 4-31

TITLE: Lloyds Maritime Information Services MARDATA

SPONSOR/AUTHOR:
Lloyds Maritime Information Service

INDUSTRY: Maritime

TIME PERIOD: 1964 - present

FREQUENCY OF UPDATE:
Ongoing

NUMBER AND TYPE OF RECORDS:
Casualty histories for more than 30,000 ships

DATA BOUNDARY: Data since 1978 for ships greater than 1000 gross tonnage
Data since 1964 for tankers greater than 4000 DWT

TAXONOMY REFERENCE: 4.1, 4.2, 4.3, 5.1, 5.2, 5.3

NATURE OF DATA: Number of Incidents, Number of Releases

DATA TYPE AND ACCESS:

Search data from database provided in hard copy or ASCII formatted diskettes.

Lloyds Maritime Information Service, Inc.
1200 Summer Street
Stamford, CT 06905

DEFINITION OF SPILL/SELECTION CRITERIA FOR INCIDENTS:

DESCRIPTION:

Contains serious casualty histories for more than 30,000 ships weighing more than 1000 gross tonnage since 1978. Also covers casualties to tankers greater than 4000 deadweight tons since 1964. User requests a search from Lloyds and the ouput is provided to them.

CAVEATS ON USE:

TRANSPORTATION DATA SOURCE 4-32

TITLE: Marine Casualty Information Reporting System (CASMAIN)

SPONSOR/AUTHOR: US Coast Guard	INDUSTRY: Maritime Transportation
TIME PERIOD: 1981-1991	FREQUENCY OF UPDATE: Superseded by MIN Module (see separate Data Source #33)

NUMBER AND TYPE OF RECORDS:
Marine Casualty Incidents

DATA BOUNDARY: US Waterways

TAXONOMY REFERENCE: 4.1.1, 4.1.2, 4.2.1, 4.2.2, 5.1.1, 5.1.2, 5.2	NATURE OF DATA: Number of Incidents,

DATA TYPE AND ACCESS:

Database available in hard copy or 9 track mainframe tape, ASCII format

Office of Information Management
U.S. Coast Guard
2100 Second Street, SW
Washington, DC 20593

DEFINITION OF SPILL/SELECTION CRITERIA FOR INCIDENTS:

Reportable incidents involve death, serious injury, structural damage greater than $25,000, loss of propulsion or steering. All groundings are also reportable.

DESCRIPTION:

Provides casualty data on US Waterways for marine transportation based on incident criteria above. The user can either obtain the whole database or have limiting searches conducted for them.

CAVEATS ON USE:

Incomplete due to the fact that some events go unreported. Cargo spills are reported as Cargo Damage and are not a reliable source of pollution events. Incidents after 1/92 are incorporated into MIN Module Data. (See separate data sheet)

TRANSPORTATION DATA SOURCE 4-33

TITLE: Marine Investigations (MIN) Module of MSIS and CASMAIN

SPONSOR/AUTHOR: US Coast Guard	INDUSTRY: Marine Transportation
TIME PERIOD: January 1992 to present	FREQUENCY OF UPDATE: Ongoing

NUMBER AND TYPE OF RECORDS:
Marine Casualty and Pollution Incidents

DATA BOUNDARY: US Waterways

TAXONOMY REFERENCE: 4.1, 4.2, 4.3, 5.1, 5.2, 5.3	NATURE OF DATA: Number of Incidents, Number of Releases

DATA TYPE AND ACCESS:

Hard copy of requested database search	Office of Marine Safety U.S. Coast Guard 2100 Second Street, SW Washington, DC 20593

DEFINITION OF SPILL/SELECTION CRITERIA FOR INCIDENTS:

See incident criteria for PIRS and CASMAIN (Transportation Data Source #'s 29 and 32)

DESCRIPTION:
Combines casualty data from CASMAIN with pollution data from Pollution Incident Reporting System (PIRS). Dates from Jan 1992 to present. Data prior to this date needs to be obtained from CASMAIN or PIRS individually.

CAVEATS ON USE:

Requests must be made to the Coast Guard and they will conduct the statistical analysis and provide a hard copy of their results.

TRANSPORTATION DATA SOURCE 4-34

TITLE: Fire/Explosion probabilities on liquid gas ships.

SPONSOR/AUTHOR: Blything, K., Edmondson, J.

INDUSTRY: marine transportation

TIME PERIOD: 1984

FREQUENCY OF UPDATE: Not Applicable

NUMBER AND TYPE OF RECORDS: Records of fires and explosions for worldwide liquified gas carrier fleet

DATA BOUNDARY: LNG & LPG Ships

TAXONOMY REFERENCE: 5.1.1.4, 5.1.2.4

NATURE OF DATA: Accident Rates,

DATA TYPE AND ACCESS:

Report No. SRD R 292

Safety and Reliability Directorate
U.K. Atomic Energy Authority
Wigshaw Lane, Culcheth Warrington
WA3 4NE
United Kingdom

DEFINITION OF SPILL/SELECTION CRITERIA FOR INCIDENTS: Fire/Explosion anywhere on an LNG or LPG ship.

DESCRIPTION:

Examination of historical accident data concerning fire/explosion incidents that have occurred on carriers used for the transport of liquid gases. For LNG Carriers, all size carriers were examined over the period 1971-1981. For LPG Carriers , incidents on carriers >5000m3 were examined over the period 1971-1981 and on carriers <5000m3 over the period 1977-1981.

CAVEATS ON USE:

TRANSPORTATION DATA SOURCE 4-35

TITLE: Analysis of Serious Casualties to Seagoing Tankers 1977-1991 and Analysis of Serious Casualties to Fishing Vessels 1982-1991.

SPONSOR/AUTHOR: International Maritime Organization (IMO)	INDUSTRY: Tanker
TIME PERIOD: 1977-1991	FREQUENCY OF UPDATE: Not Applicable

NUMBER AND TYPE OF RECORDS:
Analysis of Tanker casualty reports.

DATA BOUNDARY: Tanker incidents

TAXONOMY REFERENCE: 5.1	NATURE OF DATA: Accident Rates

DATA TYPE AND ACCESS:

Report No. W/9247e	International Maritime Organization 4 Albert Embankment London, SE1 7SR England	or	New York Nautical Instrument & Service Corp. 14 West Broadway New York, NY 10013

DEFINITION OF SPILL/SELECTION CRITERIA FOR INCIDENTS:
Serious Casualties include: A) Fire, explosion, collision, grounding, contact, heavy weather damage, ice damage, hull cracking or suspected hull defect resulting in: 1) Structural damage rendering the ship unseaworthy
2) Loss of life;
3) Pollution (regardless of quantity);
B) A breakdown necessetating towage of shore assistance;
C) A total loss

DESCRIPTION:

This report contains:

1. An analysis of serious casualties to seagoing tankers of 6,000 gross tonnage and above for 1977-1991.
2. Analysis of serious casualties to tankers of less than 6,000 gross tonnage for the period of 1978-1991.
3. Analysis of serious casualties to fishing vessels of 100 gross tonnage and above for 1982-1991.

CAVEATS ON USE: Separate statistics for oil and chemical tankers are not possible due to frequent interchangeability between trades. Casualty rates for combination carriers may be understated.

TRANSPORTATION DATA SOURCE 4-36

TITLE: Worldwide Tanker Spill Database

SPONSOR/AUTHOR: Minerals Management Service	INDUSTRY: Marine Transportation
TIME PERIOD: 1974- present	FREQUENCY OF UPDATE: Ongoing

NUMBER AND TYPE OF RECORDS:
Approx. 700 tanker spill records

DATA BOUNDARY: Oil spills worldwide

TAXONOMY REFERENCE: 4.1, 4.2, 5.1, 5.2	NATURE OF DATA: Number of Releases

DATA TYPE AND ACCESS:

	U.S. Department of Interior / Minerals Management Service
Computer Files -	Branch of Environmental Modeling
ASCII or dbase format	381 Elden Street, Mail Stop 4340
Cost = Free	Herndon, VA 22070

DEFINITION OF SPILL/SELECTION CRITERIA FOR INCIDENTS:

Oil and Petroleum spills greater than 1,000 bbls from a vessel.

DESCRIPTION:
Data available in dBase or ASCII format. Includes information on data and time of incident, vessel, cargo, spill volume, location, etc.

CAVEATS ON USE:

TRANSPORTATION DATA SOURCE 4-37

TITLE: TIGER/Line Census Files, 1990

SPONSOR/AUTHOR: US Department of Commerce

INDUSTRY: Not Applicable

TIME PERIOD: 1991

FREQUENCY OF UPDATE:

NUMBER AND TYPE OF RECORDS: Digital data from 1990 Census with geographic and cartographic information

DATA BOUNDARY: All US Counties

TAXONOMY REFERENCE: 7.1, 7.3

NATURE OF DATA: Population Data, Route Data

DATA TYPE AND ACCESS:

Computer Tape
(9-track, 1600 or 6250 bpi,
ASCII, or EBCDIC) or CD-ROM

Customer Services
Data User Services Division
Bureau of the Census
Washington, DC 20233

Cost - dependent on geographical areas represented on the files

DEFINITION OF SPILL/SELECTION CRITERIA FOR INCIDENTS: Not Applicable

DESCRIPTION: The 1990 Census TIGER/Line file is an extract of selected geographic and cartographic information from the Census Bureau's TIGER database. The files contain basic information for 1990 census geographic area codes, basic map features, and their names and address ranges.

CAVEATS ON USE:

TRANSPORTATION DATA SOURCE 4-38

TITLE: Summary Tapes (STS)

SPONSOR/AUTHOR: US Department of Commerce

INDUSTRY: not applicable

TIME PERIOD: 1991

FREQUENCY OF UPDATE:

NUMBER AND TYPE OF RECORDS: Data on population characteristics from 1990 Census

DATA BOUNDARY: All US Counties

TAXONOMY REFERENCE: 7.1

NATURE OF DATA: Population Data

DATA TYPE AND ACCESS:

CD-ROM

Cost = $150 per disk

Customer Services
Data User Services Division
Bureau of the Census
Washington, DC 20233

DEFINITION OF SPILL/SELECTION CRITERIA FOR INCIDENTS: Not Applicable

DESCRIPTION: Census data broken down into tracts and blocks as well as cities and counties. Describes population characteristics such as age, sex, race, education, etc. Tracts are areas between 2500 and 8000 people and are homogeneous with respect to population characteristics, economic status and living conditions. Blocks are smaller areas bounded by major roads, streams and invisible boundaries such as county limits.

CAVEATS ON USE:

TRANSPORTATION DATA SOURCE 4-39

TITLE: 1992 County and City Extra; Annual Metro, City and County Data Book

SPONSOR/AUTHOR: Slater, C., Hall, G.

INDUSTRY: Not Applicable

TIME PERIOD: Data from 1990 Census with local updates

FREQUENCY OF UPDATE: Annual

NUMBER AND TYPE OF RECORDS: Statistical Data on Cities, Counties, and Metropolitan areas in the US.

DATA BOUNDARY: 1990 US Census based population data

TAXONOMY REFERENCE: 7.1

NATURE OF DATA: Population Data

DATA TYPE AND ACCESS:

Book = $89.95
ISBN No. 0-89059-010-9

Bernan Press
4611-F Assembly Drive
Lanham, MD 20706

DEFINITION OF SPILL/SELECTION CRITERIA FOR INCIDENTS: Not Applicable

DESCRIPTION: This is an annual publication providing statistical information available for every state, county, metropolitan area, city and place in the United States.

CAVEATS ON USE:

TRANSPORTATION DATA SOURCE 4-40

TITLE: National Oceanic and Atmospheric Administration Environmental Data Service

SPONSOR/AUTHOR: U.S. Department of Commerce

INDUSTRY: Not Applicable

TIME PERIOD: various

FREQUENCY OF UPDATE: Annual

NUMBER AND TYPE OF RECORDS: Data on wind speed and stability class for individual weather stations

DATA BOUNDARY:

TAXONOMY REFERENCE: 7.2

NATURE OF DATA: Climactic Data

DATA TYPE AND ACCESS:

Data for specific locales at nominal cost	National Oceanic and Atmospheric Admin. National Climatic Data Center Federal Building Asheville, NC 28801	or	CEAM Bulletin Board (See Below)

DEFINITION OF SPILL/SELECTION CRITERIA FOR INCIDENTS: Not Applicable

DESCRIPTION: Data are presented by stability classes and also combined for the period indicated; first as a bivariate frequency distribution of wind direction vs. wind speed, and second as normalized values. Stability classes are based on Pasquill's class structure. Outputs are known as STAR runs.

User specifies locale desired and then can receive runs previously processed at very low cost, or have a newer run executed at a higher cost and with a longer delay.

NOAA weather information is also available on-line through the EPA's CEAM Bulletin Board for 1948-1983 [(706) 546-3402]

CAVEATS ON USE:

TRANSPORTATION DATA SOURCE 4-41

TITLE: Statistical Abstract of the United States, 112th Edition

SPONSOR/AUTHOR: U.S. Department of Commerce

INDUSTRY: Not Applicable

TIME PERIOD: 1992

FREQUENCY OF UPDATE: Annual

NUMBER AND TYPE OF RECORDS: Statistical data on population, geographic and environmental (climatic) data for areas of the U.S.

DATA BOUNDARY: United States Data

TAXONOMY REFERENCE:

NATURE OF DATA: Population Data, Climatic Data

DATA TYPE AND ACCESS:

Book
ISBN No. 0-16-038080-4

U.S. Government Printing Office
Superintendent of Documents
Mail Stop SSOP
Washington, DC 20402-9328

DEFINITION OF SPILL/SELECTION CRITERIA FOR INCIDENTS: Not Applicable

DESCRIPTION: The book provides annually updated data on population and geographic characteristics among others for areas of the U.S.. Among the geographic statistics are data on weather patterns for key U.S. cities.

CAVEATS ON USE:

5

Measurement, Calculation, and Presentation of Risk Estimates

Risk has two elements: the frequency of occurrence and the potential consequences. Transportation risk analysis concerns risk arising from the release of hazardous materials as a result of accident- and/or non–accident-initiated events. Therefore, injuries or fatalities associated with the transportation accident itself are normally excluded from consideration in order to focus on those that are attributable to the cargo being transported. General transportation injuries are more common than injuries associated with HAZMAT releases. Many accidents produce injuries while relatively few lead to releases, and even fewer lead to hazardous consequences (injuries) from those releases. The inclusion or exclusion of general transportation injuries should be determined for each database or study evaluated to ensure that data are used correctly and results are compared on a common basis.

This chapter assumes that the likelihood and the consequences of a release have been determined. Risk calculation methods for TRA are basically the same as the quantification techniques given in the *CPQRA Guidelines* (CCPS, 1989), although there are some risk measures which are unique to or more commonly associated with transportation risk. For qualitative estimates of risk, the approach is usually an integral part of the screening techniques and therefore varies from one technique to another. A brief discussion of qualitative approaches is given in Sections 5.1 and 5.2.

No one risk measure is complete or adequate for conveying all of the available information about transportation risk. Hence, the most appropriate measure or measures will depend on:

- the nature of the question being examined
- the audience for the results

- the format of previous results that may be used for comparative purposes
- the information and resources available for the study.

Section 5.1 describes some measures and also gives guidelines for selecting the measure or measures for a particular study. Calculation techniques are given in Section 5.2, and the reader is also referred to the *CPQRA Guidelines* (CCPS, 1989) for a more complete discussion of some of the measures.

All measures and techniques given in this book allow an analyst to *estimate* risk. In order to understand the results, Section 5.3 discusses the uncertainty and sensitivity of results. This will assist in communicating with both the decision-makers and any one else interested in the evaluation.

5.1. Risk Measures/Presentation

There are three basic types of risk measures which may be developed for a semiquantitative or quantitative analysis for both CPQRA and TRA:

- *risk indices,* which are single numbers or scores
- *individual risk measures,* which consider the risk to a particular person or at a particular location
- *societal risk measures*, which consider the overall risk associated with an activity to a particular population

Table 5-1 summarizes a number of risk measures defined in the *CPQRA Guidelines* (CCPS, 1989) and gives their advantages and disadvantages when used for TRA studies. Some measures such as the fatal accident rate (FAR) and the individual hazard index (IHI) are geared toward estimating risks for employees on a manufacturing site. These measures are not included as they do not generally apply to TRA.

Individual risk measures the risk to a person along a transportation route, and can be calculated for the most exposed individual, for specific locations along a route, or for an average individual in a potentially affected area. Societal risk considers the overall likelihood of different severity events occurring. It can be much higher for transportation movements than is commonly found for fixed facility operations, because of the multiplicative effect of route length and number of trips. In other cases, transportation risks can be much lower than facility risks if the route is remote from population, such as for some marine movements and pipelines. Figure 5-1 displays a typical F-N curve, which is the most common societal risk measure. There can be much greater disparity between individual and societal risk levels for TRA than CPQRA. The reader is

TABLE 5-1
Risk Measures Common to CPQRA and TRA

Measure	*Use in TRA* — *Advantages*	*Use in TRA* — *Disadvantages*
INDEX:		
Equivalent Social Cost Index (Okrent, 1981)	Takes aversion of high consequence events into account; builds on average rate of death	Index only; does not give absolute information
INDIVIDUAL RISK:		
Individual Risk Contour	Illustrates geographical distribution of individual risk; gives indication of number of people at various risk levels	Very time consuming to produce for an entire route
Maximum Individual Risk	Illustrates highest risk to any one individual from transportation activity—generally directly along the route	Does not consider how many people are at this risk level or how quickly the risk level changes with distance from the route
Average Individual Risk —exposed population	Given use of some simplifying assumptions regarding population density, this measure is often more applicable than for fixed facilities; can be compared to fixed facility and background risks and to risk criteria	Does not indicate number of individuals involved in any one accident; does not indicate maximum risk level; may be misleading when comparing risks averaged over a few people versus those averaged over many people due to large hazard distances
Average Individual Risk —total population	Allows risk to a constant population to be compared for several movements; allows for the extent of consequences to figure more heavily in risk measure	May be significantly misleading with regard to the maximum risk level if this risk is limited to a small portion of the defined community
SOCIETAL RISK:		
F-N Curve	Displays information about high likelihood/low consequence and low likelihood/high consequence events	Illustrations of total risk along route or of maximum consequence event may be threatening to management in some cases as curves are difficult to interpret
Average Rate of Death (also known as Accident Fatality Number or Annual Expected Fatalities) (Lees, 1980)	Provides indication of total risk per year	Does not differentiate beween high likelihood/low consequence and low likelihood/ high consequence events; may not be absolute criteria for tolerating or rejecting risk

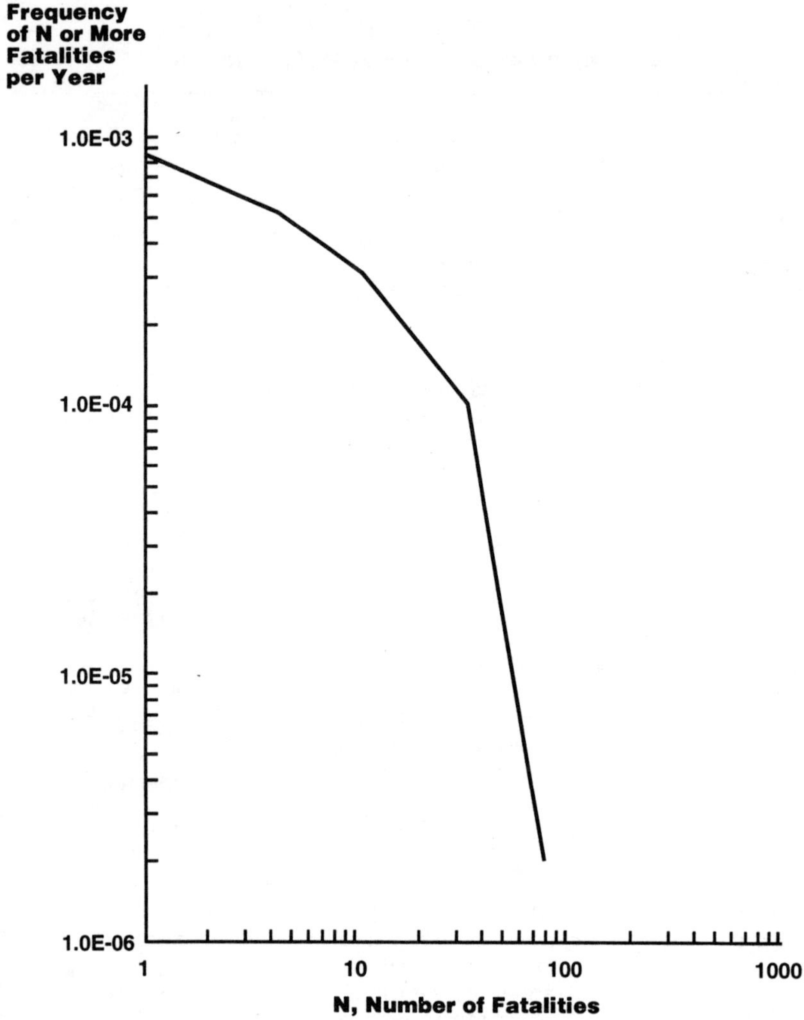

FIGURE 5-1. Typical F-N Curve.

cautioned against using internal calibrations of results that are based on past CPQRA work for TRA decisions. Care should also be taken when comparing different transportation movements if route lengths or modes vary significantly from what has previously been evaluated.

As in CPQRA, one can consider the potential for hazardous exposures, injuries, serious injuries, or fatalities. Most commonly, serious injury or fatality is considered for determination of appropriate levels of concern for overpressure, thermal radiation, and toxicity hazards.

TABLE 5-2
Additional TRA Risk Measures

Measure	*Applicability*	*Reasons for Use*
Per ton-mile	For individual or societal risk measures given in Table 5-1	Allows comparison without emphasizing differences caused by route length or number of trips
Per ton	For individual or societal risk measures given in Table 5-1	Provides risk "cost" per ton to demonstrate differences between routes and modes without specifying total annual volume
Per trip	For individual or societal risk measures given in Table 5-1	Emphasizes material and route (i.e., length) characteristics over annual volumes; allows for later incorporation of volume when known
Per community	For individual or societal risk measures given in Table 5-1	If used for a standard segment length such as 10 or 15 miles it brings results to a more comparable level with fixed facilities and recognizes public concern for the risks local to them
Per specified number of trips	For individual or societal risk measures given in Table 5-1	Allows a particular campaign (set of movements) to be evaluated and compared on an equivalent basis while still acknowledging influence of number of trips
Individual risk profile or risk transect	For an alternate display of individual risk contour information	Can basically take a cross-section of the risk contour; this can be done for several discrete locations reducing the calculation burden otherwise associated with a full set of risk contours

Table 5-2 presents several additional risk measures that may be used in a TRA. In most cases, the measures are similar to those given in Table 5-1, but use a different set of units or presentation style. For example, both individual and societal risk measures may be expressed on a per ton-mile basis rather than per year.

If one were to develop individual risk profiles (graphs of individual risk levels versus distance from the route) for a transportation movement, it might be necessary to select several locations for which this measure would be applied in order to provide examples of changes in risk along the route. There might be one or more representative cases or a worst case for a typical section of the line or route. However, the discrete portion of the risk attributable to valve or compressor stations for pipelines, railyards for rail movements, harbors for marine movements, etc. would have a different individual risk profile and might need to be shown separately.

For TRA, it is quite common to use several different risk measures in the same study. This is particularly true when considering alternate routes or modes of transport, as it is possible to unknowingly increase individual risk while decreasing societal risk, or vice versa. If one is examining risk reduction alternatives *on a given route*, then the potential for risk trade-offs may not be as much of a concern—unless the trips per year and the release sizes change significantly. An F-N curve may be sufficient to depict the differences due to these factors, but an individual risk measure may not be.

Frequency measures that may be beneficial to use in addition to the risk measure(s) include:

- the overall chance of an accident (per year or per trip)
- the overall chance of a release (per year or per trip)
- the chance of a small release (per year or per trip)
- the chance of a large release (per year or per trip)
- the chance of a release *given* an accident

These measures can be presented for a particular portion or segment of a route, or for an entire route. It may also be advantageous to express the risk in terms of the expected number of trips between incidents.

Measures that focus on consequences include:

- the maximum number of people exposed or injured per event
- the average number of people exposed or injured per event
- the area impacted by the event with the greatest consequence
- the average area impacted

As with the frequency measures, these measures add to one's understanding of the factors which are most significant in determining the overall risk level.

Screening-level or qualitative analyses typically produce risk indices, or rankings. These rankings may be very detailed, prioritized listings or may simply sort the results into broad categories. Another common qualitative/semi-quantitative risk measure is the risk matrix. There are many variations on this approach, but it basically uses an estimate of the likelihood and potential severity of an event to place it into a cell of the matrix. These cells can then be categorized as representing high, medium or low risks, for example. Figure 5-2 illustrates one possible risk matrix.

Likelihood \ Severity	*Serious* impact may cause fatality	*Extensive* Impact between serious and catastrophic	*Catastrophic* Impact ten times worse than serious
Likely release expected during period of operation	**M**	**H**	**H**
Possible release unlikely during period of operation	**L**	**M**	**H**
Unlikely release very unlikely during period of operation	**L**	**L**	**M**

FIGURE 5-2. Risk Matrix

The *HEP Guidelines* (CCPS, 1992) provide guidance on several qualitative/semiquantitative techniques that may be adapted for TRA.

The actual selection of risk measures for a particular study depends on whether a study is conducted for internal purposes or according to the specifications of a supplier, a senior management request, a corporate risk management program, or a regulatory authority. Some of the most common factors to consider when it is up to the analyst or project manager to make the choice(s) are:

- *Study Objective*—the study objectives may or may not suggest a specific risk measure, but the selection of one or more risk measures must be made when determining the study objectives due to their influence on the resources and level of effort required.
- *Required Depth of Study*—the development of quantitative measures require a semiquantitative or quantitative analysis with a risk focus. The chances of accident or release can be determined from a frequency analysis.
- *End Uses*—how the risk information is to be used in the decision-making process, and how and to whom it is communicated will play a major role in selecting the most appropriate measure. Similarly, the measure must be adequate to demonstrate the benefits of mitigation measures if that is the purpose of the study. The potential for additional end uses to be identified during or after the study should also be considered.

- *Population at Risk*—if the population is not distributed uniformly in each segment or there are particular populations or locations of concern, the selection of risk measures should consider this.
- *Risk Comparisons*—one may need to compare the TRA results to those for another TRA, to results for a fixed facility, or to a set of risk criteria. This will require comparable measures to be used, although additional measures may also be generated to better meet the study objectives.

5.2. Risk Calculations

The description of the risk measures is very similar to that for CPQRA and as such is only summarized here. The reader is referred to Chapter 4 of the *CPQRA Guidelines* (CCPS, 1989) if further information is required.

5.2.1. Individual Risk

For TRA, the risk is generally from linear, not point, sources as the product is moving from the origin to the destination. Hence, in order to calculate the risk at a specific point it is necessary to consider several combinations of wind directions and potential release points. For example, Figure 5-3 shows that a person at Point X could be affected by a release from a truck accident within Zone A that disperses to the southeast, a release in Zone B that disperses to the south, or one from Zone C that disperses to the southwest. Thus, the determination of individual risk in detail requires many more calculations for a TRA than for a CPQRA.

On the other hand the maximum individual risk can be determined more easily than described in the *CPQRA Guidelines* (CCPS, 1989), in that one can look at only points along the transportation route. If there is a particular segment with a higher accident and release rate, it will often be the one with the maximum individual risk level as well. Thus, individual risk only needs to be calculated for a limited set of points to determine the

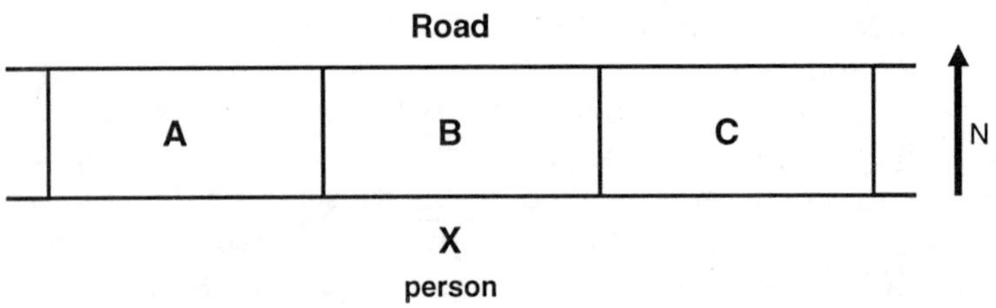

FIGURE 5-3. Contributions to Individual Risk

maximum. For example, if a train route uses the same stretch of track to travel to and from a rail yard, if any route takes a sharp turn, or if a yard or valve station are present, the potential sources of risk are increased and the maximum individual risk level may be present.

Average individual risk (for the total or the exposed population) is much easier to calculate. It can be determined by averaging the estimated individual risk levels for all the individuals in the selected area—including those in the total population case who have zero risk, or by starting with the consequence and frequency data for each incident outcome case. In the latter approach, one simply sums the product of the frequency and consequence for each incident outcome case and divides by the population of interest.

$$\text{Average Individual Risk} = \sum_{i=1}^{n} f_i \cdot N_i / P_\mathrm{T} \tag{5-1}$$

or

$$\text{Average Individual Risk} = \sum_{i=1}^{n} f_i \cdot N_i / P_\mathrm{E} \tag{5-2}$$

where

n = the number of incident outcome cases,
f_i = the frequency of incident outcome case i,
N_i = the number of fatalities for incident outcome case i,
P_T = the total population of concern and
P_E = the population within the corridor defined by the maximum hazard distance
i = incident outcome counter

This is the same as using the average rate of death index divided by the population of concern.

An approach used in TRA for sections of a route where the risks will be the same is to calculate the risks at a number of points along a line perpendicular to the route. This provides the data needed to then draw contours of constant levels of individual risk (typically, 10^{-4}/year, 10^{-5}/year, 10^{-6}/year, 10^{-7}/year and 10^{-8}/year), or to prepare a graph of individual risk level versus distance from the route (an individual risk profile).

Example

The scenario shown in Figure 5-3 can be used to demonstrate the calculation of individual risk at any given point. Assuming that this situation involved a truck transporting propane, and that only two stability classes

(D and F) are used to represent all possible meteorological conditions, the event tree given in Figure 5-4 represents the potential outcomes of a large (very rapid) release. There would be similar event trees for any other release scenarios. The event tree considers both immediate ignition and three delayed ignition cases.

The first step in the example would be to determine, for each release location A, B, and C, which incident outcomes involve (i.e., reach) Point X. Then the specific likelihood of each relevant incident outcome case would have to be determined. This can be thought of as the chance that the large release occurs in the zone of interest (A, B, or C) and the wind blows in the necessary direction (SE, S, or SW) and that the incident outcome of concern occurs.

The overall expression for annual individual risk at any specific location (such as X), assuming a constant accident rate, is thus:

$$\mathrm{IR}_{x,y} = T \cdot A \cdot \sum_{i=1}^{n} R_i \cdot \sum_{j=1}^{m} L_{i,j} \cdot W_j \cdot \sum_{k=1}^{S_i} P_{i,j,k} \tag{5-3}$$

where

$\mathrm{IR}_{x,y}$ = the total individual risk of fatality at geographical location x,y (chance of fatality per year)
T = trips per year
A = accident rate per mile
R_i = release probability for release size i
n = number of release sizes considered
$L_{i,j}$ = length of release location zone j for release size i (in miles)
m = number of release location zones and wind directions affecting location x,y
W_j = probability that wind blows in direction of concern for release location zone j (does not vary by release size)
$P_{i,j,k}$ = probability of a fatality at location x,y given that incident outcome k occurs in release location zone j with appropriate wind direction, given release size i
S_i = number of incident outcomes for release size i
i = release size counter
j = release location zone counter
k = incident outcome counter

If the accident rate varies or there are non–accident-initiated events to be considered as well, Equation (5-3) can be modified accordingly by adjusting A or adding another term for fixed sources of risk where there is no length of a release location zone to be considered.

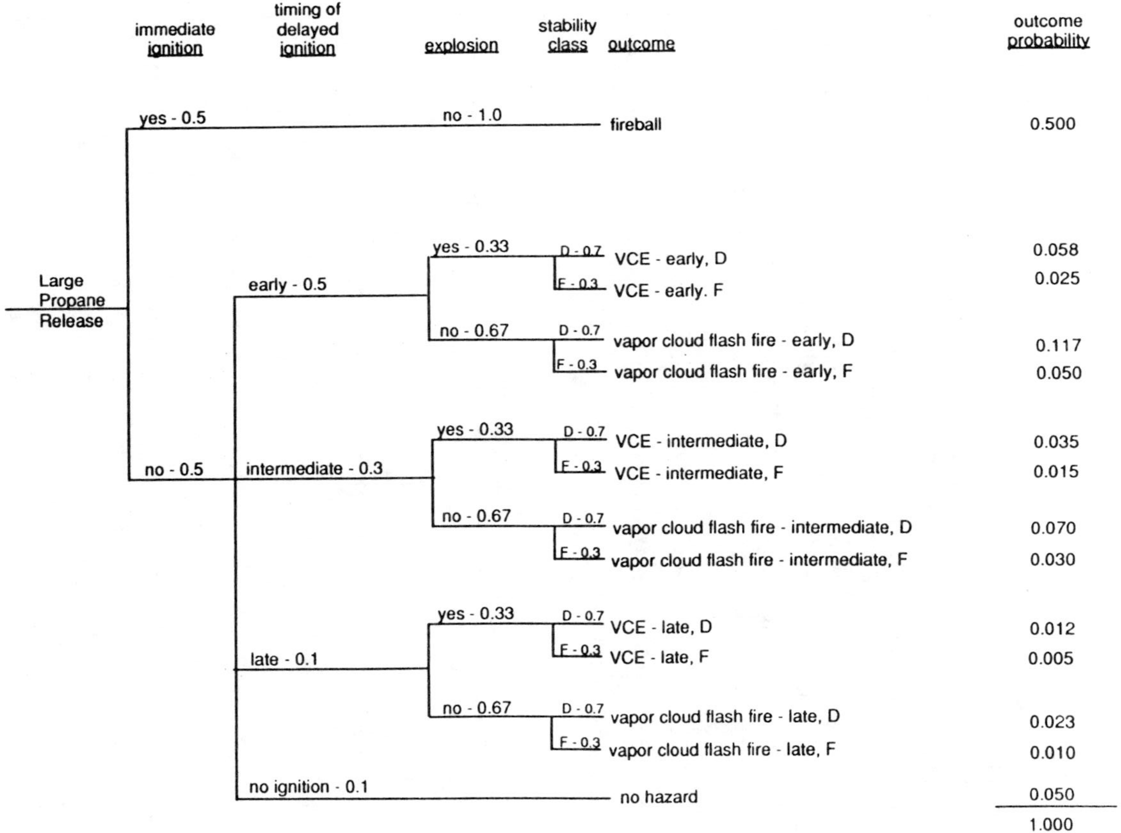

FIGURE 5-4. Sample Event Tree for a Large Release form a Propahe Tank Truck

In Equation (5-3), the term $P_{i,j,k}$ will generally equal 0 if the hazard does not reach location x,y and equal the likelihood of the incident outcome multiplied by the chance of fatality given exposure to the hazardous outcome if that hazard does reach location x,y.

Example

Table 5-3 gives a set of values for an individual risk calculation for a location sufficiently far away from a route that only a large spill dispersing due east can reach that point. A small spill or other wind directions will not pose any risk to that location. The appropriate release location zone length for small spills in this example is 300 feet and for large spills is 600 feet. The material is toxic and nonflammable, so the incident outcomes are limited to the meteorological conditions of interest—say D-stability/10 mph and F-stability/4 mph. Only the F-stability hazard zone is large enough to reach the location of interest. Equation (5-3) can be rewritten for this example as:

$$
\begin{aligned}
\mathrm{IR}_{x,y} &= T \cdot A \cdot [R_1 \cdot L_{1,1} \cdot W_1 \cdot (P_{1,1,1} + P_{1,1,2}) + R_2 \cdot L_{2,1} \cdot W_1 \cdot (P_{2,1,1} + P_{2,1,2})] \\
&= 200 \cdot 1 \times 10^{-6} \cdot [0.1 \cdot 0.06 \cdot 0.3 \cdot (0+0) + 0.02 \cdot 0.11 \cdot 0.3 \cdot (0+0.15)] \\
&= 2 \times 10^{-8} \text{ per year}
\end{aligned}
$$

TABLE 5-3
Hypothetical Data for Risk Calculation

T	=	trips per year = 200
A	=	accident rate per mile = 1×10^{-6}
R_1	=	release probability for small spills = 0.1
R_2	=	release probability for large spills = 0.02
$L_{1,1}$	=	length of release location zone 1 for small spills = 300 ft. = 0.06 mi
$L_{2,1}$	=	length of release location zone 1 for large spills = 600 ft. = 0.11 mi
W_1	=	probability wind blows to east = 0.3
$P_{1,1,1}$	=	probability of fatality for small spill in release location zone 1 dispersing to the east under D-stability = 0 (does not reach)
$P_{1,1,2}$	=	probability of fatality for small spill in release location zone 1 dispersing to the east under F-stability = 0 (does not reach)
$P_{2,1,1}$	=	probability of fatality for large spill in release location zone 1 dispersing to the east under D-stability = 0 (does not reach)
$P_{2,1,2}$	=	probability of fatality for large spill in release location zone 1 dispersing to the east under F-stability = 0.15. (0.3 for F-stability × 0.5 for likelihood of fatality given exposure to toxic concentration)

Most of the simplifications for individual risk calculations discussed in Section 4.4.1.3 of the *CPQRA Guidelines* (CCPS, 1989) can also be used for TRA. Indeed, it is often necessary to use some of the simplifications even in a detailed TRA—particularly the concepts of uniform wind and ignition source distributions.

The risks can also be calculated on a per trip basis by removing T from Equation (5-3), or simply dividing $IR_{x,y}$ by T.

5.2.2. Societal Risk

Generally, all the information needed to calculate individual risk is also needed to calculate societal risk. However, some of the simplifications typically made in a TRA regarding wind distributions, ignition probabilities, and discrete consequence effects actually make societal risk easier to calculate than individual risk. For societal risk, the risks can be calculated per segment (where all of the basic variables such as population density remain constant) and then summed.

The discussions below are for the computation of risk on an annual basis. The risks per trip can be calculated by dividing the results by T (the number of trips per year) or by setting T equal to 1 in the equation. Risks per ton or per ton-mile are most readily obtained by first calculating the risks per year or per trip and then dividing by the number of tons or ton-miles per year or per trip, as appropriate.

The main difference between calculating societal and individual risk is in determining the consequences associated with an incident outcome. Individual risk essentially determines whether or not a particular location is involved in an incident outcome case, while societal risk considers how many people are involved in an incident outcome case.

Given the length of most transportation routes, it is suggested that the route first be subdivided into segments over which the population density can be assumed to be uniform. Such segments are typically one to ten miles long in suburban and urban areas, but can be much longer in rural areas where there are no significant variations in the population density. Sections 3.2 and 3.3 discuss population densities and route segmentation in greater detail.

If the route is divided into segments, then the frequencies and consequences can be obtained for each incident outcome case in each segment. The frequency of incident outcome k for release size i on segment g is defined as:

$$F_{g,i,k} = T \cdot A \cdot R_i \cdot L_g \cdot P_{i,k} \tag{5-4}$$

where

$F_{g,i,k}$ = frequency of incident outcome k for release size i on segment g
T = trips per year
A = accident rate per mile
R_i = release probability for release size i
L_g = length of segment g in miles
$P_{i,k}$ = probability of incident outcome k for release size i
g = segment counter
i = release size counter
k = incident outcome counter

The associated consequences are given by:

$$N_{g,i,k} = \mathrm{CA}_{i,k} \cdot \mathrm{PD}_g \cdot \mathrm{PF}_{i,k} \tag{5-5}$$

where

$N_{g,i,k}$ = number of fatalities for incident outcome k for release size i on segment g
$\mathrm{CA}_{i,k}$ = consequence area associated with incident outcome k for release size i
PD_g = population density for segment g
$\mathrm{PF}_{i,k}$ = probability of fatality for incident outcome k for release size i
g = segment counter
i = release size counter
k = incident outcome counter

This calculation is repeated for all incident outcome cases. The results are then combined to construct the societal risk F-N curve as described in Section 4.4.2.1 of the *CPQRA Guidelines* (CCPS, 1989).

If there are non–accident-initiated events that can occur anywhere along the route, the likelihood of these events can be distributed along the length of the route. This effectively gives an accident rate that can be substituted into Equation (5-4) for A. More discrete or point sources of risk can be treated as their own segments, or calculated as per a CPQRA point source.

If the route has population only along one side of the route, Equation (5-4) can include an additional term for wind direction. This might simply be set equal to 0.5 or might be calculated based on wind rose data.

For the ends of segments, it is recognized that the release may originate in one segment but cause consequences in the next. This is generally insignificant as its occurrence is counteracted by releases in the previous segment which can cause consequences in the present segment. The only place that the crossing of segment boundaries is a significant concern is when the hazard length is much greater than the segment length. In such circumstances it is not possible to have the full extent of

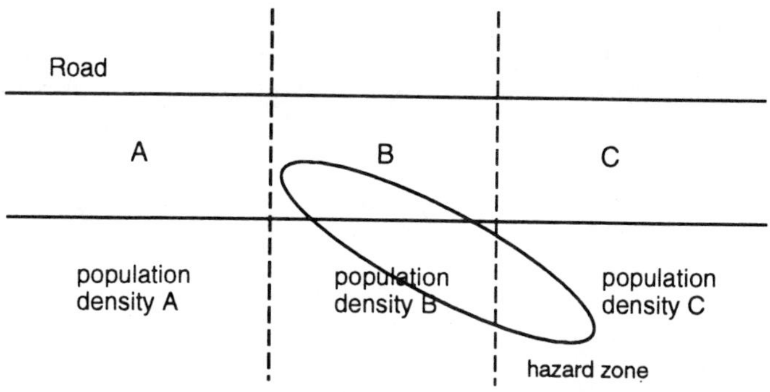

FIGURE 5-5. Hazard Length in Excess of Segment Length

the consequences realized in one segment (see Figure 5-5). Artificially modifying the population density to reflect an average with the bordering segments is one means of addressing this situation. In the example given in Figure 5-5, this would entail changing the population density for B to an average of A, B, and C or even splitting B between A and C. Short segment lengths are not recommended for use in situations with long hazard distances. Another approach is to assume that the release is perpendicular to the route.

5.2.3. Other Risk Measures

The average rate of death and equivalent social cost are both described in Section 4.4.3 of the *CPQRA Guidelines* (CCPS, 1989). The detailed formula for the average rate of death (also known as the expected number of fatalities) is:

$$\text{Average rate of death} = \sum_{g=1}^{t} \sum_{i=1}^{n} \sum_{k=1}^{S_i} F_{g,i,k} \cdot N_{g,i,k} \tag{5-6}$$

where

$F_{g,i,k}$ and $N_{g,i,k}$ are defined in Equations (5-4) and (5-5)
t = number of segments considered for route
n = number of release sizes considered
S_i = number of incident outcomes for release size i

There are other risk measures that are unique to transportation. One, a risk index, is based on the total population at risk within the corridor and the total frequency of a release (of any size). It can be expressed as:

$$\text{Risk Index} = \left(T \cdot \sum_{g=1}^{t} A_g \cdot L_g \cdot \sum_{i=1}^{n} R_i \right) \times \left(\sum_{g=1}^{t} L_g \cdot \text{PD}_g \cdot D_{max} \right) \qquad (5\text{-}7)$$

where

T = trips per year
A_g = accident rate per mile for segment g
t = number of segments considered for route
g = segment counter
L_g = length of segment g in miles
PD_g = population density for segment g
D_{max} = maximum of all hazard distances, for all release sizes
R_i = release probability for release size *i*
n = number of release sizes considered
i = release size counter

This measure may be used in simplified risk analyses where the set of potential incident outcomes and their associated consequence areas and probabilities are not considered, but where segment specific accident rates are used.

5.3. Risk Uncertainty, Sensitivity, and Importance

When conducting a quantitative TRA it is important to recognize the uncertainty, sensitivity and importance of the results. Each of these concerns have different influences on the risk measures:

- *uncertainty* deals with the effect of data and model uncertainties on the risk measure
- *sensitivity* shows the effects of varying one or more data values or methodologies
- *importance* indicates the contributions made by particular subsets of the complete analysis—such as individual route segments, large versus small spills, immediate versus delayed ignition, etc.

These topics are discussed thoroughly in Section 4.5 of the CPQRA Guidelines (CCPS, 1989), and apply directly to TRA as well. TRA has additional sources of uncertainty beyond those of concern in a CPQRA. Table 5-4 summarizes the sources of uncertainty in a TRA. The most critical

TABLE 5-4
Sources of Uncertainty in TRA

ROUTE DESCRIPTION

- Route not yet determined (new movement), unspecified (existing movement), or variable depending on certain conditions (weather, schedule, roadwork)
- Information from carriers out-of-date or incomplete
- Population data out-of-date or unavailable (rural area or some international locations)
- Condition of route unknown (e.g., track class, train speed, divided versus undivided highway)
- Weather data not representative of route conditions
- Terrain influences may vary along route

CONTAINER DESCRIPTION

- Detailed specifications not available (especially for modifications to basic container design)
- Considerable variation in shipping temperature and pressure along route or by season
- Variation in amount of cargo

HAZARD IDENTIFICATION

- Incomplete recognition of non–accident-initiated events
- Historical record inadequate to cover all major hazards
- Selection of incident outcomes to evaluate too general

CONSEQUENCE TECHNIQUES

- Uncertainties in physical modeling
  - Inappropriate model selection
  - Incorrect or inadequate physical basis for model
  - Inadequate validation
  - Inaccurate model parameters
- Uncertainties in physical model data
  - Input data (composition, temperature, pressure)
  - Source terms for dispersion and other models (size and orientation of release)
- Uncertainties in effects modeling
  - Data inappropriate for exposed populations (especially for toxicity)
  - Mitigating effects may be omitted

FREQUENCY TECHNIQUES

- Uncertainties in modeling
  - Extrapolation of historical data to present day conditions may overlook hazards introduced by new designs, conditions, or improvements
  - Limited understanding of influence of selected parameters requires simplification
  - Incompleteness in fault and event tree analysis
- Uncertainties in data
  - Data may be inaccurate, incomplete, or inappropriate
  - Data from related activities might not be directly applicable
  - Data generated by expert judgment may be inaccurate

RISK ESTIMATION

- Assumption of symmetry
  - Uniform population density may be inaccurate
  - Uniform wind rose is rare
  - Uniform ignition sources may be incorrect
  - Single release orientation for all incidents may be inaccurate
- Assumptions to reduce the depth of treatment
  - A single condition of wind speed and stability may be too restrictive
  - A limited number of ignition cases will reduce accuracy
  - A limited number of segments will reduce accuracy

among these are thought to be those that relate to release size and likelihood, and ignition probabilities and locations for flammable materials.

When addressing the sensitivity of the results to various parameters the greatest concern is typically with those parameters that have direct effects on the results. Table 5-5 highlights some of these parameters by mode. The influence of a few of the parameters is also very dependent on the risk measure being used. For example, changes in route length can have significant influences on societal risk with no or very small changes in individual risk, especially average or maximum levels. Changes in route length that are associated with major change in population density typically increase some risk measures while decreasing others. A longer truck or rail route through an area with lower population densities could decrease societal risk, hardly affect individual risk, and significantly increase the likelihood of a release. So route uncertainties can significantly influence the results.

The importance of various subsets of the risk contributors to the overall results can also vary depending on the risk measure selected. Another important consideration is whether importance is being evaluated over the entire route or for certain segments. For example, a small release in a rural area may have very little or no contribution to any risk measure but could be an important contributor in areas with high population densities.

TABLE 5-5
Typical Parameters to Evaluate in Sensitivity Analyses

Parameter	*Applicable Modes*
Route Length	Rail or road when route is unknown and must be estimated
Accident Rate	All
Probability of a Large Release	All
Large Release Size and Rate	All, but less typical on pipelines where a rupture is both credible and better defined
Ignition and Explosion Probabilities	All, for flammable materials
Population Density	All, but less typical on short pipelines where characterization is more detailed or for ocean-going vessels where concern is limited to port areas
Likelihood of Fatality Given Exposure to Incident Outcome	All
Meteorological Conditions	All

This is especially true if small spills are also much more likely than large spills.

Thus, the approaches to identifying and estimating uncertainty, sensitivity and importance given in the *CPQRA Guidelines* (CCPS, 1989) can be applied directly to quantitative TRA studies, with attention given to the way in which the techniques are applied.

5.4. References

This section lists the references cited in this chapter. Additional references that may be of interest to the reader can be found in the topical bibliography.

Center for Chemical Process Safety. *Guidelines for Chemical Process Quantitative Risk Analysis*. New York: American Institute of Chemical Engineers, 1989.

Center for Chemical Process Safety. *Guidelines for Hazard Evaluation Procedures, Second Edition with Worked Examples*. New York: American Institute of Chemical Engineers, 1992.

Lees, F. P. *Loss Prevention in the Process Industries*. (2 volumes). Stoneham, MA: Butterworths, 1980.

Okrent, D. "The Assessment and Perception of Risk." *Proceedings of the Royal Society (London) Series A 376*. (1981): 133–149.

6

TRA Application Examples

Recognizing that not all transportation studies require the same level of analysis or have the same goals, and that not all projects have the same resources available, the study cube was developed (Section 1.5.1). There are several reasons for conducting a TRA, such as selecting a mode of transport or a container, among others. Section 1.3 gives a broad but by no means complete list of reasons for conducting a TRA.

Table 6-1 lists each of the potential purposes described in Chapter 1 (see Table 1-6) and illustrates which of the study cube cells might apply to those situations.

For example, if one were to conduct a TRA in order to select a mode of transport for a particular material, certain levels of study would be more appropriate than others and some levels would not be appropriate at all. In order to decide among differing modes of transportation, many factors are important, including route lengths, container types and sizes, accident probabilities and release probabilities. All of these factors are contributing in varying degrees to the overall risk associated with that particular mode of transportation. As a result, a risk focus for this type of study would be preferred. These are also complex studies due to the number of variables being considered and a quantitative or semiquantitative approach would be preferred. A detailed set of incident outcome cases would offer the most precise answer, however, the general as opposed to detailed set of incident outcome cases may be sufficient.

If one were to focus on just the consequence portion of the TRA, there would be aspects of the problem that would not be addressed. It would be relatively easy to compare worst case scenarios among different modes based on container size, but unless accident frequencies or release probabilities, etc., are addressed, a complete picture is not provided. Similarly if the focus was on the frequency portion of the TRA, the full risk associated

TABLE 6-1
TRA Applications

Purpose of TRA	*STUDY CUBE CELL*[a]									
	Qualitative/ Conseq./ Limited Set	*Semiquant./ Conseq./ Limited Set*	*Quantitative/ Conseq./ Limited Set*	*Qualitative/ Frequency/ Limited Set*	*Qualitative/ Frequency/ Generalized*	*Semiquant./ Frequency/ Generalized*	*Quaniitative/ Frequency/ Generalized*	*Quantitative/ Risk/ Detailed*	*Semiquant/ Risk/ Generalized*	*Quantitative/ Risk/ Generalized*
Consider transportation in site selection	○	○	◉	○	○	◉	◉	◉	●	●
Demonstrate effectiveness of various alternatives	○	○	◉	○	◉	●	●	◉	◉	◉
Choose among alternate routes on relative basis	○	○	○	○	○	◉	◉	◉	●	●
Select mode of transportation	○	○	○	○	○	○	◉	◉	●	●
Control transportation risks	○	○	◉	◉	◉	●	●	○	◉	◉
Select container and/or additional protective features	○	○	○	○	◉	●	●	◉	◉	◉
Develop energency response plans	◉	●	◉	○	○	○	○	○	○	○
Understand influence of material state	◉	●	●	○	○	○	○	○	◉	◉
Examine tradeoffs in container size	○	○	○	○	○	○	◉	●	◉	●
Make risk tolerability judgments	○	○	○	○	○	◉	◉	●	◉	●
Consider benefits of trades, swaps, or toll operations	○	○	○	○	○	◉	◉	◉	●	●
Meet regulatory requirements for route determination	○	○	○	◉	◉	●	●	◉	◉	◉

[a] Complexity of Study/TRA Focus/Number of Incident Outcome Cases.
● = Very common, ◉ = Sometimes, ○ = Less common.

with each mode is not addressed and may not offer the level of detail needed to make such decisions.

This chapter provides a qualitative review of the application of TRA by considering various cells in the study cube (Figure 6-1). For the examples in this chapter, the selection of certain study cube cells is representative only and does not suggest that these particular study approaches are more appropriate or more frequently applied than those for other cells of the cube.

The examples chosen for qualitative review are:

1. Qualitative/Consequence TRA Examples
   —Application to Comparison of Different Forms (Condition or State) of One Material by Consequence Analysis
   —Application to Comparisons Against Results of More Detailed Studies
2. Semiquantitative/Frequency TRA Examples
   —Application to a Change in Supplier
   —Application to a Change in Material Container

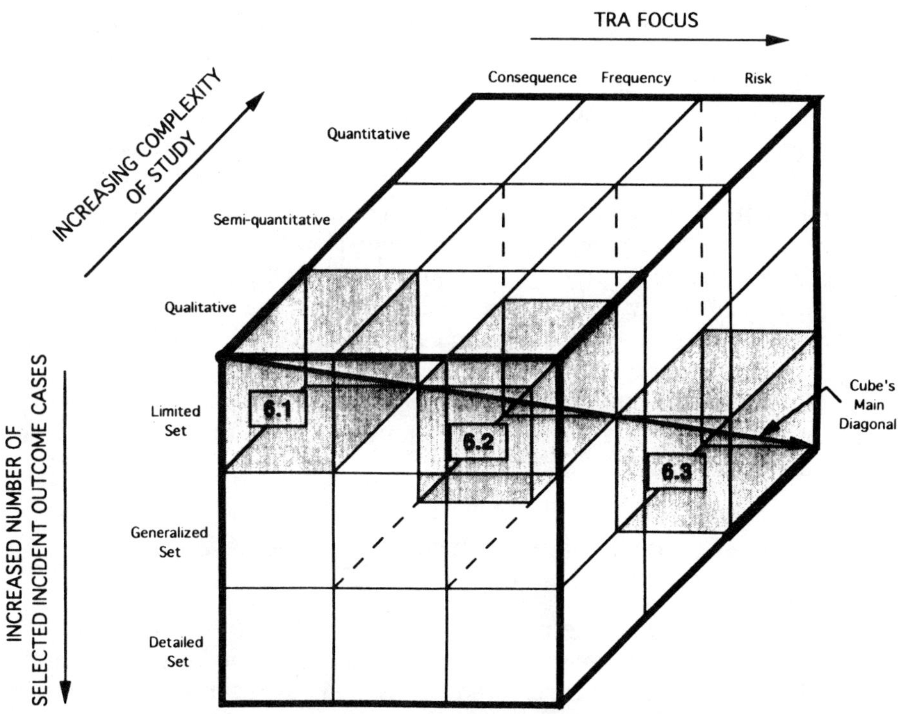

FIGURE 6-1. TRA Study Cube with Application Examples.

3. Quantitative/Risk TRA Examples
   —Application to Determination of Available Risk Reduction Options and Their Ranking
   —Application of the Evaluation of Alternative Modes for Getting From Point A to Point B

Figure 6-1 shows the study cube discussed in Chapter 1 with each of the three examples in this chapter marked in the individual cube that applies to that type of analysis.

6.1. Qualitative/Consequence TRA Examples

6.1.1. Qualitative/Consequence TRA Characterization

A Qualitative/Consequence TRA is a relatively quick study that estimates risk by focusing on consequence analysis to determine whether additional risk assessment is warranted. The broad parameters of such a study are described in Table 6-2. Such evaluations help screen movements to determine if and/or with what priority they should be evaluated in greater detail.

Qualitative/Consequence TRAs may not be appropriate for use in selecting modes of transportation or demonstrating the effectiveness of various risk reduction options, since major identified consequences often differ from major identified risks. For example, a Qualitative/Consequence TRA may highlight a particular movement due to an assumed total loss of contents (worst case release) of a container. However, more detailed risk assessment, using frequency techniques, may identify another movement to be a more likely threat, albeit with slightly reduced consequences. Further examination may also indicate that a total loss of contents has an extremely low frequency level.

Qualitative/Consequence TRAs are generally best suited for prioritizing among a set of materials, where the materials are moved in a similar fashion (mode, container size).

6.1.2. Comparison of Different Forms of One Material by Consequence Analysis

The objective of this example is to screen and bracket risks associated with transportation of the same material in different concentrations and/or physical states through estimated consequences. Estimates of consequences are based on both the size of the shipment and the chemical and physical properties of the materials being transported. The Qualita-

TABLE 6-2
Qualitative/Consequence TRA Description

Study Objective:	To determine the scale of potential consequences and the potential need for further study.
Requirement and Approach:	Straightforward approach drawing on material properties, estimated consequences, and/or consequence results of other studies to allow qualitative evaluation of movements at minimal cost.
Depth of Study:	TRA Focus—Consequence Complexity of Study—Qualitative Number of Incident Outcome Cases—Limited Set
Transportation Data:	Material transported, material specific hazards, transport conditions, shipment size (route and population data optional).
Comments:	The emphasis is on a comparison of the potential consequences posed by a release of material. This is a qualitative comparison based on the properties of the material toxicity, flash point, vapor pressure, etc.), and the amount shipped, among others. The comparison can be against other materials and/or against the results of more detailed studies. This type of study does not address the probability that accidents of this magnitude will take place or whether such scenarios are credible. These studies do not give a picture of the total risk involved with a particular movement(s), but do provide some indication of the potential consequences.

tive/Consequence TRA results may be presented as hazard zones for each incident outcome case considered.

In this example, a comparison among different forms or concentrations of one material (e.g., anhydrous ammonia versus 35 percent ammonia solutions in water) is conducted based on the individual material hazards. The worst case scenario for each material is typically examined. Using consequence estimates based on the hazards of the material (toxic, flammable, etc.), the Qualitative/Consequence TRA can be used to qualitatively compare the alternative forms of the material. The results are then compared and combined with other analyses (modal, economic, environmental, etc.) to provide input into selection decisions about the most appropriate form of the material.

These results are only one element of the decision-making process. Their purpose is to draw to the attention of management any particularly significant potential risks that may require more attention.

6.1.3. Comparisons against Results of More Detailed Studies

In many cases, detailed studies will have already been conducted for a number of hazardous materials movements. As a result, some of these results may be used for comparative purposes for other movements. In these cases, the estimated consequences for the movement of concern can be compared to detailed estimates that have already been completed in order to see if there are any potentially significant risks or if further study is warranted. This is generally done by reviewing the quantities of materials involved in each study, and then making judgmental adjustments to account for differences in quantities and hazard characteristics.

6.2. Semiquantitative/Frequency TRA Examples

6.2.1. Semiquantitative/Frequency TRA Characterization

The Semiquantitative/Frequency TRA requires a good deal more knowledge about the particular movement than the Qualitative/Consequence TRA does, as it reflects a shift towards a more quantitative approach. Consequences need not necessarily be addressed or included in this frequency-oriented study. However, if consequences are addressed, a representative set of incident outcome cases is used for Semiquantitative/Frequency TRA which covers the full range of potential incident outcomes. In contrast, a Qualitative/Consequence TRA generally uses only the largest potential or most likely incident outcomes. A range of incidents is considered for the frequency estimates as well. The broad parameters of such a study are outlined in Table 6-3.

6.2.2. Change in Supplier

This example illustrates the use of TRA to estimate community risk. In this case of a change in supplier, it is assumed that the new supplier will use a different route to transport the material(s). Thus, all variables (product, container, number of shipments, volumes, etc.) remain the same except for the route by which the material(s) is being transported. All factors associated with the new route will change, such as length and type of roadway(s).

From a frequency perspective, a change in route may have an effect on both the overall and the per mile accident rate. With all other factors constant, the risk can go up or down depending on the direction of the change in the accident rate. If consequences are also being addressed, the population characteristics along the route will also need to be ad-

TABLE 6-3
Semiquantitative/Frequency TRA Description

Study Objective:	To increase the focus of a Consequence TRA, without proceeding to a full Risk TRA, or just to examine frequency without consequence. This requires a move from a Qualitative to Semiquantitative approach.
Requirement and Approach:	The approach does not go beyond addressing frequencies. Consequences may or may not be included. Key parameters influencing frequency (and consequence) must be incorporated. Semiquantitative approach entails using estimates without getting into too much detail.
Depth of Study:	TRA Focus—Frequency and Consequence (optional) Complexity of Study—Semiquantitative Number of Incident Outcome Cases—Generalized Set
Transportation Data:	Material transported, material specific hazards, transport conditions, shipment size, annual amounts moved, number of trips per year, container type(s), route data
Comments:	An expansive list of incident outcomes is not required, but good frequency and consequence (if used) estimates are important. The primary frequency analysis is based on estimates derived from the data mentioned above and from historical incident data or other studies. These studies do not provide a complete picture of the risk since they typically address only the frequency of an accident or the probability of a release. The total risk cannot be measured unless the potential consequences are addressed as well.

dressed. For example, a decrease in population density along a particular route may offset the increased risk associated with a higher accident rate on that route—perhaps resulting from a longer route.

6.2.3. Change in Material Container

This application is similar to that discussed in Section 6.2.2 except that the route remains the same and the susceptibility of different types of containers (of basically the same size) to release of material is examined. The change in container will have no effect on the worst case release scenario if the container size is maintained. This will continue to be the total loss of contents. However, the incident frequency is modified by changing the probability of release given an accident. It is the frequency change that is affecting the overall risk level. In addition, if consequences are being addressed, the relative probabilities of large versus small releases will also influence the overall risk to the community.

6.3. Quantitative/Risk TRA Examples

6.3.1. Quantitative/Risk TRA Characterization

This is the most comprehensive depth of study and due to its more extensive resource requirements, it is generally best used sparingly and only on the toughest problems. The potential applications for a Quantitative/Risk TRA are numerous. However, to be effective, the situations for their use must be selected prudently. The broad parameters of Quantitative/Risk TRA are described in Table 6-4.

It should be noted that even with Quantitative/Risk TRA, there are still many different levels of study possible. The depth of study can vary depending on the nature and number of scenarios evaluated, the range of hazard types examined (fires, explosions, toxic vapor clouds, etc.) and the range of spill sizes considered among other factors.

TABLE 6-4
Quantitative/Risk TRA Description

Study Objective:	To extend the focus and level of a frequency TRA, relying on much more quantitative approaches.
Requirement and Approach:	A quantitative approach that applies the best available techniques to produce a risk estimate with the least uncertainty.
Depth of Study:	TRA Focus—Risk Complexity of Study—Quantitative Number of Incident Outcome Cases—Detailed Set
Transportation Data:	Material transported, material specific hazards, transport conditions, shipment size, annual amounts moved, number of trips per year, container type(s), route data, population data along route, meteorological data.
Comments:	Detailed estimation of consequences and frequencies are employed. Mitigation or risk reduction options are also addressed and evaluated. This will give a precise picture of the associated risks, however, studies of this magnitude take a significant amount of time to conduct and can be quite expensive.

6.3.2. Determination of Available Risk Reduction Options and Their Ranking

The Quantitative/Risk TRA may be undertaken due to a concern for the potential loss of contents of a container for a particular movement, as opposed to a more comparative situation. Usually, a need for this level of analysis has been determined based on the results of an initial Consequence or Frequency TRA. These studies indicate that the movement is one of particular concern and merits a detailed review of possible risk reduction measures.

The Quantitative/Risk TRA results can be used to evaluate a range of possible risk-reduction measures whose goal is to lower the frequency of release and/or reduce the consequences if a release occurs. Different risk reduction measures may affect the consequence, frequency, or both. Thus, simply looking at consequences or frequencies is insufficient. Each reduction measure would need to be addressed individually and then the resultant reduction in risk would be compared to those achieved by other risk reduction options.

For this depth of study, the primary output could be societal risk curves. Against the base case curve, additional curves can be superimposed, each specific for the implementation of a particular risk reduction option. Another possible use for the output is to compare the cost of specific risk-reduction measures with the amount of risk-reduction achieved.

6.3.3. Evaluation of Alternative Modes for Getting from Point A to Point B

The choice of transport modes is another application where Quantitative/Risk TRA studies might be commonly applied. The variations in both frequency and consequence when comparing one mode to another make the use of a common risk measure very helpful (see Section 5.1).

This option is particularly useful if the routes are also very different. Sometimes rail routes and highways travel largely in parallel, other times one is very direct and another is circuitous or involves more inner city travel. The Quantitative/Risk TRA is comprehensive enough to allow all these offsetting factors to be addressed and for the differences among modes to be identified and understood.

7

TRA Case Studies

This chapter contains five worked examples that use the TRA techniques presented earlier in the text. An example is provided for each major mode of transportation previously discussed (pipeline, highway, rail, barge, and ocean-going vessel). Figure 7-1 illustrates where each example fits in the study cube. The specific purposes for each case study are:

- *Pipeline (Section 7.1)*—to focus on and demonstrate a frequency analysis, with brief discussions of how consequence analysis and risk estimation would fit into the analysis
- *Tank truck (Section 7.2)*—to illustrate several risk measures and give guidance on the interpretation of different risk measures
- *Rail (Section 7.3)*—to provide an example of summing the risk from different portions or segments of a route
- *Barge (Section 7.4)*—to show the treatment of several transportation-specific concerns when using consequence modeling results for risk estimation
- *Ocean-going vessel (Section 7.5)*—to involve both accident- and non–accident-initiated events in a frequency analysis

All are simplified examples to illustrate the methodology and concepts of TRA and should not be applied to real situations. In reality, each study might require over 100 pages of calculations and documentation. Some of the data used come from Chapter 2, whereas other data are totally hypothetical. The data and assumptions used in an actual analysis should be situation-specific, reflecting the issues addressed in Chapters 2, 3, and 4. The examples are at least semiquantitative in nature, and focus on the selection of incidents and determination of frequency, as the use of consequence models for TRA and CPQRA is the same. Risk estimation is also reviewed as there are different assumptions than for a fixed facility analysis.

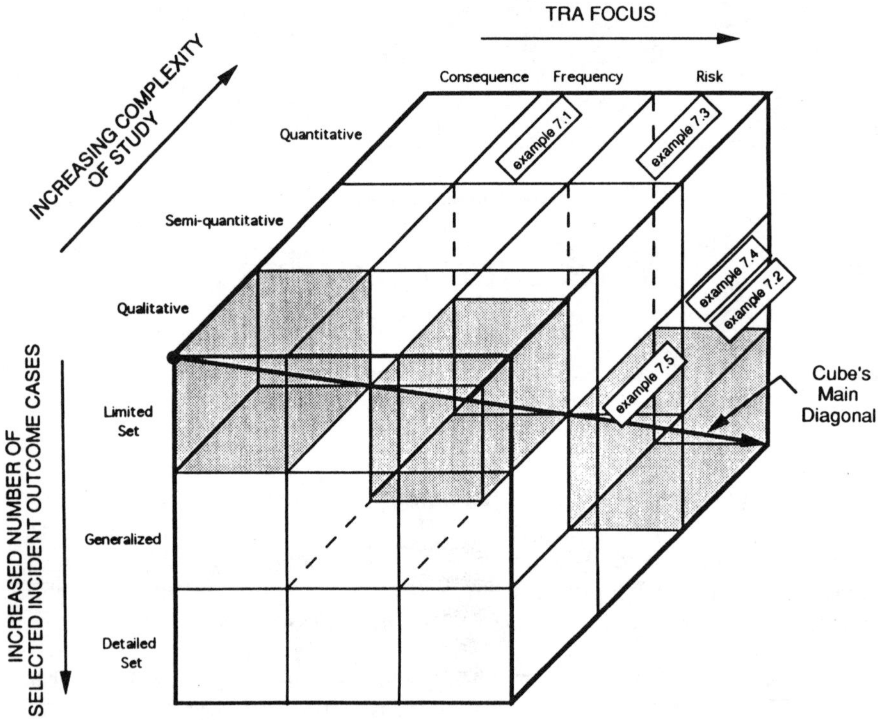

FIGURE 7.1. TRA Study Cube with Case Studies.

7.1. Pipeline Risk Assessment

7.1.1. Introduction

The situation of concern is one involving a new pipeline, currently under construction, where pipeline is the only viable means of transportation. Given this, management is concerned about how much additional protection is needed and justified. This case study focuses on and demonstrates a frequency analysis, with brief discussions of how consequence analysis and risk estimation would fit into the analysis. If necessary and appropriate, a risk analysis may be performed at a later time. The study cube cell represented by this application is the "Frequency, Quantitative, Limited Set" cell. (See Figure 7-1.) The initial frequency analysis will focus on generic hazards rather than those that may be unique to this pipeline.

7.1.2. Description

Propylene oxide will be transported from Facility A to Facility B via pipeline. Propylene oxide is a clear colorless, volatile liquid with an ether like odor. It is flammable over a wide range of concentrations with a lower flammable limit of 2.1% and is also moderately toxic by inhalation and ingestion (see Table 7-1). It is to be transported over a distance of 50 miles through areas with varying population densities (see Figure 7-2). Table 7-2 presents some of the specifics associated with this particular pipeline. The pipeline is wrapped and cathodically protected, and the line will be hydrotested before it is put into service. Given the nature of the terrain and the operations at each end of the line, leak detection and isolation are difficult and cannot be counted on to reduce the consequences of large leaks or ruptures. The pipeline comes within a mile of four towns, as shown in Figure 7-2, with one of them closer than 2000 feet.

The pipeline travels over a number of hills and valleys, with the towns located in the valleys. As a result, a pipeline failure in a low spot along the line sees not only the operating pressure but also a significant head pressure as well. This has a considerable effect on the release rates.

7.1.3. Identification, Enumeration, and Selection of Incidents

There are several types of pipeline failure incidents that may be considered. As discussed in Section 2.1.1, generic pipeline failure modes are

TABLE 7-1
Properties of Propylene Oxide

CAS Number:	75-56-9
DOT Class:	flammable liquid
ID Number:	UN 1280
Lower flammable limit:	2.1%
Upper flammable limit:	37.0%
Boiling point:	94°F
Molecular weight:	58.1
Flash point:	–35°F
Liquid specific gravity:	0.86
Solubility in water:	soluble
ERPG-3 concentration:	750 ppm
ERPG-2 concentration:	250 ppm
ERPG-1 concentration:	50 ppm

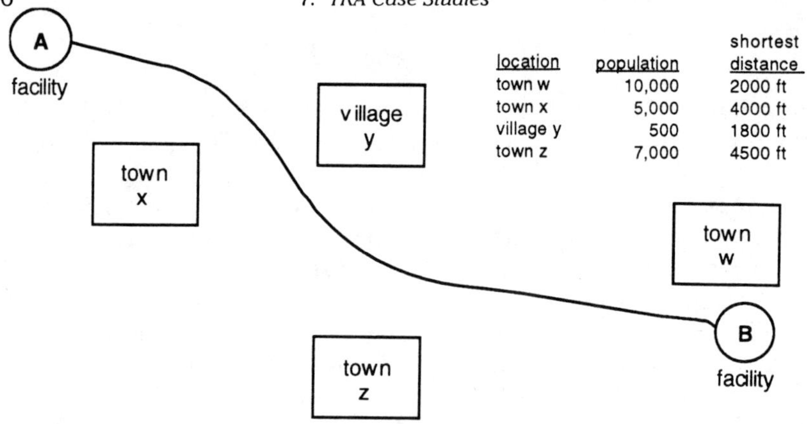

FIGURE 7-2. Description of Hypothetical Pipeline Route.

TABLE 7-2
Description of Hypothetical Pipeline

Product:	Propylene oxide
Length of pipeline:	50 miles
Diameter:	5 inches (outer diameter)
Flow rate:	33 lb/sec
Line type:	Underground (3 feet)
Normal operating pressure:	275 psi
Product temperature:	55°F
Wall thickness:	0.337 inch
Hydrostatic test pressure:	2000 psi
Valve stations:	7
Material of construction:	carbon steel

based on historical incident data and include loss of containment events resulting from causes such as:

- Corrosion leaks
- External mechanical interference (third party damage)
- Construction defects
- Material defects
- Other causes

When addressing generic failure modes in risk studies, it is common practice to assume that any pipeline will be subject to all of these kinds of events (See Section 2.1.1). The situation-specific variability comes into consideration when determining the frequency of these events, which depends on selected characteristics of the pipeline such as length, diameter, pipe wall thickness, and depth of burial. Factors such as age, maintenance, and vulnerability to third party activities should also be addressed (see Section 2.1.2). While there can be unique hazards that need to be considered in addition to the generic causes of line failure, these will not be addressed in this initial frequency analysis.

Release sizes are generally distributed by hole size. The following is an example of one type of size distribution:

Pinholes	(≈0.1 inches in diameter)
Holes	(≈1–1.5 inches in diameter)
Rupture	(line size)

Whether or not the analyst should consider pinholes depends on the characteristics of the pipeline and the commodity carried. Pinholes typically are not considered in an acute TRA when a line is underground and the material of concern is not extremely toxic. Thus for this study, the release sizes of potential concern are holes (taken to be one inch in diameter) and ruptures (taken as the pipe cross-sectional area). The 1-inch hole could also be used for a release from the valves at the valve stations; however, none of the valve stations are near the population centers. Each of the selected release sizes will be analyzed for the various generic failure causes.

7.1.4. Incident Consequence Estimation

While it is not expected that the consequences of any of this pipeline's releases can be reduced, some information regarding consequences will ultimately be needed to calculate risks. In addition, the results of a limited consequence analysis can be used to determine whether or not both holes and ruptures need to be analyzed. Table 7-3 presents some basic information that typically would need to be included as an input to a consequence modeling package. Either here or in the selection of incidents, the analyst must determine the number of meteorological conditions to be evaluated. Given that the pipeline is operational 24 hours a day, a representative daytime condition (D-stability with 11.0 mph winds) and a representative nighttime condition (F-stability with 4.5 mph winds) will be used. These conditions are assumed to occur 70 percent of the time and 30 percent of the time, respectively, based on local meteorological data.

An orientation of 10° from horizontal has been assumed for the jet release in order to maximize the potential ground level effects. Given the

TABLE 7-3
Consequence Modeling Input Parameters for Pipeline Example

Material	Propylene oxide
Release Height	0 ft
Release Orientation	10° from horizontal
Pumping Rate	33 lb/sec
Atmospheric Conditions	
—Stability Class/Wind Speed	D/11.0 mph
—Stability Class/Wind Speed	F/4.5 mph
Ambient Temperature	55°F

line and head pressures, it is assumed that the rupture will remove the overburden while the hole will not. This typically will produce very different hazard areas and distances for the hole and rupture scenarios, and the release through a hole may actually disperse farther. The rupture may produce a jet release which entrains aerosols and which disperses rapidly due to mixing with air. The leak may propagate through the soil to form a large pool which then evaporates, yielding a greater hazard distance from the resulting vapor cloud.

The output produced from the consequence analysis is driven by the hazards of concern (e.g., flammability, toxicity) and the criteria specified by the user (e.g., thermal radiation to 3200 Btu/hr-ft^2 for fires, concentration for dispersing clouds—such as the LFL and a toxicity concentration). Since propylene oxide is a flammable liquid, flammability and explosion (detonation or deflagration within the vapor cloud) hazards need to be addressed. This requires calculating dispersion to its flammable limits and the overpressure from explosions for the associated cloud masses. In addition, since this material is also toxic, dispersion to specified toxic concentrations also needs to be calculated.

A sample set of outputs is given in Table 7-4, for use in this case study. These results were obtained as part of a separate consequence analysis conducted when the pipeline was being sited. (These are hypothetical data and should not be used in an actual study.) Figure 7-3 illustrates how the data in the table can be interpreted.

Based on these results, only the hole under F-stability poses a significant hazard distance relative to the 1800-foot distance to the nearest town or village. Only East and North Towns are within the maximum hazard distances. Thus, the scenario involving a 1-inch hole will be the focus of the frequency analysis. The rupture case will also be addressed as it would have a major impact on the continuity of service.

TABLE 7-4
Hypothetical Hazard Zones for Propylene Oxide Pipeline Releases

Scenario	Stability Class/ Wind Speed	Toxic distances (ft)		Distances to LFL (ft)		Overpressure Distances (ft)
		down-wind	max. width of cloud	down-wind	max. width of cloud	radius[a]
Hole	D/11.0 mph	150	210	120	180	150
	F/ 4.5 mph	2700	5400	2550	5100	825
Rupture	D/11.0 mph	750	225	600	210	450
	F/ 4.5 mph	1050	900	900	750	750

[a]These do not include a drift factor and are thus additive to the flammable vapor cloud dispersion. One approach is to use one-half of the distance to the LFL; a more conservative approach would use the full distance to LFL.

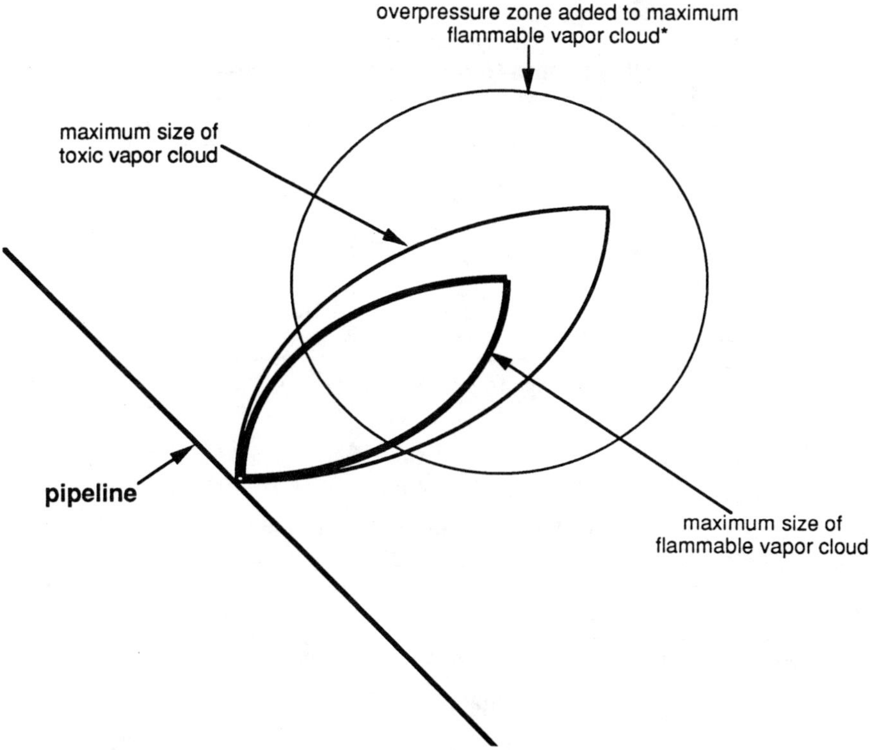

FIGURE 7-3. Hypothetical Hazard Zones for Propylene Oxide Pipeline Rupture (D-stability).

7.1.5. Incident Frequency Estimation

The base failure rates are given in Table 7-5. Adjustment factors are to be applied for the pipeline's increased wall thickness, and to the expected contribution from external mechanical interference to reflect the benefits of a surveillance system that will be put in place. Based on the data in Transportation Data Source 4-9, pipelines with a diameter of six inches or less have an overall failure rate of 2.0×10^{-3}/mile-year, which is higher than the overall rate for all diameter pipelines.

The AGA data from Transportation Data Source 4-9 are insufficient to provide a breakdown by both failure size and failure cause. A second data source (4-14) is needed if this breakdown is to be obtained. As shown in Table 2-2, the information contained within these two data sources is very consistent. Thus the additional European data from Transportation Data Source 4-14 are thought to be applicable and appropriate. Using this second source yields the values shown in Table 7-6.

TABLE 7-5
Pipeline Failure Rates (per 1000 Mile-Years)

Cause	*All Pipelines*	*Lines ≤6 in.*
External interference	0.68	1.3
Corrosion	0.20	0.15
Material/construction defects	0.26	0.34
Other	0.10	0.20
Total	**1.25**	**2.0**

Source: Transportation Data Source 4-9.

TABLE 7-6
Failure Size by Cause

Size	*External Interference*	*Corrosion*	*Material Failure/ Construction Defect*	*Other*
Rupture	23.4%	1.2%	10.4%	7.9%
Hole	50.6%	1.2%	25.0%	2.6%
Pinhole	26.0%	97.5%	64.5%	89.6%
Total	**100.0%**	**100.0%**	**100.0%**	**100.0%**

Source: Transportation Data Source 4-14.

The failure data from Table 7-5 can be combined with the data on failure size by cause from Table 7-6 to produce the failure rates by cause and size given in Table 7-7. For example, Table 7-5 gave a failure rate of 1.3×10^{-3}/mile-year for external interference on lines less than or equal to 6 inches in diameter. Using the failure size distribution for external interference from Table 7-6 gives:

Rupture:	1.3×10^{-3}/mile-year $\times$ 0.234 = 0.30×10^{-3}/mile-year
Hole:	1.3×10^{-3}/mile-year $\times$ 0.506 = 0.66×10^{-3}/mile-year
Pinhole:	1.3×10^{-3}/mile-year $\times$ 0.26 = 0.34×10^{-3}/mile-year

These are the results that appear in the external interference column in Table 7-7; the other columns are calculated in a similar manner. The failure rates in Table 7-7 allow one to focus on the likelihood of holes and ruptures. These rates now need to be adjusted for the increased wall thickness used on this pipeline as compared to the typical wall thickness for this diameter.

There are several ways to account for the benefits of increased wall thickness. One is to try and apply very limited data to the data set given in Table 7-7. Another is to assume that the increased wall thickness causes the pipeline to "behave" as though it has a larger diameter (and therefore wall thickness). Based on a combination of these two approaches, the likelihood of a rupture or hole was judged to decrease by a factor of roughly 0.7 overall. Looking at individual causes, external interference and material failure/construction defects were also found to decrease by a factor of roughly 0.7. The likelihood of a corrosion-induced failure doubled while other causes were considered only half as likely. (The data provided in Transportation Data Source 4-9 show that the corrosion failure rate for pipelines with diameters from 10 to 24 inches is roughly twice that for pipelines with diameters of 6 inches or less. This could relate to the length of field welds and the greater surface area over which wrapping may be

TABLE 7-7
Failure Rates by Cause and Size (per 1000 Mile-Years) for Pipeline Example

Size	*Total*	*External Interference*	*Corrosion*	*Material Failure/ Construction Defect*	*Other*
Rupture	0.35	0.30	0.002	0.035	0.016
Hole	0.75	0.66	0.002	0.085	0.005
Pinhole	0.89	0.34	0.15	0.22	0.18
Total	**2.0**	**1.3**	**0.15**	**0.34**	**0.20**

punctured.) Since there is no physical reason to expect that corrosion will increase, it has been left as is. This set of adjustment factors will be used in the analysis.

Given that a special surveillance system is to be put into place, whereby the pipeline right-of-way will be visually checked once a day, the likelihood of an external interference event is expected to be reduced. Since it is possible that a repair or construction crew could get set up between line surveys, and because some external interference events are due to natural causes that are not affected by the line surveys, only a 50 percent credit is given.

Applying both sets of adjustment factors to the data in Table 7-7 yields the following:

External interference

ruptures: 0.3×10^{-3}/mile-year $\cdot$ 0.7 $\cdot$ 0.5 = 1.1×10^{-4}/mile-year
holes: 0.66×10^{-3}/mile-year $\cdot$ 0.7 $\cdot$ 0.5 = 2.3×10^{-4}/mile-year

Corrosion

ruptures: 0.002×10^{-3}/mile-year $\cdot$ 1 = 2×10^{-6}/mile-year
holes: 0.002×10^{-3}/mile-year $\cdot$ 1 = 2×10^{-6}/mile-year

Material failure/construction defect

ruptures: 0.035×10^{-3}/mile-year $\cdot$ 0.7 = 2.5×10^{-5}/mile-year
holes: 0.085×10^{-3}/mile-year $\cdot$ 0.7 = 6.0×10^{-5}/mile-year

Other

ruptures: 0.016×10^{-3}/mile-year $\cdot$ 0.5 = 8.0×10^{-6}/mile-year
holes: 0.005×10^{-3}/mile-year $\cdot$ 0.5 = 2.5×10^{-6}/mile-year

These values total to overall failure rates of 1.5×10^{-4}/mile-year for ruptures and 3.0×10^{-4}/mile-year for holes, and are much lower than the corresponding values given in Tables 7-5 and 7-7. These are the final rates that will be used when estimating the risks. As other mitigation strategies are considered, they can be analyzed in a manner similar to that given above.

7.1.6. Risk Estimation

This case study does not include risk calculations. In general, the hole failure rate of 3.0×10^{-4}/mile-year would be used in conjunction with the consequence outputs from Table 7-3 to estimate the risks associated with this pipeline. This would also require consideration of the likelihood that the release takes place in a populated area, that the wind is blowing towards the population, that ignition occurs at specific locations, etc. Each specific town shown in Figure 7-2 would only be potentially vulnerable from releases which occur in the immediate vicinity of that town.

7.1.7. Discussion of Results

Once the frequencies, consequences, and risks are estimated, the results need to be evaluated. Depending on the level of the risk that is calculated, and the tolerability of that risk, there are several options for the risk analyst. These include:

- a more detailed review of the frequency of releases if it is believed that the present analysis does not account adequately for all critical factors, or if unique hazards are suspected to be significant contributors
- a more rigorous consequence analysis
- identifying and evaluating potential risk reduction measures, which would focus on reducing the likelihood of a release in this example
- develop recommendations for managing the risk and beginning operation of the pipeline

7.1.8. References

Transportation Data Sources 4-9 and 4-14 were used in this case study. These sources are described in Chapter 4.

7.2. Tank Truck Risk Assessment

7.2.1. Introduction

The following tank truck risk assessment is a simplified TRA focusing on developing and understanding the interpretation of several risk measures. There are two different routes which may be used, and each is evaluated using both societal and individual risk measures to demonstrate the differences in the information provided by each risk measure, and to aid in deciding which route to use. The study is motivated by a propane incident recently experienced by a competitor and the recognition by Super Chemicals, Inc. that they make a similar number of movements each year.

The study cube cell for this case is the "Risk, Quantitative, Generalized Set" cell. (see Figure 7-1). The choice of this cell derives from the need to consider quantitative risk estimates, and the use of generalized assumptions regarding ignition locations and probabilities.

The first step will be to identify and select the incidents of concern. The frequencies and consequences will then be analyzed and combined to calculate the overall risk for each route, and allow the analyst to compare the two routes.

7.2.2. Description

Super Chemicals, Inc. currently ships 300 tank trucks of propane each year from Here to Eternity along Route 27. Super Chemicals, Inc. is considering switching to Route 46 (see Figure 7-4). Notice the difference in lengths and distribution of population densities along the two routes. Route 46 is longer, but involves a lower population density. Route 27 is broken into three segments of uniform population density, while Route 46 has a constant population density along the entire route. Propane is commonly transported under pressure as a flammable liquid which vaporizes rapidly if released. (See Table 7-8 for some physical properties of propane.) Table 7-9 describes the specifics associated with this case.

For operational reasons, the movement is made in daylight hours, during the week. The population density along a route segment can vary with time, such as from day to night. For this example, only one set of population densities is considered for each route. If an option was to travel at night, a nighttime population distribution would need to be considered. Similarly, both on-road and off-road populations could be addressed separately. In this particular example, only the daytime, off-road population is evaluated.

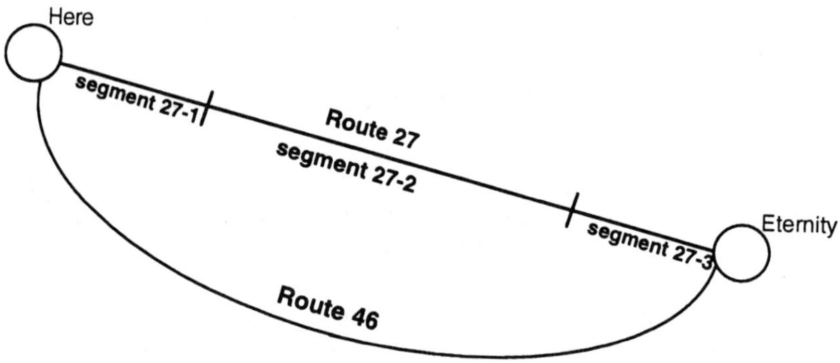

<u>Route 27</u>
Route Length: 35 miles
Population Density for Segment 27-1: 1000/mi^2 **Length:** 5 miles
Population Density for Segment 27-2: 500/mi^2 **Length:** 23 miles
Population Density for Segment 27-3: 2000/mi^2 **Length:** 7 miles

<u>Route 46</u>
Route Length: 55 miles
Population Density: 400/mi^2

FIGURE 7-4. Route Descriptions for Tank Truck Example.

TABLE 7-8
Properties of Propane

CAS Number:	74-98-6
DOT Class:	flammable gas
ID Number:	UN 1978
Lower flammable limit:	2.1%
Upper flammable limit:	9.5%
Boiling point:	–44°F
Molecular weight:	44.1
Flash point:	—
Liquid specific gravity:	—
Solubility in water:	not soluble
ERPG-3 concentration:	not available
ERPG-2 concentration:	not available
ERPG-1 concentration:	not available

TABLE 7-9
Tank Truck Example Description

Product	Propane
Amount	10,000 gallons
Container type	MC331
Trips per year	300

7.2.3. Identification, Enumeration, and Selection of Incidents

There are several types of incidents that may cause a release of propane. This case study addresses only accident-caused releases, as these are expected to be the most significant. Some typical truck accident scenarios are:

- Collisions with other vehicles, road debris, buildings, or animals
- At-grade collisions with trains
- Jackknifes or overturns

It can be assumed that all trucks are subject to these events (see Section 2.3.1) and they are therefore included in the basic accident rates. Variability comes into play when discussing how the accident frequency is affected by road type, weather conditions, traffic, and other factors (see Section 2.3.2). On the routes of concern, there are no at-grade railroad crossings.

Release sizes range from the full loss of contents to small drips, These are also affected by several different factors (see Section 2.3.3). Small drips usually result from valve and fitting leaks that are not associated with accidents. Thus, they are not considered in this case study, which focuses on accident-initiated releases.

The release sizes studied are catastrophic losses (the entire contents of the tank) and small continuous spills of 1 percent of the tank contents per minute over a time period of 10 minutes (releasing 10 percent of the tank contents), as described in Section 3.1.4. Past spill data from highway studies indicate that small spills constitute up to 10 percent of the contents, whereas large spills comprise up to 100 percent; data given in Section 3.1.4 indicate about a 65/35 split between these two categories. Given the variations of risk of concern and the expected daytime travel, only one weather condition is evaluated—D stability with 9-mph winds.

7.2.4. Incident Consequence Estimation

Consequence estimation is needed to develop risk measures. Table 7-10 presents some basic information used in the consequence modeling.

Since propane is flammable, flammability hazards (from fireballs and vapor cloud flash fires) and the potential for explosions must be addressed. As discussed in Section 3.7, a BLEVE is not a typical outcome unless there is a secondary source of fuel. For all of these hazards, the levels of concern in this study are those that may produce fatality. These have been assumed to be LFL for the maximum extent of vapor cloud dispersion, 160 kJ/m^2 for 15 seconds for fireballs, and 3 psi for explosion overpressure. Exposure to one of these levels of concern does not mean 100 percent fatalities. Protection may be offerred by buildings, rapid escape, clothing, etc. In this case study, 30 percent of exposures to levels of concern are assumed to yield fatalities.

A sample set of outputs is given in Table 7-11 for use in this study. These are hypothetical data and should not be used in an actual study.

TABLE 7-10

Consequence Modeling Input Parameters for Tank Truck Example

Material	Propane
Release size(s)	10,000 galllons and 100 gal/min for 10 min
Storage Temperature	70°F
Release Rate	10,000 gal/20 sec and 100 gal/min
Atmospheric Conditions —Stability Class/Wind Speed	D/9 mph
Ambient Temperature	70°F

TABLE 7-11
Hypothetical Hazard Zones for Releases from Propane Trucks

			Flammable Distances (ft)		*Overpressure Distances (ft)*
Scenario	*Stability Class/Wind Speed*	*Fireball Radius (ft)*	*Max. distance to LFL*	*Max. width of cloud*	*Radius from point of ignition*
Total Loss (10,000 gal)	D/9 mph	300	1600	1300	430
10% over 10 min (100 gal/min)	D/9 mph	Jet fire (localized)	300	70	Insufficient mass

The areas associated with the large releases are shown along with the outcome probabilities in Figure 7-5. The areas that have been assumed for the different ignition cases in this hypothetical example are:

- 100 percent of the fireball or overpressure (if feasible) area without a drift factor for immediate ignition.
- 20 percent of the maximum vapor cloud area for early ignition.
- 50 percent of the maximum vapor cloud area for "intermediate" ignition.
- 100 percent of the maximum vapor cloud area for late ignition.

The overpressure area for detonations or deflagrations is added to the associated vapor cloud area as a first-order approximation of the VCE area, for the various delayed ignition cases, in this example.

7.2.5. Incident Frequency Estimation

The approach taken here is to determine the appropriate accident frequency and conditional spill probability given that an accident occurs. The selected routes for this study are almost entirely four lane highways with much of the routes on interstate highways and major state highways. Given an accident, the probability of a release and its size need to be addressed.

Figure 7-5 shows an event tree (a modified version of Figure 5-4) in which the ignition timing and ignition and explosion probabilities are included and outcome probabilities are calculated. Section 3.4 suggested an immediate ignition probability of 0.75 for large propane spills. The use of 0.5 in Figure 7-5 is conservative as it increases the likelihood of delayed ignition and the associated larger hazard area.

A review of some of the data sources in Tables 2-7 and 2-8 indicate a range from 0.5 to 10 or more accidents per million vehicle miles. Since these routes are primarily on four lane highways in non-urban areas, the appropriate accident rate would be expected to be on the lower end of

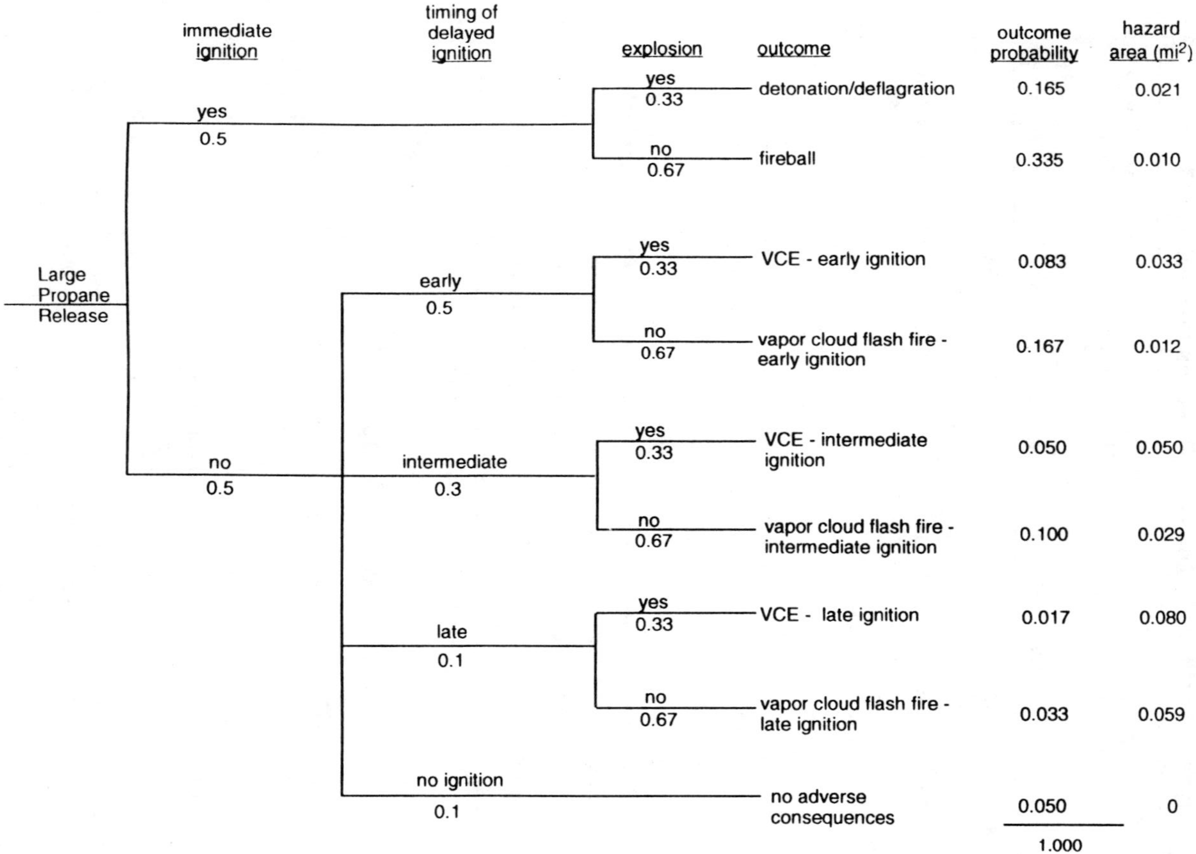

FIGURE 7-5. Sample Event Tree fpr a Large Release from a Propane Tank Truck.

the range. For this study, a reasonable yet conservative accident rate estimate is 1.5 accidents per million vehicle miles. Section 7.2.3 stated that the breakdown of small versus large spills was 65/35.

Given the similarity between the two routes, the same accident rate and release probabilities are used on each route. Since we are most concerned with consequence variations and their influence on risk levels, a more precise selection of accident rates is not of concern at this time.

The overall chance of a large spill is thus $0.35 \cdot 1.5 \times 10^{-6}$/mile or 5.3×10^{-7}/mile. For a small spill the rate is 9.8×10^{-7}/mile.

7.2.6. Risk Estimation

The objective of this step is to put all of the information on frequencies and consequences together to determine the overall risk. Table 7-12 presents the frequency and estimated consequences for each incident outcome case. The statistical consequence estimates are determined using the following expression:

Consequence = hazard zone area for incident outcome
× segment population density
× percent of exposures yielding fatality

The small spill maximum vapor cloud area is 0.00059 square miles. Even in the highest population density segment (2000 per square mile), assuming a 30 percent chance of fatality, the maximum number of calculated fatalities is less than one ($0.00059 \cdot 2000 \cdot 0.3 = 0.35$). While it is possible to have a fatality result in such cases, one is not expected to occur. Thus, small spills are not considered in the risk calculations. The frequency estimates are calculated using the following expression:

Frequency = large spill accident rate per mile × outcome probability
× miles in segment × number of trips

where the outcome probability is determined from the event tree.

Table 7-13 provides information on the risk calculated for differing fatality levels for each route. Each risk estimate is calculated by summing the frequencies for all incident outcome cases that yield the indicated number of fatalities or greater. The frequency for the lower fatality levels (i.e., lower values of N) is higher because more incident outcome cases factor into the estimate. (The decision on whether an incident is evaluated at a given level is based on the consequences.) This can be expressed as:

$$F_N = \sum_i F_i \quad \text{for all incident outcome case } i \text{ for which } N_i \geq N,$$

where

F_N = frequency of all incident outcome cases affecting N or more people

F_i = frequency of incident outcome case i

N_i = number of people affected by incident outcome case i

TABLE 7-12
Risk Results for Propane Tank Truck Movement of 300 Trips per Year

Incident Outcome Case	*Frequency (per year)*	*Consequence (fatalities)*
Route 27—Segment 1		
Explosion	1.3×10^{-4}	6.3
Fireball	2.7×10^{-4}	3.0
VCE—early ignition	6.6×10^{-5}	9.9
Flash fire—early ignition	1.3×10^{-4}	3.6
VCE—intermediate ignition	4.0×10^{-5}	15.0
Flash fire—intermediate ignition	8.0×10^{-5}	8.7
VCE—late ignition	1.4×10^{-5}	24.0
Flash fire—late ignition	2.6×10^{-5}	17.7
Route 27—Segment 2		
Explosion	6.0×10^{-4}	3.2
Fireball	1.2×10^{-3}	1.5
VCE—early ignition	3.0×10^{-4}	5.0
Flash fire—early ignition	6.1×10^{-4}	1.8
VCE—intermediate ignition	1.8×10^{-4}	7.5
Flash fire—intermediate ignition	3.7×10^{-4}	4.4
VCE—late ignition	6.2×10^{-5}	12.0
Flash fire—late ignition	1.2×10^{-4}	8.9
Route 27—Segment 3		
Explosion	1.8×10^{-4}	12.6
Fireball	3.7×10^{-4}	6.0
VCE—early ignition	9.2×10^{-5}	19.8
Flash fire—early ignition	1.9×10^{-4}	7.2
VCE—intermediate ignition	5.6×10^{-5}	30.0
Flash fire—intermediate ignition	1.1×10^{-4}	17.4
VCE—late ignition	1.9×10^{-5}	48.0
Flash fire—late ignition	3.7×10^{-5}	35.4
Route 46		
Explosion	1.4×10^{-3}	2.5
Fireball	2.9×10^{-3}	1.2
VCE—early ignition	7.3×10^{-4}	4.0
Flash fire—early ignition	1.5×10^{-3}	1.4
VCE—intermediate ignition	4.4×10^{-4}	6.0
Flash fire—intermediate ignition	8.7×10^{-4}	3.5
VCE—late ignition	1.5×10^{-4}	9.6
Flash fire—late ignition	2.9×10^{-4}	7.1

TABLE 7-13
Data for F-N Curve for Tank Truck Example

	Frequency of Fatalities Greater Than or Equal to N (per year)				
	$N = 1$	$N = 3$	$N = 10$	$N = 30$	$N = 48$
Route 27	5.3×10^{-3}	3.4×10^{-3}	7.0×10^{-4}	1.1×10^{-4}	1.9×10^{-5}
Route 46	8.3×10^{-3}	2.5×10^{-3}	1.5×10^{-4}	—	—

Minor rounding off of the consequences is applied in developing Table 7-13. This information can also be presented graphically. Figure 7-6 uses an F-N curve to present the risk. This graph plots cumulative frequency versus the number of fatalities and gives a picture of the overall risk along each route for this study. The F-N curves could also be presented on a per trip or per ton basis. However, since the same number of trips and quantity per trip occurs for each route, nothing different would be learned about the F-N curves with these alternate presentations.

The data given in Table 7-12 can also be used to estimate the average individual risk level as defined in Equation (5-1), which summed the products of each frequency/consequence pair and then divided by the total population. On both routes, 0.38 miles is the maximum hazard distance, 1600 feet for vapor cloud dispersion plus 430 feet for the overpressure distance. The factor of two accounts for the fact that this distance can extend to either side of the route.

$$\text{Average Individual Risk} = \sum_{i=1}^{n} f_i \cdot N_i / P_T \tag{5-1}$$

where

n = the number of incident outcome cases,
f_i = the frequency of incident outcome case i,
N_i = the number of fatalities for incident outcome case i,
P_T = the total population of concern and
i = incident outcome counter

The total population at risk on Route 27 is thus:

$$\begin{aligned} &\ 5 \text{ miles} \times 2 \times 0.38 \text{ miles} \times 1000 \text{ people/mi}^2 \\ &+ 23 \text{ miles} \times 2 \times 0.38 \text{ miles} \times 500 \text{ people/mi}^2 \\ &+ 7 \text{ miles} \times 2 \times 0.38 \text{ miles} \times 2000 \text{ people/mi}^2 \\ &= 23{,}180 \text{ people} \end{aligned}$$

On Route 46, the total population at risk is:

$$55 \text{ miles} \times 2 \times 0.38 \text{ miles} \times 400 \text{ people/mi}^2 = 16{,}720 \text{ people}$$

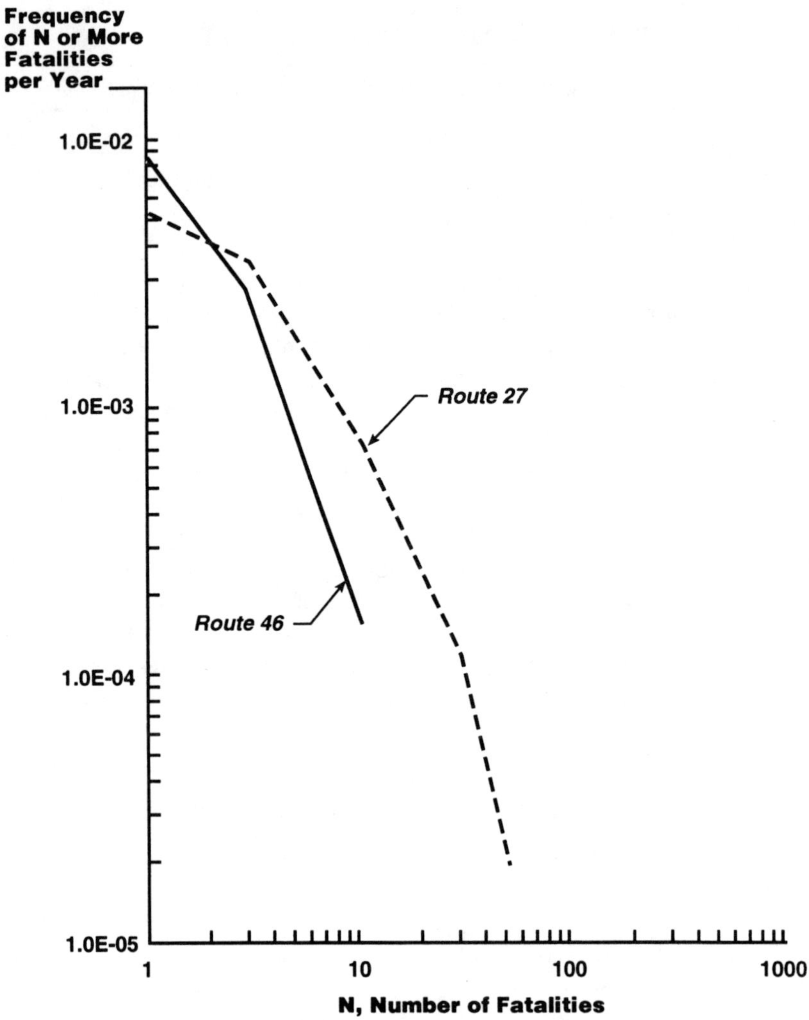

FIGURE 7-6. F-N Curves for Propane by Tank Truck Example.

Using these figures and the data in Table 7-12 to determine the sum of the products of each frequency and consequence gives:

	Route 27	*Route 46*
Average Individual Risk	1.3×10^{-6}/yr	1.3×10^{-6}/yr

Thus, this risk measure does not distinguish between the two routes, but may still provide a comparison to other movements.

7.2.7. Discussion of Results

This analysis provides input to the overall decision-making process. The risks along each route can be compared and a decision made on which route to use based solely on fatality risk, or by taking other factors such as cost, uncertainty, environmental impacts, and delivery time into consideration as well. If neither set of results is tolerable, mitigation or a more rigorous analysis can be considered.

Societal risks are used to emphasize the difference between the routes. As shown in Figure 7-6, the likelihood of one or more fatalities is greater for Route 46, however the risks quickly become less than those for Route 27. The maximum number of fatalities is limited to roughly 10, while on Route 27 there can be up to 48 fatalities predicted. Thus, if the goal is to reduce consequences, Route 46 is the clear choice. If the chance of one or more fatalities is the greatest concern, the choice is less obvious.

Super Chemicals, Inc. might consider the potential benefits of transporting the propane at night, increasing driver training, monitoring and controlling speeds, or even going to the use of a pipeline.

7.3. Rail Risk Assessment

7.3.1. Introduction

The following case study is one in which two different suppliers are being considered as a source of chlorine. Each supplier uses a different size and type of tank car and route. Consequences were estimated in a previous study and those results are applicable to this study. The objective of this case study is to evaluate the risks of transporting this material on the different routes and to use these risk estimates as one factor in considering which supplier to choose. The case study itself provides an example of summing the risk from different portions or segments of a route.

The study cube cell for this case is the "Risk, Quantitative, Limited Set" cell (see Figure 7-1), as both the frequencies and consequences are sufficiently different that risk estimates are required. The steps in this case study are:

- Identify and select the incidents of concern.
- Estimate the frequency of each incident.
- Combine the frequency analysis with the previous information on consequences to estimate the overall risk for each route.
- Compare the results for the two routes.

7.3.2. Description

Chlorine may be transported using two different routes (see Figure 7-7) in two different sizes of railcars. Chlorine is a toxic gas that has a pungent suffocating odor. It is extremely toxic by inhalation and is heavier than air. Table 7-14 provides some of the physical properties of chlorine. It is currently being transported on Route 1 from Present Supplier, Inc. The customer is considering switching to another supplier near by. Table 7-15 summarizes the shipment information for the two options. The route lengths and associated population densities are different for each option. In addition, there are differences in track quality along the routes. Route 1 involves three yard visits, while Route 2 has four yard visits. Table 7-16 provides the details for each route.

The return of the "empty" rail cars is not included in the present analysis. Such cars can contain up to 10,000 pounds of material.

7.3.3. Identification, Enumeration, and Selection of Incidents

There are several types of incidents that may yield a release of the chlorine. This case study focuses only on accident-caused releases. Some typical accident scenarios are listed below:

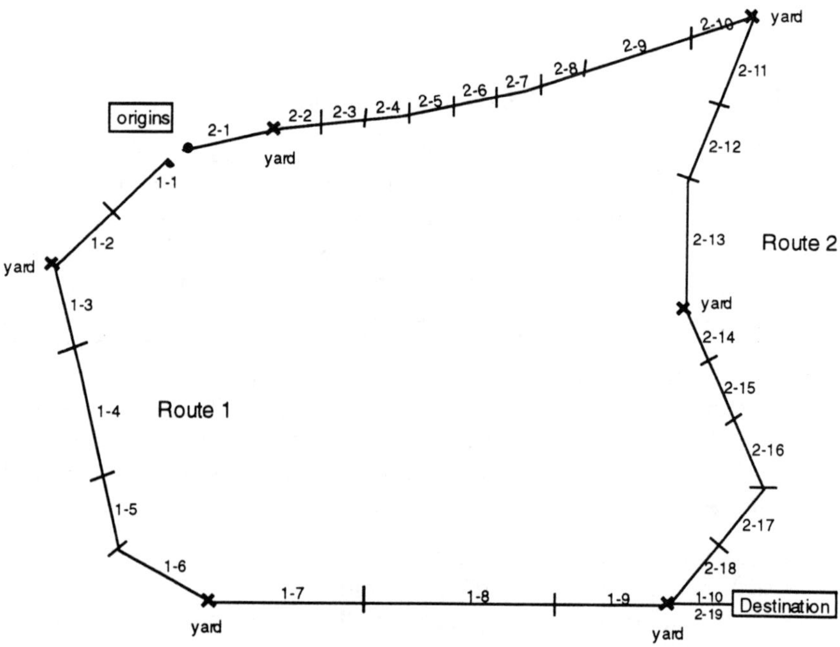

FIGURE 7-7. Route Descriptions for Rail Example.

TABLE 7-14
Properties of Chlorine

CAS Number:	7782-50-5
DOT Class:	poison gas
ID Number:	UN 1017
Lower flammable limit:	—
Upper flammable limit:	—
Boiling point:	–29°F
Molecular weight:	70.9
Flash point:	—
Liquid specific gravity:	1.57 at 93°F
Solubility in water:	not soluble
ERPG-3 concentration:	20 ppm
ERPG-2 concentration:	3 ppm
ERPG-1 concentration:	1 ppm

TABLE 7-15
Description of Example Rail Shipment Options

	Route 1	*Route 2*
Material	Chlorine	Chlorine
Amount	90 Tons	55 Tons
Car Type	105 A 500W; insulated	105 A 300W; insulated[a]
Temperature	70°F	70°F
Trips/Year	50	82

[a]An older, slightly thinner-walled tank car that is still in use up to this size.

- Collisions with vehicles at grade crossings
- Collisions with other trains
- Derailments

It can be assumed that all rail cars are subject to these events (see Section 2.2.1). Frequencies of these events are affected by speed and track class among other things. In general, derailment rates are high on low class track and reduce rapidly as quality improves (see Section 2.2.2). Release sizes can range from pinholes to a catastrophic loss of the entire contents. Small drips are not usually associated with accidents and will not be

TABLE 7-16
Route Characterizations for Rail Example

Route and Segment Number	*Population Density (per mi.^{2})*	*Track Class*	*Typical Speed (mph)*	*Length (miles)*
ROUTE 1				**717**
1-1	780	4	50	56
1-2	490	4	50	53
1-3	375	4	60	56
1-4	480	4	60	88
1-5	1226	4	60	46
1-6	435	4	60	74
1-7	562	3	40	102
1-8	218	3	40	126
1-9	454	3	40	74
1-10	675	3	40	42
ROUTE 2				**870**
2-1	900	4	60	66
2-2	602	5	80	32
2-3	441	5	80	28
2-4	712	5	80	28
2-5	380	5	80	30
2-6	190	5	80	28
2-7	65	5	80	32
2-8	112	5	80	32
2-9	89	5	80	74
2-10	220	5	80	42
2-11	176	4	60	63
2-12	120	5	80	53
2-13	237	5	80	88
2-14	237	4	60	39
2-15	154	4	60	42
2-16	118	4	60	49
2-17	327	4	60	49
2-18	520	4	60	53
2-19	675	3	40	42

considered in this study. Given the hazards posed by chlorine, anything larger than a small drip can be very significant.

The analyst selects the following release cases as representative of the complete range of potential releases: (1) catastrophic loss of the entire contents of one car and (2) continuous release of contents at a rate of 1 percent per minute over 60 minutes. This selection is based on past analyses of spill data as discussed in Chapter 2 and is quite reasonable given the pressurized nature of the cargo. In this case study multiple car spills are not considered, as only one car is shipped at a time.

7.3.4. Incident Consequence Estimation

The consequence estimation and hazard distances were estimated in a previous study that looked at releases which could occur on site due to accidents while loading. A sophisticated consequence modeling package was used, with a concentration of concern of 140 ppm. This value was selected based on discussions with the company's toxicologists. The results are listed in Table 7-17. They were derived for D-stability 10 mph and F-stability 4 mph. D-stability is representative of the prevalent conditions 70 percent of the time, while F-stability covers the other 30 percent of the time, based on STAR data (see Transportation Data Source 4-40 in Chapter 4) for several locations along the route. (These data are hypothetical and should not be used in an actual study.) For the total loss scenarios, the cloud shapes are very different under D and F stabilities. The D-stability scenario produces a longer, thinner cloud, while under F-stability the shape is more triangular. As shown in Figures 7-8 to 7-11, the total areas covered under the two stabilities are similar.

TABLE 7-17
Hypothetical Hazard Zones for Chlorine Releases from Rail Cars

			Toxic Dispersion (ft)	
Car Size	*Scenario*	*Stability Class/ Wind Speed*	*Downwind distance*	*Maximum width of cloud*
90 ton	1. Total loss	D/10 mph	30,000	5,000
		F/ 4 mph	18,000	11,000
	2. 1% per min for	D/10 mph	5,000	600
	60 min	F/ 4 mph	20,000	1,800
55 ton	1. Total loss	D/10 mph	18,500	4,000
		F/ 4 mph	11,000	8,000
	2. 1% per min for	D/10 mph	3,000	500
	60 min	F/ 4 mph	11,500	1,500

7.3.5. Incident Frequency Estimation

This section estimates the accident frequency and conditional spill probability given an accident occurs. Route 1 has both FRA class 3 and class 4 track, while the majority of Route 2 is on class 4 and class 5 track. The following have been used for accident rates:

	Mainline	*Yard*
Class 4, 5	2×10^{-7} per loaded car mile	6×10^{-6} per car visit
Class 3	5×10^{-7} per loaded car mile	6×10^{-6} per car visit

These mainline rates are taken from Table 2-4. The yard rate was determined after discussions with the railroads involved. These rates include the chance that the car derails. Since only one car is assumed to be shipped at a time a very simple approach to estimate the spill frequency can be used. The following spill probabilities are used and differ due to the car types used by the two suppliers. The values were obtained from a previous study conducted by the analyst involving the same car types, and were determined considering such factors as head and shell thicknesses, protective features such as head shields, etc.

	Conditional Probability of Release Given Derailment			
	105 A 300W/Route 2		*105 A 500W/Route 1*	
	Mainline	*Yard*	*Mainline*	*Yard*
Total Loss	0.18	0.03	0.15	0.01
1% per min. for 60 min.	0.18	0.06	0.15	0.04

In addition, event trees similar to that discussed in the previous case study on tank trucks have been developed and are presented as Figures 7-8 to 7-11. These trees combine the information discussed above for release size and weather conditions. Separate trees are provided for both types and sizes of rail cars, as well as for mainlines versus yards.

7.3.6. Risk Estimation

The objective of this step is to combine the frequency and consequence information in order to estimate the overall risk. Table 7-18 illustrates the set of frequency and consequence results for one yard and two mainline segments. The frequency estimates are calculated as:

$$\begin{aligned}\text{Frequency} &= \text{accident rate} \\ &\quad \times \text{outcome probability} \\ &\quad \times \text{miles in segment} \\ &\quad \times \text{trips}\end{aligned}$$

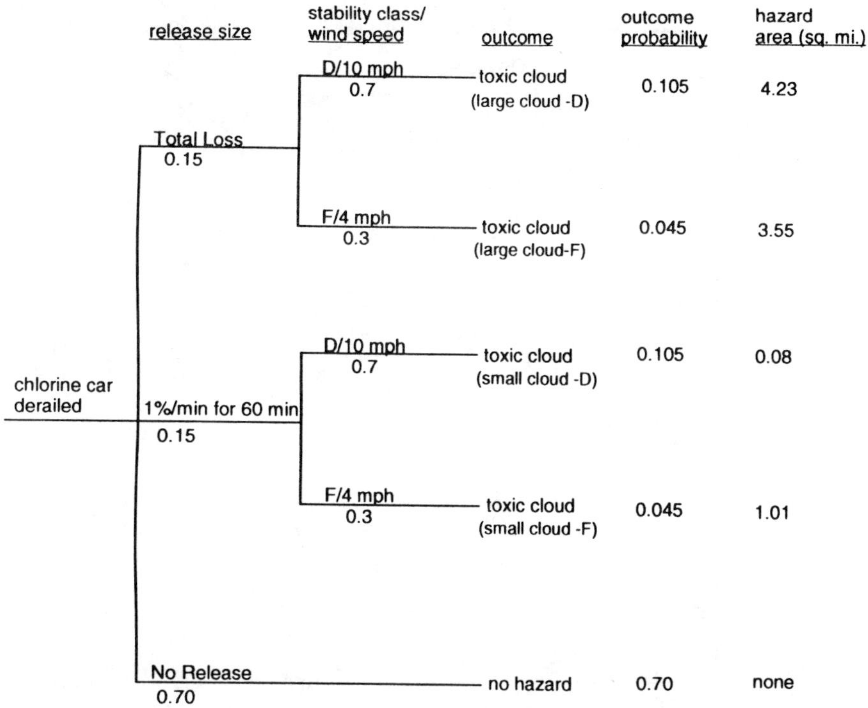

FIGURE 7-8. Event Tree for Chlorine Rail Transportation Example—Route 1, mainline.

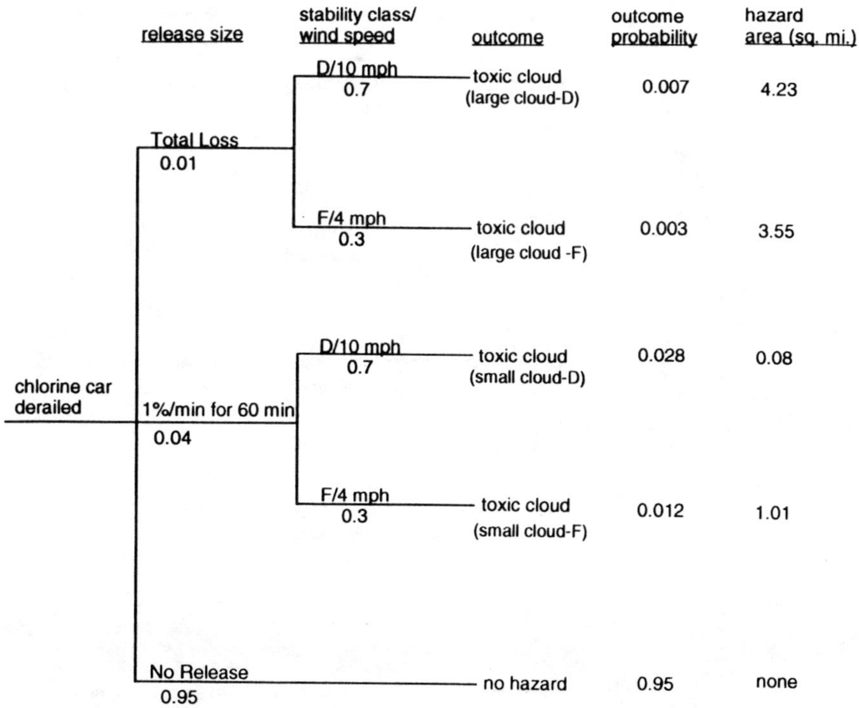

FIGURE 7-9. Event Tree for Chlorine Rail Transportation Example—Route 1, yard.

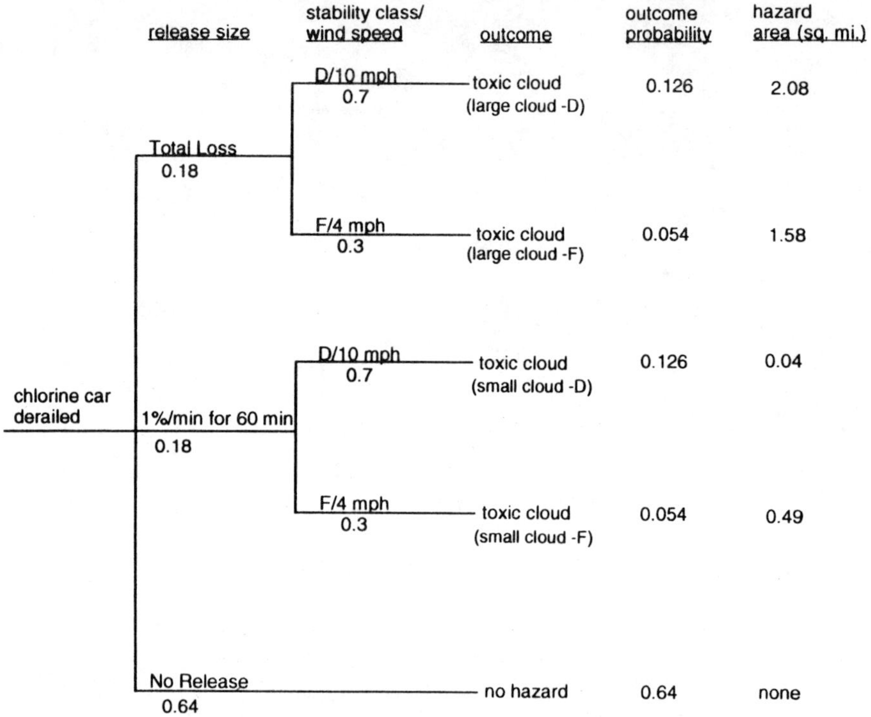

FIGURE 7-10. Event Tree for Chlorine Rail Transportation Example—Route 2, mainline.

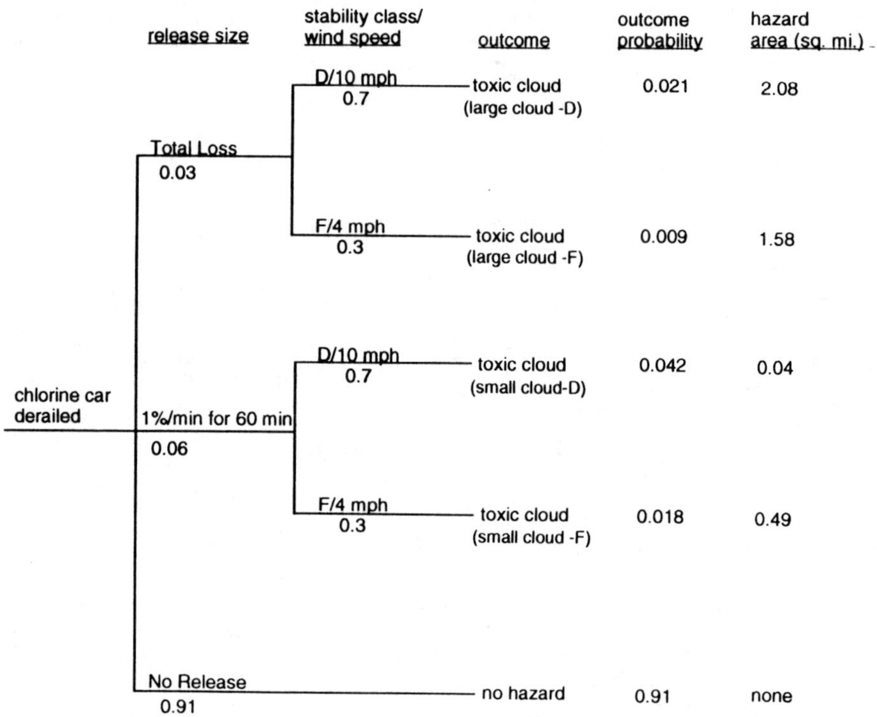

FIGURE 7-11. Event Tree for Chlorine Rail Transportation Example—Route 2, yard.

TABLE 7-18
Representative Risk Results for Three Segments on Route 1

Incident Outcome Case	*Frequency (per year)*	*Consequence (fatalities)*
Segment 1-1		
Large cloud—D	5.9×10^{-5}	660
Large cloud—F	2.5×10^{-5}	554
Small cloud—D	5.9×10^{-5}	12
Small cloud—F	2.5×10^{-5}	158
Segment 1-2		
Large cloud—D	5.6×10^{-5}	415
Large cloud—F	2.4×10^{-5}	348
Small cloud—D	5.6×10^{-5}	8
Small cloud—F	2.4×10^{-5}	99
Yard		
Large cloud—D	2.1×10^{-6}	366
Large cloud—F	9.0×10^{-7}	307
Small cloud—D	8.4×10^{-6}	7
Small cloud—F	3.6×10^{-6}	87

The consequence estimates are determined using the following expression:

$$\begin{aligned}\text{Consequence} &= \text{hazard footprint area}\\ &\quad \times \text{population density}\\ &\quad \times \text{percent of exposures yielding fatality}\end{aligned}$$

The population density for the yard is taken as the average for the segments on either side. In transportation risk, the use of a constant population density along a segment makes the analysis simpler. In this case study, the percent of exposures yielding a fatality is taken as 20 percent, since the population is expected to be indoors a significant portion of the time. Prugh (1985) discusses the effectiveness of evacuation, sheltering in place, etc. as does the *CPQRA Guidelines* (CCPS, 1989).

Sample calculations for the "large cloud-D" frequency for each segment are:

Segment 1-1: 2×10^{-7}/mile · 0.105 · 56 miles · 50/year = 5.9×10^{-5}/year

Segment 1-2: 2×10^{-7}/mile · 0.105 · 53 miles · 50/year = 5.6×10^{-5}/year

Yard: 6×10^{-6}/car visit · 0.007 · 50/year = 2.1×10^{-6}/year

Consequences for the same incident outcome cases are:

Segment 1-1: 4.23 sq. mi. · 780 people/sq. mi. · 0.2 = 660 people

Segment 1-2: 4.23 sq. mi. · 490 people/sq. mi. · 0.2 = 415 people

Yard: 4.23 sq. mi. · 432.5 people/sq. mi. · 0.2 = 366 people

Table 7-19 provides information on the risk calculated for differing fatality levels for each route, when this process is carried out for all segments. These results were obtained using a spreadsheet computer program.

Figure 7-12 displays these results graphically. The chance of one or more fatalities is lower on Route 1. However, Route 1 poses higher risks of incidents resulting in 100 or more fatalities. Another risk comparison is the average rate of death and/or maximum consequence for each route. The overall chance of a spill and likelihoods of particular spill sizes may also be used. Table 7-20 presents these comparisons. All frequencies and the average rate of death are given on an annual (or per year) basis. The average rate of death is calculated by summing the product of the frequency and consequence for each incident outcome case for each segment, as stated in Equation (5-6).

$$\text{Average rate of death} = \sum_{g=1}^{t} \sum_{i=1}^{n} \sum_{k=1}^{S_i} F_{g,i,k} \cdot N_{g,i,k} \tag{5-6}$$

where

$F_{g,i,k}$ and $N_{g,i,k}$ are defined in Equations (5-4) and (5-5)
t = number of segments considered for route
n = number of release sizes considered
S_i = number of incident outcomes for release size i

TABLE 7-19
F-N Curve Data for Rail Example

	Frequency of Fatalities Greater Than or Equal to N (per year)						
	$N = 1$	$N = 3$	$N = 10$	$N = 30$	$N = 100$	$N = 300$	$N = 1000$
Route 1	3.7×10^{-3}	3.7×10^{-3}	2.6×10^{-3}	2.4×10^{-3}	2.1×10^{-3}	1.4×10^{-3}	4.8×10^{-5}
Route 2	5.2×10^{-3}	4.4×10^{-3}	3.6×10^{-3}	3.0×10^{-3}	1.2×10^{-3}	1.5×10^{-4}	—

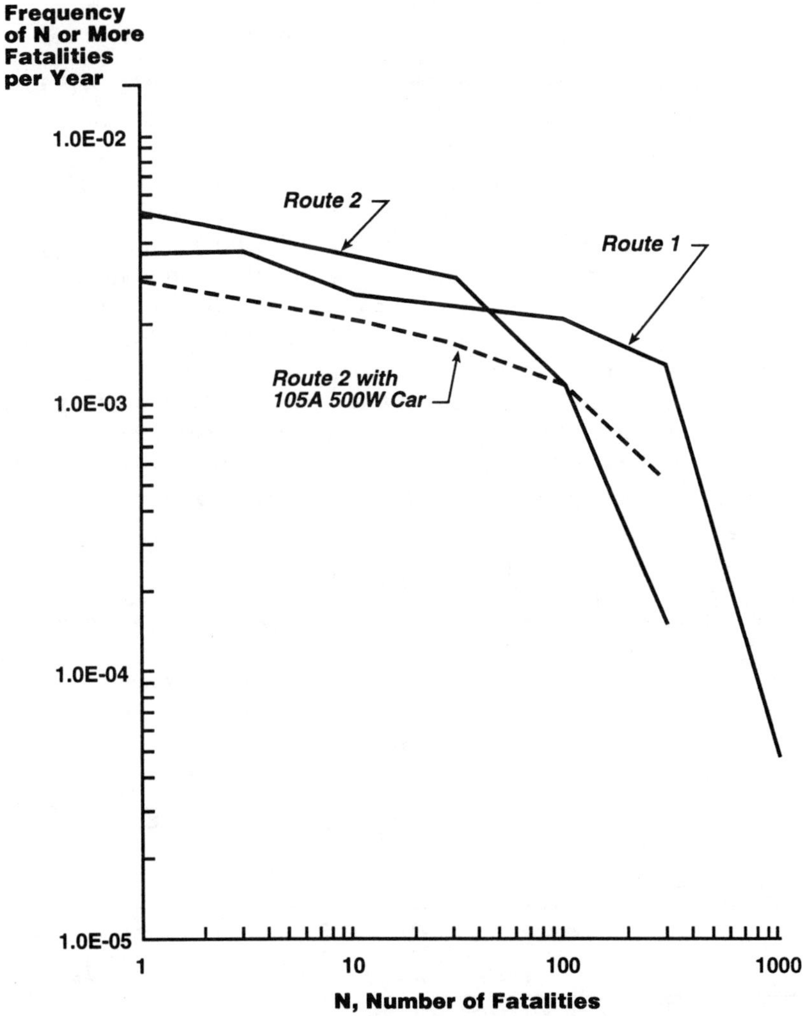

FIGURE 7-12. F-N Curves for Chlorine by Rail Example.

TABLE 7-20
Other Risk Results for Rail Example

	Rate of Death (fatalities/year)	*Overall Spill Frequency (per year)*	*Chance of Large Spill (per year)*	*Chance of Small Spill (per year)*	*Maximum Consequences (fatalities)*
Route 1	0.8	3.7×10^{-3}	1.9×10^{-3}	1.9×10^{-3}	1037
Route 2	0.4	5.7×10^{-3}	2.8×10^{-3}	2.9×10^{-3}	374

7.3.7. Discussion of Results

Quantitative risk assessment identifies the significant contributors to risk and helps suggest potential mitigation strategies. Based on the similarity in the results for the two routes, it may be desirable to focus on the more significant difference (almost a factor of three) in the maximum predicted consequences. This difference is due both to the lower population densities and lower quantity per container on Route 2. One obvious potential mitigation strategy to consider is using a 105A 500W car on Route 2. This reduces the frequencies while increasing the consequences slightly, due to the larger quantity per container. The various results for Route 2 with this alternate car type are shown in Table 7-21, and on Figure 7-12. From these results, one might conclude that there was a reasonable trade-off between the changes in frequencies and consequences. However, no decision is likely to be based on risk alone. Costs, reliability of supply, and other factors will be important additional considerations.

7.3.8. References

Prugh, R.W. "Mitigation of Vapor Cloud Hazards." *Plant/Operations Progress* 4 (2, April): 95-104, 1985.

Center for Chemical Process Safety. *Guidelines for Chemical Process Quantitative Risk Analysis*. New York: American Institute of Chemical Engineers. 1989.

Transportation Data Source 4-40 was also used in this case study.

TABLE 7-21
Example Results for Route 2 with Mitigation

F-N CURVE DATA

	Frequency of Fatalities Greater Than or Equal to N (per year)					
	N = 1	*N* = 3	*N* = 10	*N* = 30	*N* = 100	*N* = 300
Route 2	5.2×10^{-3}	4.4×10^{-3}	3.6×10^{-3}	3.0×10^{-3}	1.2×10^{-3}	1.5×10^{-4}
Route 2 with 105A 500W car	2.9×10^{-3}	2.5×10^{-3}	2.1×10^{-3}	1.7×10^{-3}	1.2×10^{-3}	5.1×10^{-4}

OTHER RISK RESULTS

	Rate of Death (fatalities/year)	*Overall Spill Frequency (per year)*	*Chance of Large Spill (per year)*	*Chance of Small Spill (per year)*	*Maximum Consequences (fatalities)*
Route 2	0.4	5.7×10^{-3}	2.8×10^{-3}	2.9×10^{-3}	374
Route 2 with 105A 500W car	0.4	2.9×10^{-3}	1.4×10^{-3}	1.5×10^{-3}	761

7.4. Barge Risk Assessment

7.4.1. Introduction

This case study addresses population characteristics versus release location. One factor unique to barge (and other marine) studies is the location of an accident, for example, mid-river or on one of the banks. The magnitude of the effect on surrounding populations will be influenced by the buffer distances that the river provides for some combinations of accident location and wind direction. The objective of this study is to estimate the risks associated with the transportation of butadiene which is flammable, by barge along a selected route. Worldwide Plastics could continue to use rail transportation, if the barge option poses too high a level of risk.

The study cube cell representing this case study is the "Risk, Quantitative, Generalized Set" cell (see Figure 7-1), as the example focuses on the risk estimation process, particularly the treatment of several transportation-specific concerns when using consequence modeling results.

7.4.2. Description

Butadiene is currently being transported from Rock, Louisiana to Hardplace, Mississippi by rail. The access to navigable waterways at both plants make barge an obvious alternative to evaluate. Table 7-22 lists some of the

TABLE 7-22
Properties of Butadiene

CAS Number:	106-99-0
DOT Class:	flammable gas
ID Number:	UN 1010
Lower flammable limit:	2.0%
Upper flammable limit:	11.5%
Boiling point:	24°F
Molecular weight:	54.1
Flash point:	–105°F
Liquid specific gravity:	0.65 at 21°F
Solubility in water:	slightly soluble
ERPG-3 concentration:	5000 ppm
ERPG-2 concentration:	50 ppm
ERPG-1 concentration:	10 ppm

physical properties of butadiene. The total length of the barge route is 300 miles with 100 miles on the Intracoastal Waterway and 200 miles on the Mississippi River. There would be 13 shipments per year along this route in a hopper-type barge, similar to chlorine barge designs. The barge has four pressurized tanks each holding about 500,000 pounds of butadiene. Each tank is positioned in a watertight hold—separated from the other tanks by a longitudinal bulkhead along the center of the barge and by fore and aft cross-wise bulkhead structures. The barge has both a double-hull and a double bottom. Table 7-23 summarizes the facts for this movement. The population densities range from urban to rural with the majority of the route being through rural areas (see Figure 7-13).

The population density along the route was determined using U.S. Census data, and the route was subdivided into segments with uniform population densities by per segment. Table 7-24 gives the details of this characterization. If the population density for a given segment varied significantly for each side of the waterway, two approaches were reviewed. If the difference in densities was greater than a factor of three, the segment was divided into two subsegments of equal length and each was assigned the density for a different side of the waterway. If the difference was less than a factor of three, the densities were averaged and applied to the original segment.

7.4.3. Identification, Enumeration, and Selection of Incidents

There are several types of incidents that may cause a release of the cargo. This case study focuses on accident-caused releases. Incidents resulting from contamination, overfilling, etc. are not included. Some typical barge accident scenarios are as follows:

- Collisions
- Rammings

TABLE 7-23
Example Barge Description

Product:	Butadiene
Amount:	4 Tanks at 500,000 lb each
Container type:	Pressurized tanks
Storage Temperature:	70°F
Number of Shipments:	13
Route Length:	300 Miles

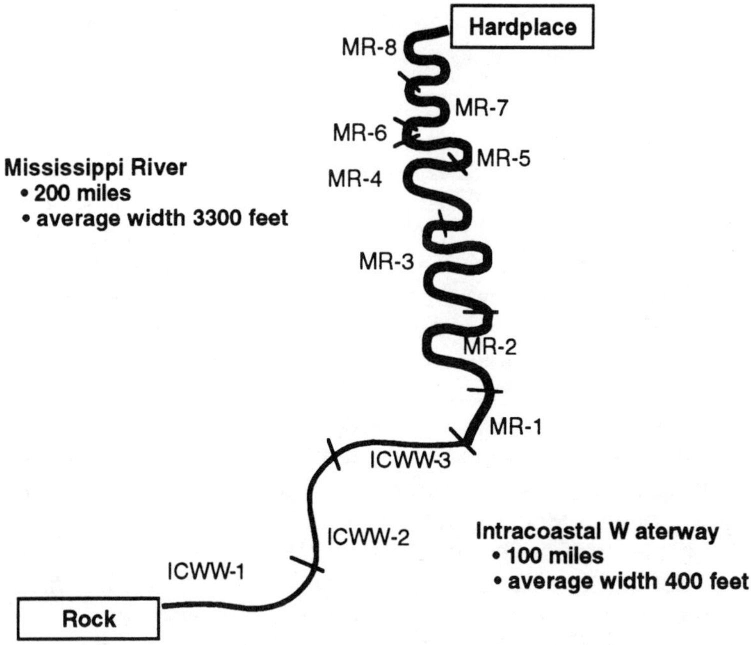

FIGURE 7-13. Route Description for Barge Example.

TABLE 7-24
Route Characterization for Barge Example

Segment	*Length (miles)*	*Population Density (per square mile)*
ICWW-1	45	7
ICWW-2	29	44
ICWW-3	26	105
MR-1	13	857
MR-2	27	109
MR-3	56	53
MR-4	32	199
MR-5	11	254
MR-6	3	467
MR-7	22	194
MR-8	36	62

- Allisions
- Fires and Explosions
- Groundings
- Other

It can be assumed that all barges are potentially subject to these events (see Section 2.4.1). Data sources on the number and frequency of these types of events can be found in Chapter 4. An example of how this data may be used to generate frequency estimates is given in Section 7.4.5. Frequencies are affected by parameters such as vessel speed, and width and depth of the waterway (see Section 2.4.2). Potential release sizes range from very small to the contents of one or more cargo tanks.

Impact-type marine casualties of sufficient severity to cause a spill generally result in damage to localized areas of a barge structure, such that a credible spill size would be limited to a maximum of the contents carried in one cargo tank. Some other impact-type casualties, such as collisions at or above the waterline, may result in spills much smaller than the total contents of the breached cargo tank. Other types of casualties such as material or structural connection failures, or breaks in piping, valves, fittings, etc. may result in relatively small release quantities, up to as much as 10 percent of the total cargo tank contents. A credible estimate for a large spill case would be the loss of the entire contents of one cargo tank in the barge. The smaller release could be modeled as a loss of 1 percent per minute over 10 minutes from one cargo tank. (See Section 3.1.4.)

The selection of incidents for consideration may need to rely on the consequence analysis already having been conducted. Since buffer distances (the distance between where the accident/release occurs and the shoreline where the affected populations start) are taken into account, some consequences may not need to be addressed. This is due to the fact that the downwind and crosswind distances used to define the hazard zones might not extend all the way to the shoreline.

7.4.4. Incident Consequence Estimation

Table 7-25 presents the results of the consequence analysis estimates for butadiene. These results were obtained using two sets of meteorological conditions: D-stability with 9 mph winds and F-stability with 4 mph winds. These were selected based on STAR data for Baton Rouge, which also suggested a split of 65 percent for D and 35 percent for F. (These data are hypothetical and should not be used in an actual study.)

Spills on water can pose modeling challenges, as one must consider solubility, smoother surfaces, temperature exchange, unconfined (effec-

TABLE 7-25
Hypothetical Hazard Zones for Barge Releases of Butadiene

Scenario	Stability Class/ Wind Speed	Flammable Distance (ft) Max. distance to LFL	Flammable Distance (ft) Max. width of cloud	Overpressure radius (ft)	Pool Fire radius (ft)	Fireball radius (ft)
Large	D/ 9 mph	1700	510	850	260	2500
	F/ 4 mph	5500	470	1200	200	2500
Small	D/ 9 mph	190	240	insuff. mass	130	—
	F/ 4 mph	1030	300	300	90	—

tively) spills, etc. Section 3.7 identifies several references that may be of assistance in modeling releases on water. Some releases may also be directly into rather than onto water if the bottom of a tank is penetrated.

The effective area of a hazard zone needs to take into account the buffer distance provided by the river. Some scenarios, especially for the small (continuous) release case, may not reach shore and therefore may not need further evaluation. The calculation of the effective (land-based) hazard areas needs to be done separately for different parts of the route as not all portions of the route have the same width (e.g., the Mississippi River versus the Intracoastal Waterway).

Several different assumptions were made on the buffer distance between a spill event and a populated area. First the actual river widths were analyzed for the segments of concern, giving:

	Average Width	*Maximum Width*
Mississippi River	3300 ft	9800 ft
Intracoastal Waterway	400 ft	500 ft

Three basic cases can be considered. These are shown in Figure 7-14 and described below:

- *No buffer*: This is representative of very narrow sections of the waterways, and cases where the accident/release occurs at the shoreline and the wind is blowing onshore. No credit is taken for there being some distance between the event and populated areas.
- *Mid-waterway*: This case assumes releases occur in the middle of the waterway, giving 1650 ft of buffer zone on the Mississippi River and 200 ft on the Intracoastal Waterway.
- *Waterway edge*: Assumes release occurs at waterway's edge and 50 percent of the time the wind blows onshore (no buffer) and 50 percent of the time it blows across the full width of the waterway

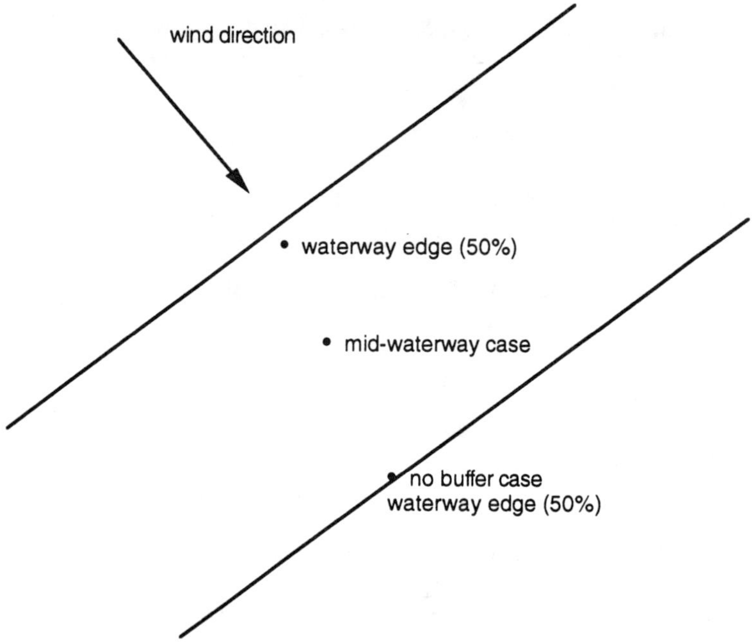

FIGURE 7-14. Barge Example Release Cases.

(3300 ft buffer on the Mississippi River and 400 ft buffer on the Intracoastal Waterway). Given the symmetry implied by assuming uniform population densities, this case covers the barge being on either edge of the waterway.

This latter case is believed to be most representative of reality as accidents which occur mid-waterway will often result in the barge drifting toward shore. In addition, groundings and rammings are more likely near shore. Last, shipping lanes and channels are generally not in the middle of the waterway. It is computed by taking an equal mix of the no buffer and the full buffer cases, where the full buffer case should recognize that at least half of a fireball hazard zone will be onshore. This means that the combined results are slightly conservative because they assume that the full fireball hazard is onshore in the no buffer case.

7.4.5. Incident Frequency Estimation

This section determines the appropriate accident frequency and conditional spill probability given that an accident occurs, as well as the incident outcome probabilities. Using a combination of CASMAIN accident data and Corps of Engineers data (see Transportation Data Sources 4-28, 4-32, and 4-33) on vessel traffic on the route of interest, an accident rate of 1.5 $\times 10^{-3}$ accidents per transit and a starting spill probability of 20 percent will be used in the analysis. Given a distance of 300 miles, this produces an overall spill rate of 1×10^{-6} per mile. The release probability is believed to be unduly conservative because the barge has both a double hull and a double bottom. For this case study, the chance of a spill is judged to be a factor of five lower, giving an overall rate of 2×10^{-7} per mile. The analyst believes this is a conservative estimate, based on the U.S. Coast Guard evaluation discussed in Section 2.5.3.

Figure 7-15. shows an event tree, similar to those presented earlier. Given the severity of the event needed to penetrate both hulls, a greater fraction of the resulting releases are expected to be large releases. Therefore, the spills have been distributed as 50 percent large (defined as a total loss of contents of one tank) and 50 percent small (taken as 1 percent per minute for 10 minutes, from one tank).

The likelihood of immediate ignition is judged to be higher in the events that are capable of causing a large spill, and this is reflected in the event tree. The chance of a release not igniting immediately, but then igniting as it encounters additional ignition sources, was determined to be lower than for the propane truck case. This is because the lower population densities on the barge route imply fewer ignition sources. It was also determined that only two stages of delayed ignition would be considered.

The event tree only covers the initial outcome of an accident. For simplicity in this case study, the influences of one tank's release on the other tanks are not addressed. Such effects might include BLEVEs resulting from an initial fire, explosion-induced (overpressure) failure of a second tank, etc.

7.4.6. Risk Estimation

The overall calculation of risk combines the estimates of release frequency with the incident outcome probabilities, the route characterization, and the consequences. The first step is to determine what the hazard areas are for various incident outcomes and release locations—both in terms of the Intracoastal Waterway versus the Mississippi River and in terms of mid-waterway versus the near shore (no buffer) versus the far shore (full buffer). Table 7-26 summarizes all the combinations that could be required, depending on which release locations are to be evaluated.

As stated earlier, the most realistic case is considered to be the *waterway edge* case. This requires that hazard areas be developed for

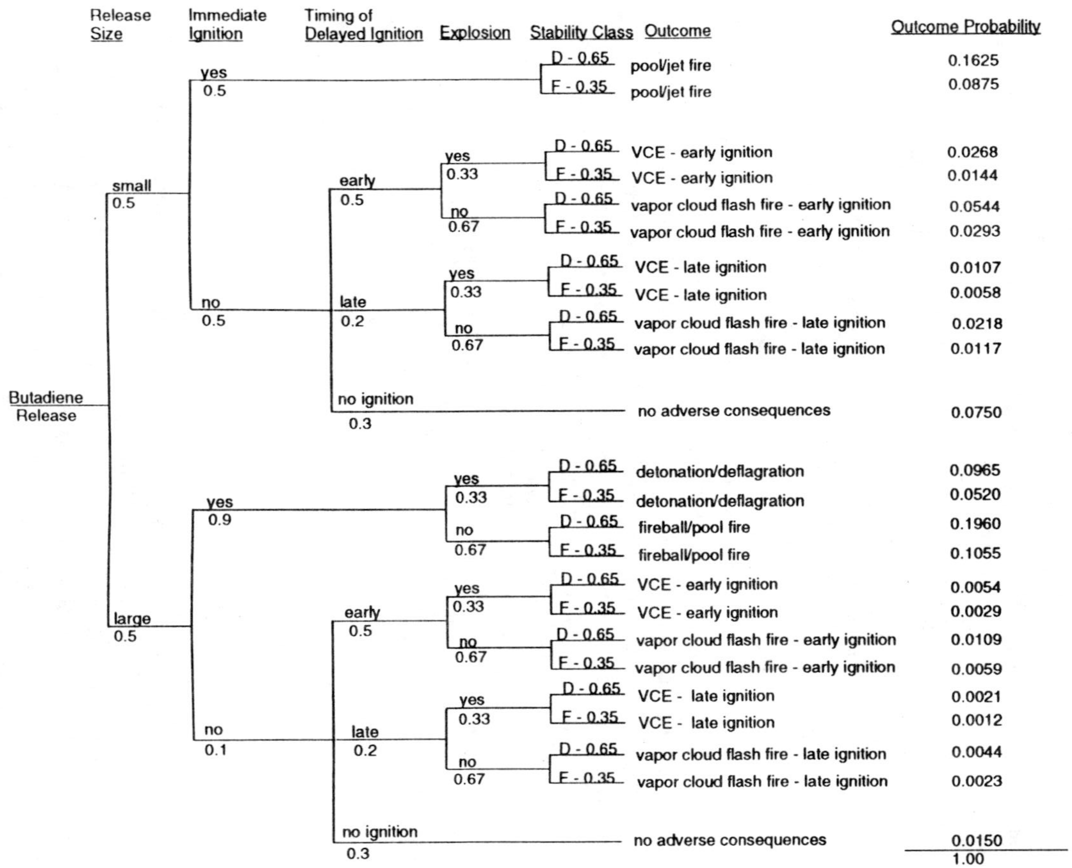

FIGURE 7-15. Event Tree for Butadiene Transport by Barge Example.

both the *no buffer* and *full buffer* situations, but not for the *mid-waterway* case. The direct results of the consequence modeling provide the needed data for the *no buffer* case, which is unaffected by the width of the waterway. For the *full buffer* situations, a different set of areas will be determined for the Intracoastal Waterway than for the Mississippi River. A number of the hazard distances given in Table 7-25 are less than 400 feet and many more are less than 3300 feet. In these circumstances, no onshore population would be impacted. Therefore, no hazard area would be reported as being of interest in Table 7-26.

Another consideration in determining the appropriate hazard areas is the treatment of early ignition—with or without explosions. The analyst must select representative points in the cloud's development where ignition is assumed to occur. For the propane truck case study, early ignition

TABLE 7-26
Hazard Areas of Potential Interest for Barge Example

		Full Buffer		*Mid-Waterway*	
Incident Outcome Case	*No Buffer*	*Intracoastal Waterway*	*Mississippi River*	*Intracoastal Waterway*	*Mississippi River*
Small Release					
Pool/jet fire:D	Yes	No	No	**NOT INCLUDED IN EVALUATION (see text)**	
Pool/jet fire:F	Yes	No	No		
VCE—early ignition: D	Yes	No	No		
VCE—early ignition: F	Yes	Yes	No		
Flash fire—early ignition: D	Yes	No	No		
Flash fire—early ignition: F	Yes	No	No		
VCE—late ignition: D	Yes	No	No		
VCE—late ignition: F	Yes	Yes	No		
Flash fire—late ignition: D	Yes	No	No		
Flash fire—late ignition: F	Yes	Yes	No		
Large Release					
Detonation/Deflagration:D	Yes	Yes	No		
Detonation/Deflagration:F	Yes	Yes	No		
Fireball/Pool fire:D	Yes	Yes	No		
Fireball/Pool fire:F	Yes	Yes	No		
VCE—early ignition: D	Yes	Yes	No		
VCE—early ignition: F	Yes	Yes	No		
Flash fire—early ignition: D	Yes	Yes	No		
Flash fire—early ignition: F	Yes	Yes	No		
VCE—late ignition: D	Yes	Yes	No		
VCE—late ignition: F	Yes	Yes	Yes		
Flash fire—late ignition: D	Yes	Yes	No		
Flash fire—late ignition: F	Yes	Yes	Yes		

was associated with 20 percent of the maximum vapor cloud area, while intermediate ignition was associated with 50 percent of the maximum area. In this study, only early ignition is considered. However, the analyst might wish to use different area assumptions for the *full buffer* and *no buffer* cases, or even for the *full buffer* on the Intracoastal Waterway versus the Mississippi River.

It has been assumed that early ignition occurs at one-third of the maximum downwind distance. In the case of a vapor cloud explosion resulting from early ignition, the buffer distance is compared to the sum of the overpressure distance and one-third of the downwind vapor cloud distance. The maximum downwind distance for each incident outcome case is then compared to the buffer distance for each waterway.

The analyst would then have to calculate the required hazard areas. Each segment is analyzed twice. The frequencies are the same in both evaluations, and are given by:

$$\underset{\text{(per year)}}{\text{Frequency}} = \underset{\text{(per mile)}}{2 \times 10^{-7}} \cdot \underset{\text{(miles)}}{\text{segment length}} \cdot \underset{\text{(per year)}}{\text{13 trips}} \cdot \text{outcome probability}$$

One frequency is obtained for each of the 22 incident outcome cases. The two corresponding sets of consequences are for the *no buffer* case and the *full buffer* case which applies to the waterway for that particular segment. In this example, based on the data in Table 7-24, three segments would have the 400-foot buffer associated with the Intracoastal Waterway and eight would have a 3300-foot buffer on the Mississippi River. The consequences are calculated by the same formula used elsewhere:

$$\underset{\text{(people)}}{\text{Consequence}} = \underset{\text{(sq. miles)}}{\text{effective hazard area}} \times \underset{\text{(people per sq mile)}}{\text{population density}} \times \begin{array}{l}\text{percent of exposures yielding}\\ \text{consequences of concern}\end{array}$$

The consequence of concern might be injuries or fatalities, but must be determined in advance so that the consequence models are run for the appropriate damage criteria (e.g., 1 psi versus 3 psi).

The analyst can combine the resulting data on all the segments in one of two ways. The suggested approach is to calculate the risk measures separately for the *no buffer* and *full buffer* cases and then to average the results. This provides the analyst with an immediate evaluation of the sensitivity of the results to assumptions about buffer distances.

The second approach is to divide the frequency of each incident outcome case by two and to use both sets of consequences—creating 44 pairs of frequency/consequence data for each segment. These can then be used to calculate the desired risk measures.

7.4.7. References

Transportation Data Sources 4-28, 4-32, and 4-33 were used in this case study.

7.5. Ocean-Going Vessel Risk Assessment

7.5.1. Introduction

The following ocean going vessel risk assessment is part of a larger study that also includes land-based transportation by truck or rail, as well as linkages with facility based QRAs that address loading and unloading risks. Hydrogen fluoride is being transported in an intermodal container. The objective of this case study is to calculate the risks associated with the transportation of hydrogen fluoride into two alternate ports. This case study focuses on the port transits as well as in-port buffer storage and transfers.

The study cube cell for this case is the "Risk, Semiquantitative, Generalized Set" cell (see Figure 7-1), as the risks will only be coarsely estimated at this time. The case study is designed to show the inclusion of both accident- and non–accident-initiated events.

The first step will be to identify and select the incidents of concern. Then their frequencies and consequences will be estimated. Consequences are measured as number of pounds released. These data will be synthesized to come up with estimates of the overall risk for the two port options.

7.5.2. Description

Hydrogen fluoride (see Table 7-27) can be brought into either Port A or Port B via ocean-going vessel. It is being transported in an intermodal container on a container ship. Accident data for both ports are available from each Captain of the Port's (COTP) office. As a result, the analysis is broken into three sections: Port A, Port B and the harbor transits. Table 7-28 describes the facts surrounding the scenario.

The ISO containers are a DOT-approved design, and have a single-wall pressure vessel construction. They are designed for a working pressure of roughly 100 psi. Given the limited number of shipments, they are expected to be transported individually.

Both Port A and Port B have designated hazardous materials storage areas within the container facility.

TABLE 7-27
Properties of Hydrogen Fluoride

CAS Number:	7664-39-3
DOT Class:	corrosive, poison
ID Number:	UN 1052
Lower flammable limit:	—
Upper flammable limit:	—
Boiling point:	67°F
Molecular weight:	20.0
Flash point:	—
Liquid specific gravity:	0.99
Solubility in water:	soluble
ERPG-3 concentration:	50 ppm
ERPG-2 concentration:	20 ppm
ERPG-1 concentration:	5 ppm

TABLE 7-28
Description of Example Container Movement

Product	Hydrogen Fluoride (anhydrous)
Amount	35,000 lb
Container Type	Intermodal (ISO container)
Storage Temperature	70°F
Number of Shipments	30

7.5.3. Identification, Enumeration, and Selection of Incidents

Considering the scope of this case study, there are several types of incidents that may yield an HF release (Section 2.6.1). Typical accident- and non–accident-initiated scenarios include:

- ship accidents in the harbor
- dropping another container on the HF container during unloading operations (while aboard the ship)
- dropping the container during its unloading (transfer) from the ship
- dropping a container from a top lifter
- releases from valves while the container is sitting aboard the ship or in interim storage

The typical sequence of operations at both ports involves a harbor transit followed by a mooring of the container ship. Once the vessel is moored, containers are off-loaded by a computerized yard crane from the vessel to a truck. The truck then drives the container to the intended storage area where a wheeled top loader lifts the container and places it in storage. To the extent possible, multiple containers with the same destination/customer are grouped together to minimize the need for any interim lifts. Depending on the status of the customs paperwork, the container might be in storage for one to five days before being picked up. The average duration is three days.

Containers are stacked with the bottom container at ground level, in part to limit the potential for drivers coming in and taking containers away without involving the staff and equipment of the port. Based on a detailed review of the operations, it has been determined that the transfer operations within the container facilities will not involve sufficient speeds or potential drop heights from the top lifter to yield releases in the event of an accident, with the exception of loading and unloading the ship (Section 2.6.3). This yields a reduced set of scenarios of concern:

- dropping the container from the unloading crane
- dropping another container on the HF container aboard ship
- ship accidents in the harbor
- releases from valves while the container is sitting aboard the ship or in interim storage

To cover this spectrum of releases, the analyst determines that there are three representative release sizes:

1. a full loss of cargo (35,000 pounds)
2. a 10 percent cargo loss (3500 pounds)
3. a 1 percent cargo loss (350 pounds)

Only scenarios involving loaded containers are addressed. Scenarios involving contamination or reactions of the HF are not addressed. Such events result from problems initiated during filling or emptying operations, which are being addressed separately. The likelihoods of any such release can be added to the results of this transportation study.

7.5.4. Incident Consequence Estimation

Since consequences are being represented by the quantity of material lost in this semiquantitative study, there is no need to select meteorological conditions or to estimate hazard distances. Similarly, the need to address spills on land versus those on water is not of concern at this time. If it were, most of the releases would be treated as "on land" since even releases aboard the vessel are likely to be onto a vessel surface or directly to atmosphere.

7.5.5. Incident Frequency Estimation

Given the range of accident- and non–accident-initiated scenarios, frequency data must be gathered from a number of different sources, and visits to the port facilities may be required. The sources of data to review include:

- published sources
- past analyses
- contacts with the U.S. Coast Guard, various ports, and other parties
- data gathered for other studies
- equipment reliability data for valves

Statistical information on accidents involving ISO-containers during port operations is not maintained by any government or industry agency. Agencies contacted by the analyst include several port authorities, the Department of Transportation, the Coast Guard, the Maritime Administration, Lloyd's of London, the IMO, trade associations such as represented by the Journal of Commerce, and larger shipping lines.

Each of these organizations claimed that they do not maintain a data file on container accidents. The one exception to this search was that the Coast Guard provided a list of 28 incidents that occurred in the time period from 1982 to 1989 in which cargo was lost or damaged while aboard a ship, either underway or at berth. This listing does not include accidents which might occur portside, however, such as loading or unloading containers, nor does it include details of the number of containers involved in each incident or whether spills or breaking of the container occurred.

Container Dropped from Unloading Crane

Sources of accident data involving containers during port-side operations, primarily the loading and unloading of the containers using container gantry cranes, include the container port operators and terminal manager as well as the Captain of the Port's office. For example, in Port A the terminal operators for two berths have experienced three drop accidents in five years. Based on a total of 315,000 lifts during this time period, this accident history leads to a drop rate of 10 per million lifts. According to the Operations Manager of Port B's container terminal, they have had no drops during their limited period of operations.

In previous risk studies, data were obtained on container accidents from a major shipping line and from five container port facilities. These sources generally maintain a file of specific accident reports, primarily for insurance and litigative purposes, but not an accident data base in a form useful for risk evaluation studies. Furthermore, most such sources consider their accident reports proprietary. Based on this limited information, it was estimated that the likelihood of accidents during port activities in U.S. ports was 10 per million lifts. Note that Port A's accident history is consistent with the previously estimated accident rate for U.S. ports.

Given the lack of history for Port B, but the advanced design of its equipment and rigorous inspection policies, a rate of 5 per million lifts is judged to be appropriate.

Thus, for the scenario involving a container drop during unloading the frequency estimates are:

Port A: 30 lifts/year $\cdot$ 10×10^{-6}/lift = 3×10^{-4}/year
Port B: 30 lifts/year $\cdot$ 5×10^{-6}/lift = 1.5×10^{-4}/year

Given that a container is dropped, it is estimated that a spill will occur 80 percent of the time, and that such spills will be distributed evenly between small spills (up to 10 percent of contents) and large spills (100 percent of contents). The small spills would be evenly distributed into releases of 10 percent of the cargo and smaller releases, such as from leaking fittings, equal to one percent of the cargo. This gives the following spill frequencies by release size:

	Port A	*Port B*
100% loss	1.2×10^{-4}/yr	6×10^{-5}/yr
10% loss	6×10^{-5}/yr	3×10^{-5}/yr
1% loss	6×10^{-5}/yr	3×10^{-5}/yr

Another Container Dropped on HF Container

ISO containers are normally not moved on board the ship until they are unloaded. However, they may be exposed to container lifts or cargo movements which pass overhead. If one of these containers or cargo packages is dropped, it may impact the ISO container and cause it to release its contents.

The chance of a container dropping is taken to be the same on a per lift basis as for the ISO container itself. Given the nature of the ISO container's cargo, most shipping companies would try to limit the number of overhead lifts above it. Based on discussions with the terminal operators, it has been determined that a conservative estimate of the number of overhead lifts is 10 for Port A and 15 for Port B. Given that the container or cargo package could be dropped anywhere in its lift path, it is estimated that there is a one in ten chance that the HF container would be impacted.

Since the potential drop heights for the cargo passing overhead will be lower than for the case where the HF container is dropped during unloading, the conditional spill probabilities will also be lower. A total conditional spill probability has been assumed as:

- 8% yielding a loss of 1% of the cargo,
- 8% yielding a 10% loss, and
- 4% yielding a total loss of contents.

20% overall

This produces the following spill frequencies by release size:

Port A Spill Frequencies

100% loss:

30 HF containers/yr · 10 lifts/HF container · 10×10^{-6}/lift · 0.1 · 0.04
$= 1.2 \times 10^{-5}$/yr

10% loss:

30 HF containers/yr · 10 lifts/HF container · 10×10^{-6}/lift · 0.1 · 0.08
$= 2.4 \times 10^{-5}$/yr

1% loss:

30 HF containers/yr · 10 lifts/HF container · 10×10^{-6}/lift · 0.1 · 0.08
$= 2.4 \times 10^{-5}$/yr

Port B Spill Frequencies

100% loss:

30 HF containers/yr · 15 lifts/HF container · 5×10^{-6}/lift · 0.1 · 0.04
$= 9 \times 10^{-6}$/yr

10% loss:

30 HF containers/yr · 15 lifts/HF container · 5×10^{-6}/lift · 0.1 · 0.08
$= 1.8 \times 10^{-5}$/yr

1% loss:

30 HF containers/yr · 15 lifts/HF container · 5×10^{-6}/lift · 0.1 · 0.08
$= 1.8 \times 10^{-5}$/yr

Accidents during Harbor Transits

One data source (J.J. Henry, 1973) suggests that worldwide harbor and dock casualties occur with a frequency of 1.2 per thousand transits. For container ships of modern design and with advanced navigation equipment, the casualty rate is expected to be substantially lower than this average rate. An estimate of 0.25 per thousand transits was used by the analyst in a prior study.

Anecdotal information about past container ship accidents suggests that, on average, 10 of the 2000 containers onboard are involved. This leads to a chance of involvement of 10 in 2000 or 0.005. Thus, the rate of container involvement per transit is 1.25×10^{-6} per transit. (Since the container is to be off loaded, only a one-way transit and not a full port call is of interest.)

The same conditional spill probability and spill size distribution as were used for dropped containers are applied here. This gives the same results for both Ports A and B:

Spill frequencies for accidents during harbor transits

100% loss: $30 \text{ transits/yr} \cdot 1.25 \times 10^{-6}\text{/transit} \cdot 0.5 \cdot 0.50 = 9.4 \times 10^{-6}\text{/yr}$

10% loss: $30 \text{ transits/yr} \cdot 1.25 \times 10^{-6}\text{/transit} \cdot 0.5 \cdot 0.25 = 4.7 \times 10^{-6}\text{/yr}$

1% loss: $30 \text{ transits/yr} \cdot 1.25 \times 10^{-6}\text{/transit} \cdot 0.5 \cdot 0.25 = 4.7 \times 10^{-6}\text{/yr}$

Releases from Valves

All valve releases are expected to be leaks, and should not release more than 1 percent of the cargo before detection and repair. Even if slightly more cargo is lost, it will be at a very low rate of release and the smallest spill size is still believed to be the most appropriate. If the ship is at sea, the container may be able to be pushed overboard if the leak cannot be controlled. As stated in Section 7.5.3, the container is in port for an average of 3 days. The harbor transit and time at dock is kept to a minimum for schedule purposes, and two days is a conservative estimate according to operators at both ports. Thus, there are roughly 150 days per year (30 containers per year at a total of 5 days per container) when a leaking valve would be a problem to Port A or Port B. There are two valves per container. Using a failure rate for external leakage of valves of 1×10^{-6}/hr from the *Offshore Reliability Data Handbook* (1992), assuming that 10 percent of external leaks are significant, gives:

$$150 \text{ days/year} \cdot 24 \text{ hours/day} \cdot 1 \times 10^{-6}\text{/hr} \cdot 0.1 \cdot 2 = 7.2 \times 10^{-4}\text{/year}$$

for a release frequency.

7.5.6. Risk Estimation

Table 7-29 presents the likelihoods of various size releases for each port, including the harbor transit associated with each port, based on the results for each of the scenarios of interest.

7.5.7. Discussion of Results

While the total chance of a release of any size is very similar for both ports, the relative likelihood of large losses varies more significantly. This might

TABLE 7-29

Summary of Release Frequencies by Release Size for Hypothetical Intermodal Container Movements at Port A and Port B (per year)

Scenario	*1% Loss (350 lb)*	*10% Loss (3500 lb)*	*100% Loss (35,000 lb)*	*All Sizes*
PORT A				
Container dropped	6.0×10^{-5}	6.0×10^{-5}	1.2×10^{-4}	2.4×10^{-4}
Drop onto container	2.4×10^{-5}	2.4×10^{-5}	1.2×10^{-5}	6.0×10^{-5}
Accident in harbor	4.7×10^{-6}	4.7×10^{-6}	9.4×10^{-6}	1.9×10^{-5}
Valve release	7.2×10^{-4}	—	—	7.2×10^{-4}
Port A Total	8.1×10^{-4}	8.9×10^{-5}	1.4×10^{-4}	1.0×10^{-3}
PORT B				
Container dropped	3.0×10^{-5}	3.0×10^{-5}	6.0×10^{-5}	1.2×10^{-4}
Drop onto container	1.8×10^{-5}	1.8×10^{-5}	9.0×10^{-6}	4.5×10^{-5}
Accident in harbor	4.7×10^{-6}	4.7×10^{-6}	9.4×10^{-6}	1.9×10^{-5}
Valve release	7.2×10^{-4}	—	—	7.2×10^{-4}
Port B Total	7.7×10^{-4}	5.3×10^{-5}	7.8×10^{-5}	9.0×10^{-4}

be an important factor in discriminating between the two alternatives, depending on the results of the other risk studies for land-based transportation and loading/unloading activities.

The risk of large releases is dominated by the scenario where an HF container is dropped. Thus a strategy to protect the HF container from overhead activities will not be as effective in reducing risk as one might otherwise think—especially for Port A. Valve releases dominate the small spill frequency as well as the overall results. Hence, one potential next step would be to attempt to get more specific estimates of significant external leaks from valves.

7.5.8. References

J. J. Henry Co. *An Analysis of Oil Outflows Due to Tanker Accidents*. Prepared for the U.S. Coast Guard. AD-780-315. 1973.

Offshore Reliability Data Handbook, 2nd Edition. Hovik, Norway: Published by the OREDA participants; distributed by DNV Technica. 1992.

8

Future Developments

This chapter describes a number of areas of current research or where future developments in TRA are expected, and suggests areas for future research. The following topics are addressed in this chapter:

- General Risk Assessment Advances/Trends (Section 8.1)
- Release Rates and Volumes (Section 8.2)
- Release Probabilities for Specific Containers (Section 8.3)
- Accident Rates (Section 8.4)
- Integration with Network Models (Section 8.5)
- Comparisons with Actual Performance (Section 8.6)
- Risk Reduction Strategies (Section 8.7)
- Environmental Risk (Section 8.8)
- Increased Cooperation (Section 8.9)

8.1. General Risk Assessment Advances/Trends

Risk assessment is an evolving technology in terms of both qualitative and quantitative techniques. Numerous approaches have been developed over the years and show significant variances. This is particularly true for qualitative approaches, which have often been developed on a company by company basis. Many companies are beginning to try and integrate their fixed facility and transportation assessments to prioritize resources.

In general, quantitative transportation risk assessment models generate a numerical estimate of risk using mathematical combinations of the following factors:

- movement characteristics (length, number)
- accident rate

- release probability
- release size and/or rate
- hazard zone
- consequences to exposed populations, property, and the environment

Since the majority of these factors also apply to CPQRA studies, most improvements and future developments for CPQRA will also apply to TRA. In particular, this includes better information on acute (short-term exposure) toxicology, enhanced consequence models, new means of manipulating population and environmental characteristics data (such as GIS networks), and improved computer tools for performing large numbers of risk calculations.

This said, it is also known that the various data sources, risk estimation methods, and models available for TRA do not always produce results that are consistent with one another. In general, this results from approaches that are inconsistent and/or the use of different risk measures (see Chapter 5). To date, there has been no consensus on a best practice, particularly in terms of the appropriate risk measure to use. It is expected that with time and experience there will be one or two preferred measures, but that these will vary by user. There can also be problems with data that are not well defined or that are incomplete. Ongoing data collection efforts that have been designed with risk assessment as one of the intended applications are expected to improve this situation in the future.

Several international standard-setting bodies, such as the International Electrotechnical Commission are preparing standards or guidelines for risk assessment that may also help to improve the consistency of such studies.

Even the scope of HAZMAT transportation studies is quite variable. In the United States, the risk study typically looks at a full route or a part of a route for a given movement or set of movements. In the United Kingdom, one recent study (Advisory Committee, 1991) attempted to evaluate the entire set of risks in the country for particular products. In general, studies in the United States are initiated by chemical companies and address their own products, albeit with an increasing level of concern for purchased materials. In most other countries, studies seem to be initiated by government requests or the U.S. portion of an international company.

Transportation risks are also beginning to be more fully addressed in the regulatory environment through an increased emphasis on data collection and the application of risk assessment techniques—either to establish or comply with regulations. Recent changes by the U.S.DOT in their hazardous materials regulation are indicative of this, as is the Oil Pollution Act of 1990 (OPA '90) and the potential future application of that act to hazardous substances besides oil.

Enhancements and advances in consequence modeling will also benefit TRA. However, some of the models have become so specific that there is a mismatch between the selection of representative transportation-specific releases and the detailed inputs required by the consequence models. The challenge is to pick one or more representative releases and then to revisit these situations as models advance even further, while keeping the analysis relatively simple.

More information on incidents which occur are needed in order to make more realistic assumptions about ignition locations and probabilities. Similarly, since all the parameters needed to model a vapor cloud fire or explosion cannot be precisely specified in a TRA, a better method for estimating hazard areas is needed.

8.2. Release Rates and Volumes

There are uncertainties present in the estimation of release rates and volumes due to a lack of hard data. The need for more information in this area has been recognized and is an area of ongoing interest and data collection. In addition, as more TRAs are being conducted, more people are recognizing the importance of such information, and more data are being collected. This is coming largely from private industrial sources and not all data are available to the public. Primarily, the difficulty arises because whereas the volume released is reported, the following items are not:

- release rate
- duration
- hole size
- initial quantity in container

8.3. Release Probabilities for Specific Containers

The range of possible release quantities can be represented by a broad spectrum of distinct values, each with associated conditional probabilities of occurrence in an accident. This is complicated further by the different types of containers in use and their respective release probabilities. The statistical basis for estimating the probability of large releases is very limited, even without taking into account the specific container type.

Smaller releases are generally more common and have somewhat more data available. For certain types of containers, virtually no release data may exist.

In attempts to be conservative, one may overestimate the probability of large releases. This overestimation may shift the emphasis from one population at risk to another, which may skew the conclusions. This information could also have a large impact on the outcome of the TRA.

Unfortunately, this is still an area where there is not much information currently available. As with estimating release rates and volumes (Section 8.2), the need for more data in this area has been recognized and there are currently several industry panels compiling and analyzing data of this nature. As many different industries are involved in these projects (e.g., container manufacturers, chemical producers, transporters, etc.), some amount of consensus will need to be obtained. This is especially a problem in the estimation of mitigation effectiveness. In addition, other things could be done to facilitate the compilation of data such as including the container specifications in accident reports so data are not lost. This will also require support on the part of the federal and state governments who compile these reports (see Table 4-4, pages 168–169).

8.4. Accident Rates

Accident rates are based on reported information from various sources. As a result, there are limitations on the quantity and availability of accident data from which one can statistically estimate accident probabilities. In particular, some national databases may be incorrect by as much as a factor of three, largely due to underreporting. Existing individual sources of accident data and their limitations are discussed in depth in Chapter 4.

For specific routes, the smaller number of incidents, especially at the level of individual segments, incorporates large statistical uncertainty. In addition, relevant accident information such as time of day, day of week, and weather conditions can only be considered in a general way due to the lack of prior valid statistical studies of such factors. This presents a need for better reporting of incidents as well as management of data and communications between authorities handling such data (e.g., between state agencies or state and federal agencies).

The total population of vehicles, rail cars or trains, and vessels experiencing reported incidents and the total miles traveled (denominator data) are also not well known. There are currently several data collection efforts underway to compile this type of data. Most of these projects are being conducted by different industry trade groups with interests in particular modes of transportation (e.g., CMA's OPA '90 task group and the Inter Industry Rail Safety Task Force). In addition, the federal government is also advocating the conduct of an economic cost/benefit analysis prior to promulgating new regulations. Since the total costs of investment in new safety equipment will be derived based on this type of data, industry is being pushed to collect better data to get a more accurate picture of the costs associated with such changes.

The future is expected to bring many more data collection efforts which may be focused on individual carriers, may incorporate sampling techniques, or may involve obtaining data gathered by Local Emergency Planning Committees (LEPC) to understand the volumes of materials transported through their communities. While some data collection efforts are underway today, complaining about data limitations is more common than is finding companies which are willing to fund data collection efforts.

8.5. Integration with Network Models

As computer models become more sophisticated, they rely on more input parameters to generate output. While this may be advancing the science of risk analysis, it is also making some of these models more difficult to use. This requires either a sufficient degree of proficiency on the part of the user or an amount of user interface so minimal as to make the model very inflexible and dependent on the designers' assumptions. This also may mean that those who are not knowledgeable about TRA can suddenly conduct such analyses, without any ability to check their work or the credibility of an answer. A decrease in user data needs can, however, lead to a corresponding decrease in human error when using computer models. More databanks will be directly accessed by computer models and those databanks will supply the models with information that previously had to be entered by the user. An example of this is the built-in database of chemical properties that some consequence modeling packages automatically access to generate the output for a given scenario. In other cases, such as for meteorological data, more models will be linked to a Station of Atmospheric Measurements (SAM) and be fed real-time weather data. This data will be automatically processed by the computer application without any further user involvement.

Population data and route information are also emerging areas of development. With the advent of Geographic Information Systems (GIS), applications have been developed in the areas of mapping routes and analyzing population trends. These applications have been used for several years by such organizations as the U.S. Postal Service. GIS data are now available in CD-ROM format for use on personal computer systems. Many applications have incorporated mapping capabilities and population data (U.S. Census based) into their features. (See Section 3.10 on available software.) Although it is currently possible to link population data and mapping capabilities into selected software applications and to allow these applications to provide details on where the routes are and the surrounding population characteristics, not all route data needs are satisfied. Elements such as track class for railroads and speed limits on highways are currently not part of the route information generated from GIS data and as a result, these factors cannot be incorporated automatically and must be addressed externally by the user. Additional areas of research are also being conducted in incorporating accident data into the models as well. However, this effort is limited to a large extent by the quality of accident data that is published (see Chapter 4).

Another area that has been evolving separately but has a significant amount of overlap with hazardous materials transportation is the area of emergency response. Chemical databases are routinely accessed by responders to provide information on the hazards posed by a particular material, recommended response procedures and recommended personal protective equipment, among other things. There is an opportunity for this material to be used in transportation studies instead of during a response. More information will also become available on response capabilities and availability along a particular route.

8.6. Comparisons with Actual Performance

As the science of risk analysis evolves, TRA will also continue to develop. As existing data become more refined and more data become available, risk estimates presented in future TRAs will become more certain and realistic.

The question always remains, however, as to how accurate the risk estimates really are. The best way to measure this is to compare them to actual performance. This is easier said than done. Comparison of risk estimates to actual performance would require much more comprehensive and consistent record keeping, including collecting data on variables that are not now recorded.

For instance, in the past, risk has been measured on an annual basis. One potential comparison is to the risk on a per-ton mile basis. To do this, data are needed that provide information on the total amounts of material being transported and the distances traveled, by mode of transport. Data of this sort typically have not been collected in the past and future data collection efforts would have to be enhanced to provide the information necessary for this type of comparison.

Comparisons can be made based on risk, or on the total number of accidents, or on the frequency of spills of various sizes. The latter types of comparison should have the most available data and offer the best calibration of the results in the foreseeable future.

As risk estimates improve in terms of their accuracy and ability to be directly compared to actual experience, it is expected that the use of TRA for insurance purposes will increase. There are two primary forms that such developments may take, and the two are not exclusive. The first is for a company to use the results of risk assessments to make decisions regarding deductibles, self-insurance levels, or total liabilities. The second form is the use of risk assessments by insurers to influence the setting of insurance premiums. This is an application that seems logical and straight forward, but it has not met with much success in the past. Future competitiveness and the increased tendency to self-insure may yield a break through in this area, however.

8.7. Risk Reduction Strategies

A TRA is conducted to identify the risks associated with the transportation of a hazardous material over a route(s) via a certain mode(s). Once the risks are identified, an evaluation needs to be made as to their tolerability. If the risks are deemed intolerable, then strategies need to be explored that will reduce the risk to a level which is tolerable to an organization. There are several types of activities that may bring risks down to a more tolerable level and some are discussed below.

Some obvious choices to test for feasibility include changing the container (e.g., going from a MC306/406 to a MC307/407) in which the material is shipped or using another route through a less densely populated area. However, many of these alternatives will offer limited benefits and only incremental reductions in the overall risk.

A concept that has been taking hold in recent years has been the "partnership" concept, and this trend is expected to be accelerated in the future. This involves carriers, suppliers and customers working together to look at transportation from an overall perspective. This outlook not only may affect the risks associated with a particular hazardous material movement but also the manner in which that company(s) does business. For instance, a customer may be considering the risks of barge versus rail transportation for one of their materials. Normally this customer would look individually at the risks associated with each mode and route. However, the supplier of the material is also the supplier for another site further up river. By sharing information with the supplier, the two may be able to combine deliveries, thereby reducing the overall number of shipments on that segment of the river. This may turn out to have a significant effect on the overall risk that may not have otherwise been achieved by switching modes. These partnership issues open up opportunities for more significant reduction results (i.e., approximately an order of magnitude) as opposed to just incremental reductions, and also allow greater cost options to be practical.

One risk reduction strategy common in the past that is expected to be less common in the future is risk shifting or transferring. In the past, one company might decide that a risk was too high and stop moving a product themselves—letting someone else make the movement instead. This other organization may not have even controlled the risk as well as the first company originally did. In the future, more companies are expected to follow the course of action that minimizes the risk and to acknowledge that doing so may have a short-term financial disadvantage. However, the longer-term view recognizes that this is the best way to maintain the integrity of the supply and to minimize reactionary regulations that may have adverse cost implications for the entire industry.

Other risk reduction strategies evolve from technological advances. Satellite tracking of shipments and enhanced vessel navigation systems can have major effects on existing risk levels by lowering accident rates through collision avoidance, either directly or by assuring that drivers take breaks and correct routes are followed. Other high technology solutions include simulator training, rail flaw detection, etc.

On the strategic front, transportation requirements should also be considered. Siting facilities to eliminate HAZMAT transportation altogether should be considered when applicable, as well as whether it is better to transport raw materials, intermediates, or products.

8.8. Environmental Risk

Environmental risks have been an area of increasing concern, and additional discussions of environmental risks as they pertain to TRA are given in Section 3.9. While long term hazards to the environment resulting from transportation incidents have not been traditionally addressed in a TRA, this trend is changing. This is largely a result of recent transportation incidents, mostly oil spills, which have increased public concern regarding environmental risks.

The railroads are currently working with industry to examine the environmental effects of several commonly transported materials. Environmental impacts are also being addressed for other modes of transportation (e.g., the extension of the OPA '90 requirements to other hazardous substances). While the initial list of materials considered is often relatively short, future developments are expected to include a much greater set of materials. This parallels the development of fixed facility regulations which started with selected materials and has quite an extensive list in EPA's Risk Management Program (RMP) rule.

Natural Resource Damage Assessment (NRDA) is a recently developed method of dealing with transportation risks (Grigalunas et al., 1988). It is used currently for actual oil spills as a method to calculate the associated costs. Costs are calculated for such items as cleanup of the spill, loss of cargo, and habitat restoration. Some transporters have used these models predictively to get an idea of the potential costs associated with an oil spill. Operations may then be altered based on this information.

In the future, this approach is expected to have an increased level of application, both for oil and other chemicals. More data on the impacts of spills of other materials will be needed in order to develop the cost models appropriate for different geographic areas and habitats.

8.9. Increased Cooperation

Partnerships between suppliers, carriers and customers are becoming more common not only to deal with hazardous material transportation risks through voluntary efforts such as product stewardship and Responsible Care®, but also to address such issues as costs and logistics. Interindustry task forces are currently active in rail and marine transportation, and those parties involved in bulk liquid truck transport are beginning to form an interindustry group to consider transportation risks.

By forming these interindustry groups, more comprehensive data collection efforts and analyses can be conducted, and the "true" benefits of various approaches to mitigation can be determined more sensibly. The joint efforts can lead to overcoming some biases about the types of changes that should be made, can motivate changes in equipment and infrastructure that are beyond the influence of individual chemical companies, and can even guide regulatory changes.

The increased awareness of risk issues that such partnerships are bringing about also seems to be broadening the scope of risk that companies consider. Thus companies who formerly focused on risk to members of the public will begin to address environmental risks and vice versa.

Section 8.7 provides a further discussion of partnership issues as they relate to TRA.

8.10. References

This section lists the references cited in this chapter. Additional references that may be of interest to the reader can be found in the topical bibliography.

Advisory Committee on Dangerous Substances, Health and Safety Commission. *Major Hazard Aspects of the Transport of Dangerous Goods; Report and Appendices*. London: HMSO, 1991.

Grigalunas, T., J. Opaluch, D. French, and M. Reed. "The Natural Resource Damage Assessment Model for Coastal and Marine Environments (NRDAM/CME)." *Proceedings of 1987 Oil Spill Conference*. Washington, D.C., 1988.

APPENDIX A

Standard Container Illustrations and Specifications

Many types of highway cargo tanks, rail cars, drums and bulk storage tanks exist, but each type is designed for specific materials or groups of products. The design of highway cargo tanks, rail cars, drums, and bulk storage tanks is standardized. Their design and performance is regulated or controlled in the United States by several federal agencies or design recommendations made by various trade organizations (e.g., ASME). The U.S. Department of Transportation (DOT) regulated the design of highway cargo tanks, rail cars, cylinders, and drums, among others.

I. Highway Cargo Tanks

Currently, the U.S. Department of Transportation specifies five different designs for highway cargo tanks, the Motor Carrier (MC) series. Descriptions of these and the materials which they are designed to contain are provided in Tables 1 and 2. The tanks listed in Table 1 are no longer manufactured but may still be in service if properly maintained. Those designs have been superseded by those listed in Table 2.

Highway cargo tanks built to current specifications will vary considerably in construction detail, shape and configuration. For example, MC306 highway cargo tanks fabricated by two different manufacturers are most likely not identical, but the general design and features will be consistent.

Department of Transportation (DOT) specifications for all of these tank trucks can be found in 49 CFR 178.

TABLE 1
Early Motor Carrier Series Cargo Tanks

Design Number	*Material(s) Transported*
MC300	Primarily for flammable liquids or poisonous liquids (B)
MC301	Primarily for flammable liquids or poisonous liquids (B)
MC302	Primarily for flammable liquids or poisonous liquids (B)
MC303	Primarily for flammable liquids or poisonous liquids (B)
MC304	Flammable liquids or poisonous liquids, vapor pressure of 18 psi+
MC305	Primarily for flammable liquids or poisonous liquids (B)

TABLE 2
Current Motor Carrier Series Cargo Tanks

Design Number	*Material(s) Transported*
MC306/406	Flammable and combustible liquids and Class B poisons
MC307/407	Flammable and combustible liquids and Class B poisons and chemicals with a vapor pressure of 18 psi at 100°F or greater but not more than 40 psi at 70°F
MC312/412	Corrosive liquids
MC331	Liquefied gases such as LPG, chlorine, and anhydrous ammonia
MC338	Cryogenic gases

MC306/406

MC306/406 tanks (Figure 1) carry flammable and combustible liquids such as gasoline and diesel fuel, and poisonous materials such as magnesium arsenate, mercuric arsenate, calcium arsenate, and copper cyanide, or "white water" solvents. They are considered nonpressure tanks and are hydrostatically tested to 3 psi. The maximum allowable working pressure (MAWP) must be between 2.65 and 4 psig. Typically, they are constructed of aluminum. Other important design features include:

- The capacity to carry from 2000 to 9500 gallons of product.
- A multicompartmented design feature to carry different commodities, which if combined will cause dangerous conditions. The compartments are separated by double bulkheads, each double

bulkhead is separated by an air space with a vent and drain provided.

- The ability to be unloaded from the bottom.
- Springloaded valves that remain closed during transport. These valves are either mechanically, pneumatically or hydraulically operated and must be equipped with an automatic heat activated closure system (usually fusible links, nuts or bolts) which operates at a temperature not more than 250°F.
- In addition to normal means of closure, each internal or external self-closing stop valve must be fitted with a remotely activated means of closure located more than 10 feet from the stop valve. These remotes are generally found toward the front or rear of the tanks opposite the side of the outlet piping.
- Piping must be protected from breakage by accident protection devices (structural) or shear sections located no more than 4 inches from stop valves.
- A rear bumper to protect piping and valving in event of a collision.
- Manhole covers that make a secure closure and include a safety device that prevents the cover from opening fully when internal pressure is present.
- Both pressure and vacuum relief systems designed to operate and have sufficient capacity to prevent tank rupture or collapse due to over-pressurization or vacuum resulting from tank heating, cooling, loading or unloading. Each pressure relief system must be designed to prevent loss of loading from the system in case of pressure surges, vehicle upset or accident, regardless of vehicle orientation.

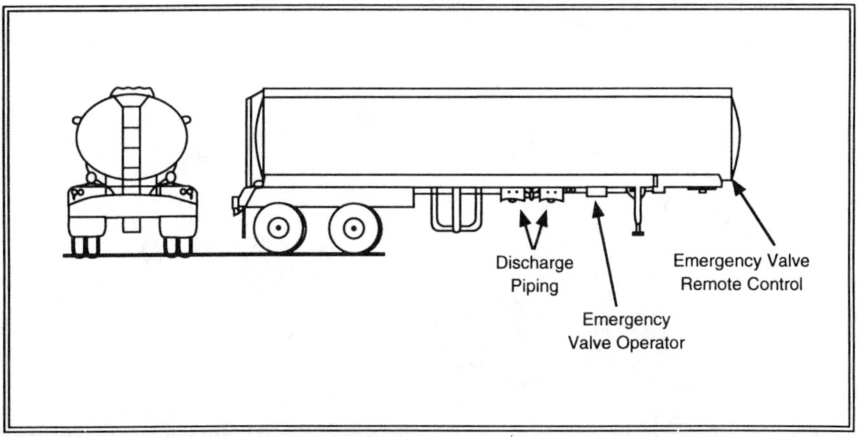

FIGURE 1. MC 306/406 Cargo Tanks. Description: Oval or round cross-section, smooth lines (may have external ribs), with what appears to be a catwalk on top.

- Overturn protection for fittings, manway covers, and vacuum or pressure vents. If these devices are mounted on the surface of the container, overturn protection may be accomplished by providing guards that protect these devices, or these devices may be inherently protected if they are located within the body of the container.

MC307/407

MC307/407 tanks (Figure 2) carry products with vapor pressures not more than 40 psi at 70°F. Common products carried include flammable liquids and mild corrosives such as acrylic acid, liquid coal, tar dye and monoethanolamine solutions. MC307/407 tanks are considered low pressure tanks with a minimum design pressure of 25 psig.Typically, they are constructed of steel. Other important features include:

- The capacity to carry from 2000 to 8000 gallons of product.
- The tank may be multicompartmental with double bulkheads, air spaces, vents, and drains as required.
- The ability to be unloaded from the bottom and springloaded mechanically, pneumatically or hydraulically operated valves equipped with an automatic heat activated closure system. This feature is identical to MC306/406.
- Internal and external self-closing stop valves have a secondary closure system located at least 10 feet from the stop valves. The remote device will usually be located at either end of the cargo tank opposite the side of the tank outlet piping.

FIGURE 2. MC307/407 Cargo Tank. Description: The MC307/407 tank has a horseshoe-shaped tank with a ladder and platform located along the top of the container.

- A rear bumper to protect the tank and piping in the event of a collision. Manhole covers that meet the same requirements of MC406 except that each manhole must be capable of withstanding internal fluid pressure of 40 psi or the estimated pressure of the tank, whichever is greater.
- Vents that are designed to limit internal pressure to 150 percent of the maximum allowable working pressure. These vents must be pressure activated (springloaded). Also, fusible and/or frangible (breakable) venting may be provided with fusible vents operating at 250 °F and frangible discs bursting at not less that 110 percent or more than 120 percent of the maximum allowable working pressure.
- Overturn protection for fittings, manway covers and vents.
- Discharge piping that is designed to break away from emergency valves or be protected by guards.
- The insulation provided by heater coils in a jacket that allows loading and unloading of viscous products.

MC312/412

MC312/412 tanks (Figure 3) carry high density liquids and corrosives such as acetyl chloride, hydrochloric acid and sodium hydroxide solutions. Typically, they are constructed of steel, stainless steel or aluminum and are often lined with a material suited to resist degradation or reaction to its contents. An MC312/412 has a design pressure of at least the pressure required for unloading and is hydrostatically tested at 1.5 times the maximum allowable working pressure, but not less than 5 psig. Other important design features include:

- A multicompartmental design feature that allows for separate compartments.
- The ability to be unloaded from the top, frequently by air pressure. Top discharging valves must be located as close as practical to the discharge point. Piping from the top discharge with an outlet below the top liquid level in the container requires an additional valve to blank flange or sealing cap to prevent siphoning in event of top discharge valve failure.
- Bottom outlets located with the valve seat inside the container or immediately adjacent to the outlet if bottom discharge is provided.
- An additional or secondary closure system that does **not** require a heat activated system.
- A rear bumper to protect the tank and piping in the event of a collision.

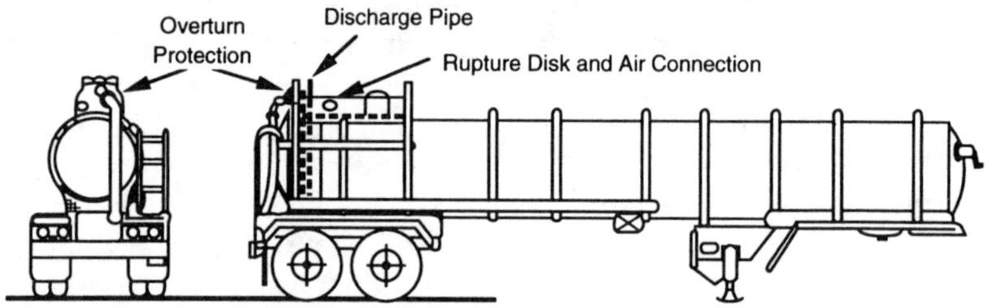

FIGURE 3. MC312/412 Cargo Tanks. Description: Cylindrical rear cross section, narrow diameter, external ribs with ladder and platform located along the top of the container. Because of the materials that an MC312/412 carries, it is usually smaller than MC306/406 and MC307/407 containers.

- Manhole covers that are structurally different in design, but must meet the same requirements of the MC306, MC307, MC406 and MC407 cargo tanks.
- Vents that are either mechanical pressure relief vents or frangible discs.
- Overturn protection for fittings, manway covers and vents that is the same as featured for MC306, MC307, MC406 and MC407 cargo tanks.
- Discharge piping that is designed to either break away from emergency valves or be protected by guards. This is the same as featured for MC306, MC307, MC406 and MC407 cargo tanks.
- The insulation with heater coils in a jacket that allows loading and unloading of viscous products. This is a feature that is associated with the MC306 and MC307.

MC331

MC331 tanks (Figure 4) carry gases which are liquefied by the application of pressure. Products usually transported are compressed gases and some very hazardous liquids such as anhydrous ammonia, chlorine, liquefied petroleum gas (LPG), liquefied carbon dioxide, and liquid methyl parathion. Since many hazardous materials now classified as Poison Inhalation Hazard (PIH) are also gases or extremely volatile liquids at ambient temperature and pressure, they are also shipped in these tanks (e.g.,

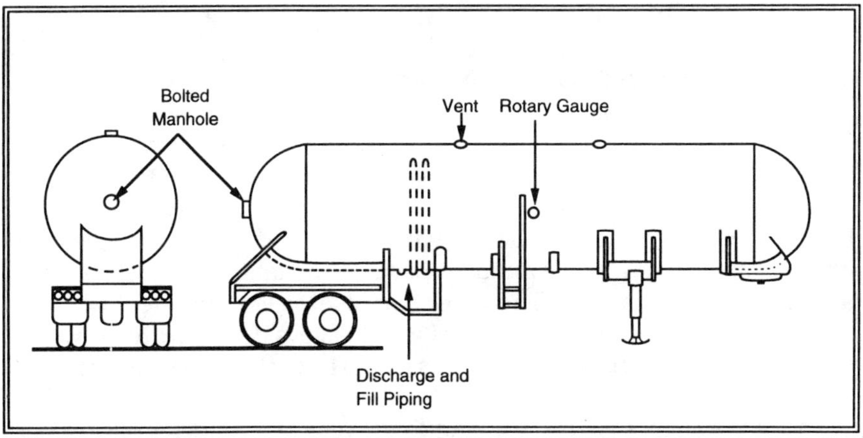

FIGURE 4. MC331 Tanks. Description: Cylindrical rear cross-section (usually larger diameter than MC312/412), hemispherical heads, and smooth surface without external ribs.

ethylene oxide, chlorine). Typically, an MC331 is constructed of steel and has a design pressure of at least 100 psi, but not more than 500 psi. MC331 containers are never filled completely. Some outage or vapor space is necessary for product expansion during transit and overfilling can cause hydrostatic failure. Also venting is provided, but is only effective for venting vapor. In an overturn situation, venting of liquid will not lower internal pressure. Other important features include:

- An internal pressure gauge, thermometer for product temperature and gauging device to determine the liquid level.
- Insulation or a jacket of aluminum or stainless steel, or a coating of white or aluminum paint on at least the upper two thirds of the container to prevent heating by the sun.
- Excess flow shutoff valves on product discharge openings which are designed to stop product flow in the event of a failure in downstream piping or hoses.
- Internally seated valves that provide for discharges.
- An additional or secondary closure system (emergency controls) and a heat activated system. On semi-trailers 30 feet in length or more, emergency controls will be located at both ends of the trailer.
- A rear bumper.
- A bolted manhole, usually located in the rear head of the container.
- Vents that are either mechanical pressure relief vents or frangible discs.
- Overturn protection for fittings, manway covers and vents.
- Discharge piping that is designed to break away from emergency valves or be protected by guards.

MC338

MC338 tanks (Figure 5) carry cryogenic gases (cryogenic gases are liquefied by refrigeration as opposed to pressurization) such as liquid helium (i.e., liquid at -425 F). From the outside, these containers have the same basic design as other tank containers, for example, the MC331 container. The MC338 tank, however, is constructed as a container within a container, similar to a thermos bottle, with annular (totally surrounding the inner container) space. The annular space is evacuated and insulated with a multilayered mylar film. The outer container is typically steel with the inner container constructed of special steel alloys due to extremely cold temperatures. They have a design pressure of at least 25.3 psi, but not more than 500 psi.

MC338 containers are designed to prevent heat transfer of product, but are not refrigerated. Therefore, over time heat transfer will occur and pressure in the tank will rise. This will cause the relief valve to operate. Due to the properties of cryogenic gases, venting refrigerates the load, thus lowering the internal pressure and subsequently allowing closing of the relief valve once pressure returns to normal. Other important features include:

- Annular space with a safety relief device, often a frangible disc, to prevent tank rupture from overpressure in the event of a leak in the internal tank.
- The same safety devices as the previously described containers - safety relief valves, protected discharge valves, collision protection, automatic heat activated valve closures, remotely located emergency valve operating system, etc.

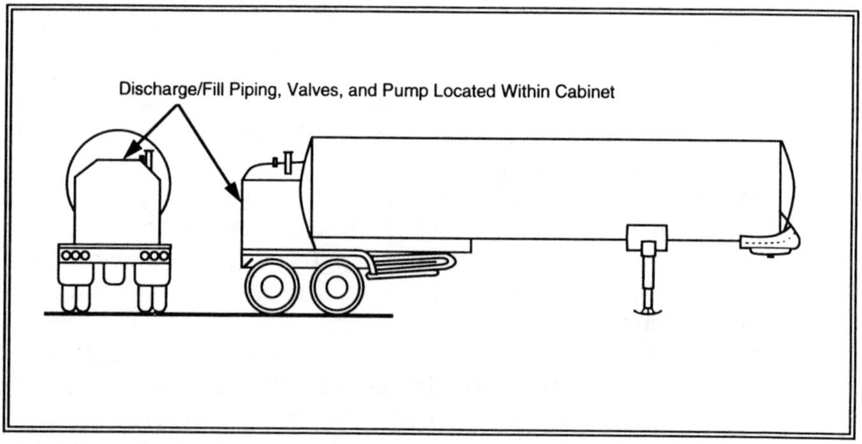

FIGURE 5. MC338 Tanks. Description: Cylindrical rear cross-section, hemispherical heads, and smooth surface without external ribs.

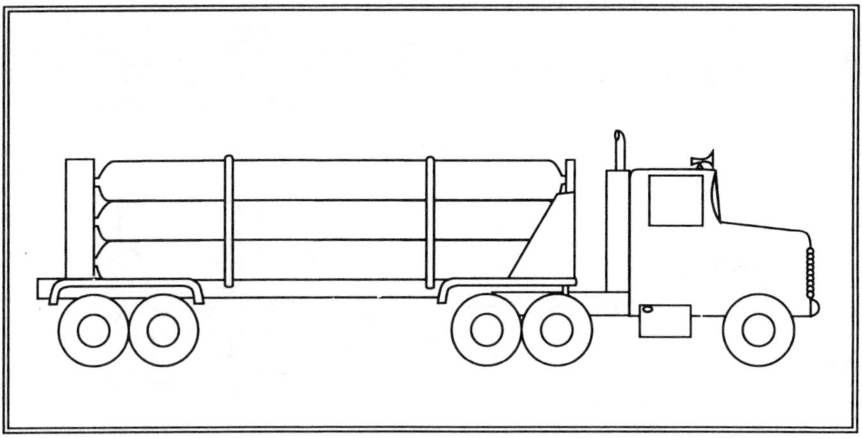

FIGURE 6. Tube Trailer. Description: Tube trailer is the common name for a semi-trailer that transports bulk nonliquified compressed gases.

Tube Trailers

Tube trailers (Figure 6) consist of a group of seamless cylinders, nine to forty eight inches in diameter, permanently mounted on a semi-trailer. The tube trailer may have as few as three larger cylinders or over twenty of the smaller ones. Cylinder service pressures range from 3000 to 5000 psi. Examples of materials shipped in tube trailers are helium, hydrogen, nitrogen and oxygen.

II. Intermodal Tanks

Intermodal tank containers (Figure 7) may carry a wide variety of hazardous materials and are composed of both a tank and its integrated frame. They may be nonpressure, pressure, and cryogenic. Intermodal tanks do not normally exceed 6300 gallons. The tanks are anchored by pins in each corner. This type of tank is used so that it may be transported by more than one mode of transportation (rail, road, and marine).

The International Maritime Organization (IMO) publishes a code for transporting dangerous goods entitled the International Maritime Dangerous Goods (IMDG) code. Section 13 of the IMO code lists the specifications for 7 types of IMO tanks. In general the tank types are:

Types 1, 2, and 4	Liquids
Types 5 and 6	Compressed Gases
Types 7 and 8	Cryogenic Liquids

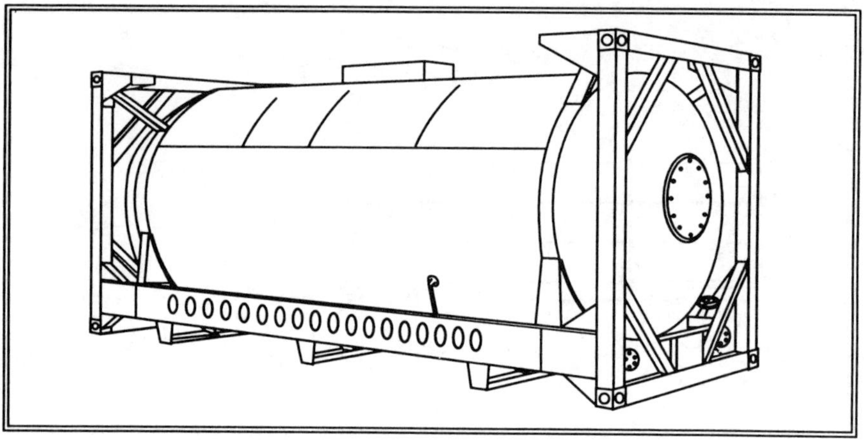

FIGURE 7. Intermodal Tank Container. Description: Intermodal tank containers may carry a wide variety of hazardous materials. They may be nonpressure, pressure, and cryogenic.

Note: There is no Type 3 container

IMO tanks are also usually designed to fit in a 20-foot cell space (i.e., the space occupied by a 20 foot freight container on a ship). This would make the frame dimensions 20 ft. × 8 ft. × 8 ft. Off-size containers are stowed above deck and are more expensive to ship.

III. Railroad Tank Cars

There are three categories or classes of railroad tank cars: (1) nonpressure tank cars; (2) pressure tank cars; and (3) miscellaneous tank cars. The miscellaneous class includes cryogenic liquid tank cars, multi unit tank car tanks, high pressure tank cars, pneumatically unloaded covered hopper cars and wooden tank cars.

DOT Specifications for the tank cars can be found in 49 CFR 179. In addition, various handbooks on railcars (e.g., the GATX tank car book) can also be used as references.

Nonpressurized Tank Cars (USDOT-103,104,111,and 115 and AAR-201,203,206 and 211)

Nonpressurized tank cars (Figure 8) carry flammable and combustible liquids, flammable solids, oxidizers and organic peroxides, poisonous and corrosive materials. Materials posing inhalation hazards cannot be transported in 111 cars. Solids are often loaded in the molten state. Also, certain nonhazardous materials such as edible and inedible tallow, fruit and vegetable juices, tomato paste, and caramel are transported in nonpressurized railroad tank cars. They are considered nonpressurized although test pressures range from 35 to 100 psi. They are constructed of steel, aluminum, or alloy steel (stainless or nickel steel). Other important design features include:

- The capacity to carry 4000 to 45,000 gallons of product. (In the US, since 1970, the maximum capacity of new tank cars for regulated commodities—hazardous materials—has been limited to 34,500 gallons.)
- The division into compartments, which may be of differing capacity and may transport different commodities.
- One manway (newer nonpressure tank cars) or one expansion dome (older nonpressure tank cars). A manway allows access to the interior.
- External fittings as necessary for loading and emptying the container (optional bottom outlet or bottom washout).
- Appropriate safety devices (either a safety relief valve, safety vent, or both).
- A container within a container for one type (U.S. DOT-115 or AAR-206). The tank containing the product is supported by foam within a heavier outer tank.

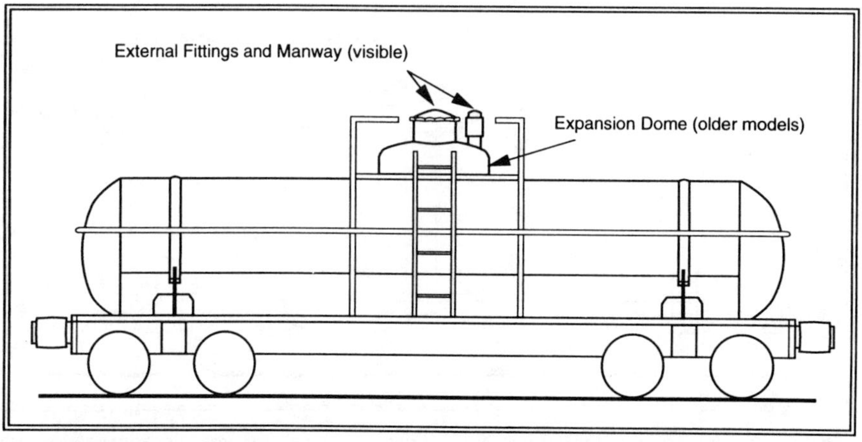

FIGURE 8. Nonpressurized Tank Cars. Description: Cylindrical rear cross-section, convex heads and visible fittings or an expansion dome.

Pressure Tank Cars (U.S. DOT-105, 109, 112, 114 and 120)

Pressure tank cars (Figure 9) carry nonflammable, flammable, and poisonous gases (materials designated as PIH by DOT). They are also used to transport products such as pyrophoric liquids, sodium metal, motor fuel antiknock compound, bromine, anhydrous hydrogen fluoride, and acrolein, inhibited. Tank test pressures range from 100 to 600 psi. They are constructed of steel or aluminum. Other important design features include:

- The capacity to carry from 4000 to 45,000 gallons.
- Top loading.
- Various combinations of bottom opening and washout (e.g., a bottom opening and washout, only bottom opening, only bottom washout or none).
- A manway of sufficient size to permit access to the interior.
- Manway cover plate designed for the mounting of all valves and gauging and sampling devices (covered by protective housing).
- Insulation and/or thermal protection.
- At least top two-thirds of container painted white, if not insulated or protected.
- Head puncture resistance (head shields) for containers that are used to transport flammable gases and anhydrous ammonia. These shields protect the lower portion of the heads against punctures and can be built into a jacket while others are visible. If visible, they appear as "half head" or trapezoidal shaped plates mounted on both ends of the container.

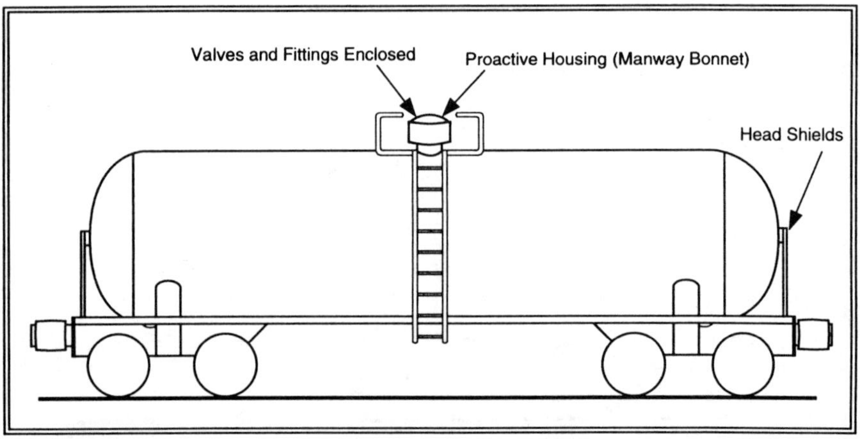

FIGURE 9. Pressurized Tank Cars. Description: Cylindrical rear cross-section, convex heads and a protective housing (called a manway bonnet) 18 to 24 inches high and 30 to 36 inches in diameter located at top center of the container.

Miscellaneous Tank Cars (U.S. DOT-113, 106, 107, 110 and AAR-207, 208 and 204)

There are a number of rail cars listed under this class. Three examples are described below.

Cryogenic Liquid Tank Cars

Cryogenic liquid tank cars (Figure 10) are used to transport low pressure, very cold refrigerated liquefied gases. Common products that are carried include hydrogen, ethylene, argon, nitrogen, and oxygen. These tanks are usually constructed as a container within a container, with the space between the inner container and outer container filled with insulation and evacuated, in order to keep the product cool for up to 30 days. The outer container is made of steel and the inner container is usually made of steel alloy (stainless or nickel steel). They are tested to pressures ranging from 60 to 175 psi. Other important features include:

- Loading and unloading fittings that are located in cabinets at diagonal corners of the railroad car or on the end of the car at ground level or center.
- Safety relief devices—including safety relief valves and safety vents.

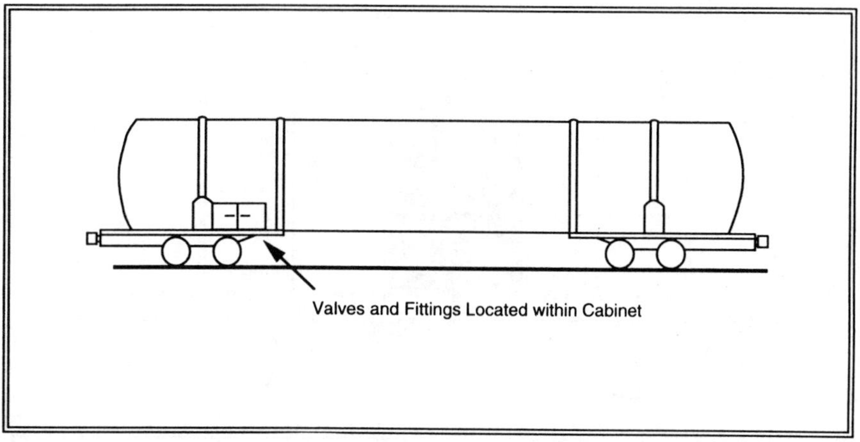

FIGURE 10. Cryogenic Liquid Tank Cars. Description: Cylindrical rear cross-section, ellipsoidal heads and fittings on sides or ends.

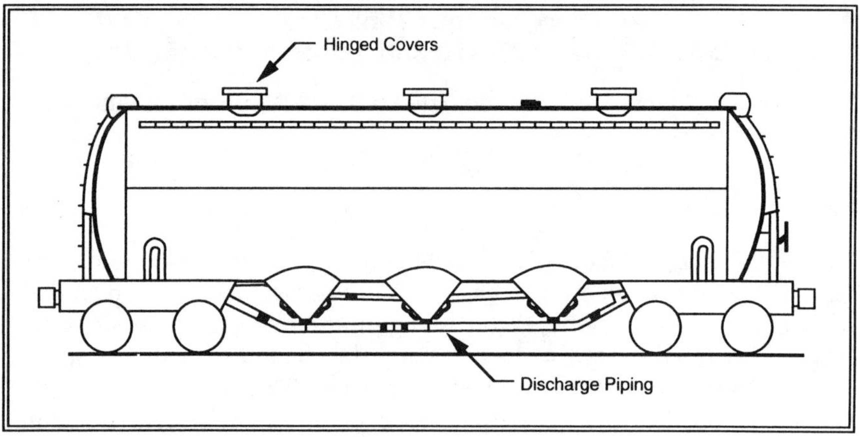

FIGURE 11. Tank Type Covered Hopper. Description: Covered hopper car.

Tank Type Covered Hoppers

Tank type covered hoppers (Figure 11) are used to transport granular products such as caustic soda. These rail cars are considered nonpressure containers although pressure is used during unloading. Tank test pressures range from 20 to 80 psi.

Rail Box Cars

Rail box cars (Figure 12) are enclosed rail freight cars usually with doors in the middle of both sides. Box cars are built primarily of steel although some have wooden interior linings. Box cars range in size up to 85 feet

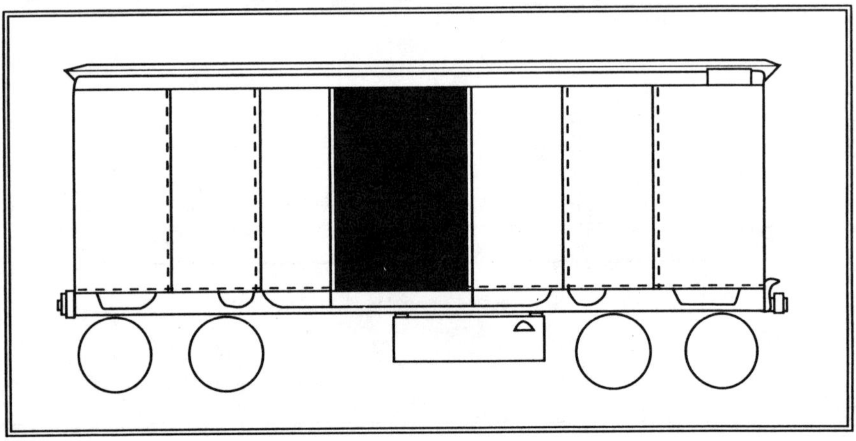

FIGURE 12. Rail Box Car. Description: Rail box cars are enclosed rail freight cars usually with doors in the middle of both sides.

long with capacities up to 8000 cubic feet. Box cars may be refrigerated, and refrigerated cars may carry as much as 360 gallons of diesel fuel in a tank underneath the car.

IV. Drums

The contents of a drum may be identified from the drum type. DOT Specifications for drums and other nonbulk packages can be found in 49 CFR 178. In addition to drums, specifications for other types of nonbulk packaging including cylinders, boxes, barrels, bags, and jerricans are also included in Part 178. Special drum types have associated hazards as illustrated below:

- *Polyethylene or polyvinyl chloride lined, metal drums*—Often contain strong acids or bases. If the lining is punctured, the substance usually quickly corrodes the steel, resulting in a significant spill or leak.
- *Exotic metal drums (aluminum, nickel, stainless steel or other unusual metals)*—Very expensive drums usually contain extremely dangerous materials.
- *Single-walled drums used as a pressure vessel*—These drums have fittings for both product filling and placement of an inert gas, such as nitrogen. May contain reactive, flammable or explosive materials.
- *Laboratory packs*—Used for disposal of expired chemicals and process samples from university laboratories, hospitals, and similar institutions. Individual containers within the lab pack are often not packed in absorbent material. They may contain incompatible materials, radioisotopes, shock sensitive, highly volatile, highly corrosive or very toxic chemicals.
- *Plastic drums*—Often contain acids or bases. They are resistant to the corrosive properties that are typical of these materials.

Information about the contents of a drum may also be provided by the drumhead configuration as illustrated in Table 3.

TABLE 3
Drum Head Configuration

Configuration	*Information*
Whole lid removable	Designed to contain solid material
Has a bung or removable-lid type bung	Designed to contain a liquid
Contains a liner	May contain a highly corrosive or otherwise hazardous material

V. Barges

There are currently proposed changes to the U.S. Coast Guard standards for Type I and Type II barges that transport hazardous materials. These changes have to do with requiring the barges to move tanks containing hazardous materials further from the ship's hull walls. This is designed to prevent punctures in bulk hazardous materials containers that could occur in barge collisions.

- Type I barges carry "very severe environmental and safety hazards." These will have their cargo stowed four and one-half feet from the hull's side and two feet from the hull's bottom.
- Type II barges carry "appreciably severe environmental and safety hazards." These will have their cargo stowed three feet from the hull's side and two feet from the hull's bottom.

In addition, all barges must be double-hulled. Figure 13 illustrates a pushed barge.

VI. Vessels

Hazardous materials are shipped via the marine mode in several types of vessels. Characteristics of the vessel are dependent on the properties and hazards (e.g., gases, liquids, etc.) of the specific material to be transported.

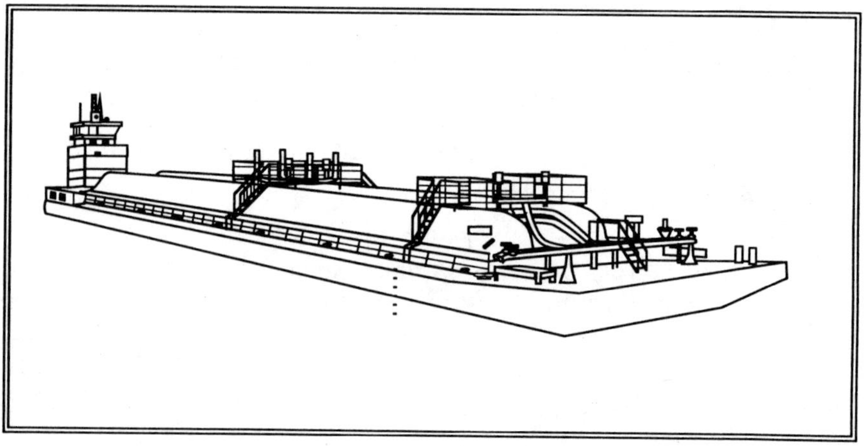

FIGURE 13. A Pushed Barge.

Liquids

Tanker is a term generally used to describe vessels used to carry bulk liquid cargoes. The term includes both self-propelled vessels (tankships) and non–self-propelled vessels (tank barges). Tankers are used for all types of bulk liquid cargoes including petroleum, hazardous chemical cargoes and nonhazardous liquid cargoes. They are classified by several variables including:

- Type of vessel (tankship, tank barge, etc.)
- Crew (manned or unmanned)
- Size
- Cargo

Gases

The U.S. Coast Guard has specified and enforces design standards for US ships and foreign ships carrying bulk liquefied gases into US ports. These standards are contained in Title 46 of the CFR. In addition Lloyd's Register approves the different containment systems for the carriage of liquefied gases in ships.

The standards cover the design of ships used to transport materials such as LNG and LPG. LNG is refrigerated and shipped in vessels with capacities of up to 130,000 cubic meters. LPG is shipped as a liquefied gas in ships with capacities of up to 15,000 cubic meters. Larger capacity ships are also refrigerated.

APPENDIX B

Information Needs for a Study and Sources of Data

The checklist given in Table B-1 covers the basic data needs for a TRA. Some of the items will be unnecessary in a qualitative TRA or less important in a semiquantitative TRA. Their significance will also be influenced by whether the TRA focus is on consequences, frequencies, or risks. Potential data sources (from Chapter 4) are also identified.

TABLE B-1
Information Checklist

SITUATIONAL FACTS

Materials of concern

Container specifications
- ❑ type
- ❑ capacity (or diameter for pipelines)
- ❑ wall thickness
- ❑ operating temperature
- ❑ operating pressure
- ❑ number of compartments or tanks (all but pipelines)
- ❑ depth of burial (pipelines)
- ❑ age (pipelines in particular)
- ❑ hull type (marine)
- ❑ protective features (e.g., headshields, excess flow valves, cathodic protection)
- ❑ insulation

Other equipment of concern (e.g., pig launchers, valves, pumps, . . .)

Failure history (if available)

Route specifications
- ❑ length
- ❑ number of port calls (marine)
- ❑ number of yard visits (rail)
- ❑ details of route
- ❑ carrier(s) used
- ❑ track classes and speeds (rail)
- ❑ type of highways (truck)
- ❑ type of waterways (marine)
- ❑ railcar, barge, or container placement
- ❑ train length (number of cars)
- ❑ special traffic control systems

Operating procedures (e.g., uses of pilots or tugs, backup drivers, etc.)

Number of trips per year (rail, truck, marine)

Flow rate (pipeline)

Number of cars or barges of concern per movement

DATA NEEDS

Accident rates (Transportation Data Sources 4-1 to 4-15, 4-17 to 4-27, 4-29 to 4-35)

Release probabilities (4-1, 4-7 to 4-9, 4-23, 4-25 to 4-27, 4-29 to 4-31, 4-33, 4-36)

Ignition and explosion probabilities (Section 3.4)

Meteorological conditions (4-40, 4-41)

Population densities (4-37, 4-38, 4-39, 4-41)

Traffic levels (marine in particular) (4-28)

Currents (marine)

Likelihood of fatality given exposure to incident outcomes

APPENDIX C

Sample Outline for TRA Reports

The report outline presented in Table C-1 will help (1) present the background and results of the TRA in a way that satisfies the needs of the user while withstanding scrutiny and question and (2) preserve the assessment in a retrievable form. This outline will be most suitable for semiquantitative or quantitative TRAs. Not all chapters will always be needed, particularly if the analysis does not involve a consequence or a frequency analysis.

The final report is usually the most retrievable document from a TRA, and, thus, should contain as many details as possible regarding the analytical calculations, assumptions, and results (within the bounds of practicality). Ideally, it would be possible to regenerate or expand the assessment from this report; however, other files are often needed as well.

TABLE C-1
Sample Outline for a TRA Report

PREFACE
EXECUTIVE SUMMARY

1. INTRODUCTION AND BACKGROUND
   - 1.1. Description of the Transportation Movement(s)
   - 1.2. Purpose and Scope of Study
   - 1.3. General Approach, Methods, and Analytical Techniques Used
   - 1.4. Organization of Report
2. HAZARDS IDENTIFICATION
   - 2.1. Route Description(s)
   - 2.2. Container Description(s)
   - 2.3. Accident-initiated Events
   - 2.4. Non–accident-initiated Events
   - 2.5. Selection of Incident Outcomes
3. DATABASE DEVELOPMENT
   - 3.1. Accident Rates
   - 3.2. Release Probabilities
   - 3.3. Non–accident-initiated Event Data
   - 3.4. Demographic Data
   - 3.5. Meteorological Data
   - 3.6. Summary of Data Used for the Analysis
4. FREQUENCY ANALYSIS
   - 4.1. Boundaries of Analysis
   - 4.2. Methodology
   - 4.3. Specific Assumptions
   - 4.4. Results and Identified Problems
5. CONSEQUENCE ANALYSIS
   - 5.1. Summary of Analytical Methods and Limitations
   - 5.2. Specific Assumptions, Demography
   - 5.3. Results
6. RISK ESTIMATION
   - 6.1. Individual Risk
   - 6.2. Societal Risk
7. RESULTS AND RECOMMENDATIONS
   - 7.1. Results of Sensitivity Studies
   - 7.2. Summary of Major Recommendations (container selection, route selection, training, procedures, etc.)
   - 7.3. Conclusions

APPENDICES
- A. Movement Documentation (e.g., routes, route characterizations, MSDSs, operating procedures, container specifications)
- B. General Assumptions and Boundary Conditions for the Analyses
- C. Supplemental Details Regarding Analytical Methodology/Computer Risk Vulnerability, Dispersion Models, etc.
- D. Copies of Detailed Frequency Calculations
- E. Supplemental Calculations
- F. References

APPENDIX D

Conversion Factors

Four-Significant-Figure Conversion Factors between Common Chemical Engineering Units and SI Units.

Multiply the numerical value of units in the first column by the conversion factor in the second column to obtain the standard SI units given at the head of each individual section.

(Engineering units)(conversion factor) = (SI units above)

L, Length, m	
Å	1.0×10^{-10}
in.	0.02540
ft	0.3048
yd	0.9144
mi	1609
A, Area, m^2	
barn	1.0×10^{-28}
$in.^2$	6.452×10^{-4}
ft^2	0.09290
yd^2	0.8361
acre	4047
hectare	10,000
mi^2	2.590×10^{6}
V, Volume, m^3	
$in.^3$	1.639×10^{-5}
U.S. fluid ounce	2.957×10^{-5}
quart (liq)	9.464×10^{-4}
L, liter	1.0×10^{-3}

U.S. gallon (liq)	3.785×10^{-3}
ft^3	0.02832
barrel (42 gal)	0.1590
yd^3	0.7646
m, Mass, kg	
oz (av.)	0.02835
lb. (av.)	0.4536
ton (2000 lb)	907.2
ton (metric), t	1000
ton (2240 lb)	1016
w, Mass flow rate, kg/s	
lb/yr	1.438×10^{-8}
lb/h	1.260×10^{-4}
lb/min	7.560×10^{-3}
ton*/day *2000 lb.	0.01050
lb/s	0.4536
q, Volumetric flow rate, m^3/s	
ft^3/h	7.87×10^{-6}
gpm (gal/min)	6.309×10^{-5}
ft^3/min	4.72×10^{-4}
ft^3/s	0.02832
T, Temperature, K	
°F	5/9(°F+459.7)
°C	°C + 273.2
R	5/9
Δ°F, ΔR	5/9
Δ°C, ΔK	1.0
u, Velocity, m/s	
ft/min	0.005080
km/h	0.2778
ft/s	0.3048
mi/h, mph	0.4470
1 mph=	1.609 km/h
f, Frequency, Hz	
cycles per second	1.0
F, Force, N	
dyne	1.0×10^{-5}
lb_f	4.448

p, Pressure, Pa	
dyne/cm^2	0.1000
mm of Hg	133.3
torr	133.3
lb$_f$/ft^2	47.88
in. of water	248.8
ft. of water	2989
in. of Hg	3377
lb$_f$/in.^{2}, psi	6895
bar	100,000
std atm	101,325
E, Energy; W, work: J	
erg	1.0×10^{-7}
dyne · cm	1.0×10^{-7}
W · s	1.0
ft · lb$_f$	1.356
cal	4.184
Btu	1055
kcal	4187
kW · h	3.6×10^{6}
P, Power; q, heat transfer rate: W	
ft · lb$_f$/min	0.02260
Btu/h	0.2931
cal/s	4.184
Btu/m	17.57
hp	745.7
ton of refrig	3517
Radioactivity	
1 röntgen =	2.58×10^{-4} C/kg
1 rd =	1.0×10^{-2} J/kg
1 curie =	3.7×10^{10}/s
α, υ, D, Diffusivity, m^2/s	
ft^2/h	2.581×10^{-5}
cm^2/s	1.0×10^{-4}
ft^2/s	0.09290
C_p, Heat capacity, J/kg · K	
Btu/lb · °F	4187
cal/g · °C	4187

k, Thermal conductivity, W/m · K	
Btu/h · ft^2 · °F/in.	0.1442
Btu/h · ft^2 · °F/ft	1.731
h, Heat transfer coefficient, W/m^2 · K	
Btu/min · ft^2 · °F	0.0946
Btu/h · ft^2 · °F	5.678
μ, Newtonian viscosity, Pa · s	
lb/hr · ft	4.134×10^{-4}
centipoise	0.0010
poise, g/cm · s	0.1000
lb/ft · s	1.488
p, Density, mass concentration, kg/m^3	
grain/gal	0.01712
g/L	1.0
lb/ft^3	16.02
lb/gal	119.8
g/cm^3	1000
lb/in.^{3}	27,680
q/A, Thermal flux, W/m^2	
Btu/hr · ft^2	3.155
Btu/s · ft^2	11,356
cal/s · cm^2	41,840
a, Acceleration, m/s^2	
cm/s^2	0.010
in./s^2	0.0254
ft/s^2	0.3048
std accel of grav, g	9.807 m/s^2
σ, Surface tension, N/m	
lb_f/ft	14.59
lb_f/in	175.1
I, Moment, torque, N · m	
dyne · cm	1.0×10^{-7}
in. · oz_f	7.062×10^{-3}
in. · lb_f	0.1130
ft · lb_f	1.356

Note: Miles are statute miles. Btu's and calories are from the thermochemical scale.

APPENDIX E

Qualitative Evaluations

This appendix contains a listing of appropriate topics for inclusion in a checklist that a company may use to audit or review the operations of a carrier for a particular movement or set of movements. As the causes of many transportation related accidents and releases can be traced to operational or maintenance inconsistencies, evaluation of carriers and their practices may be a valuable tool in reducing the risks associated with the transportation of hazardous materials.

The following material is presented as an outline of topics that one might consider when evaluating the performance of a particular carrier. The majority of the topics discussed are most applicable to road and rail transportation, although they can be applied to marine transportation as well. Where pipeline transportation is concerned, additional considerations may need to be addressed and some of those presented will not apply at all.

- **Carrier Name**
- **Address**
- **Contact Person**
- **Telephone Number**
- **Type of Carrier**
- **Accident History**
  - —*Number of Accidents:* These data should be obtained for several years and can be used to establish accident trends over time.
  - —*Number of Preventable Accidents:* This will point out areas where future accidents can be avoided.
  - —*Accident Frequency per million miles:* These data can be used as an accident rate in conducting a risk analysis, or to compare with the rates of other carriers.
  - —*Miles Traveled:* This can be combined with the number of accidents to derive accident rates.

- **Hazardous Materials Transported**
  —*Materials Handled*
  —*Quantities per year and Quantities per trip:* These are both indicators of the amount and type of experience the carrier has with hazardous materials.
  —*Hazards:* This information helps to understand the degree of awareness on the part of the carrier as to the nature of the materials that are being handled. They can also be compared to emergency response capabilities.
    - Toxic
      - Inhalation
      - Ingestion
      - Dermal
    - Fire (note: flammable materials may give off products of combustion that are also hazardous)
    - Corrosive
    - Reactive
      - Incompatible materials
    - Explosive
    - Radioactive
    - Environmental Hazards

- **Employee Training and Qualification**—There are specific training requirements mandated by the government. In addition, effective training programs may correlate with better safety records.
  —General Awareness and Familiarization
  —Function Specific
  —Safety Training
  —Driver Training/Qualification
  —Vehicle Inspection
  —Loading/Unloading
  —Blocking/Bracing
  —Hazard Communication

- **Program Implementation and Recordkeeping**—In addition to regulatory requirements, these are an indicator of the state of the safety program and the quality of these programs may correlate to the accident history of that carrier.
  —Controlled Substance Testing
  —Logs of Safety Meetings
  —Written Procedures for Inspection, Repair and Maintenance
    - Vehicles
    - Trailers
    - Cargo Tanks

Fittings
Hoses
—Written Programs
Safety Policy
Maintenance/Inspection/Testing
Training
Loading/Unloading
Driver Handbook
Incident Reporting
Examination, Certification, and Testing of Tanks
Emergency Procedures
Vehicle Checklist for Drivers

- **Emergency Response**—A good emergency response program may be of assistance in mitigating the consequences of a particular release.
  —Capabilities
  —Notification Procedures
  —Contractor Arrangements in Place
  —24 Hour Phone Number on Shipping Papers
  —Exercises (dates and scope of)

- **Incident Reporting**—These are extremely useful for evaluating potential improvements to reduce accident rates.
  —Maintain file of reported incidents
  —Reports sent to DOT
  —Provisions for incident review/analysis

- **DOT Audits**—Lack of compliance with DOT standards is an indirect indicator of the state of the safety program and may correspond to the carrier's accident history.
  —Results
  —Fines/Citations

- **Routing**—This is important information for input into the risk calculations for evaluations of particular movements, and for tracking shipments.
  —Trip length (in miles)
  —Trip duration (in hours or days)
  —Physical condition/terrain of route
  —Specifics of routes used
  —Individual driver versus team
  —Enroute inspection requirements

Transportation Risk Analysis Topical Bibliography

SECTION	ENTRIES
1.0. Highway	1– 54
1.1. Data	1– 23
1.2. Methodology	24– 47
1.2.1. Qualitative	24– 31
1.2.2. Quantitative	32– 47
1.3. Other	48– 54
2.0. Marine	55–129
2.1. Data	55– 88
2.2. Methodology	89–120
2.2.1. Qualitative	89– 99
2.2.2. Quantitative	100–120
2.3. Other	121–129
3.0. Multiple Modes	130–246
3.1. Data	130–166
3.2. Methodology	167–236
3.2.1. Qualitative	167–191
3.2.2. Quantitative	192–236
3.3. Other	237–246
4.0. Pipeline	247–292
4.1. Data	247–262
4.2. Methodology	263–275
4.2.1. Qualitative	263–269
4.2.2. Quantitative	270–275
4.3. Other	276–292
5.0. Rail	293–343
5.1. Data	293–322
5.2. Methodology	323–340
5.2.1. Qualitative	323–327
5.2.2. Quantitative	328–340
5.3. Other	341–343
6.0. Related Topics	344–371

1.0. Highway

1.1. Data

1. American Petroleum Institute; Summary of Motor Vehicle Accidents in the Petroleum Industry for 1988; (1988).
2. Australian Bureau of Statistics; Survey of Motor Vehicle Use; Victoria, Australia. 9208.0. (1992).
3. Croce, P. A. et al.; A Feasibility Study of a Sealed Safety Monitor for Trucks Carrying LNG and Other Hazardous Materials; Prepared by Arthur D. Little, Inc. for the U.S. Department of Energy, Office of Operational Safety Programs DOE/EV/10502-1. (1982).
4. Davies, P. Lees, F.; Impact Speed of Heavy Goods Vehicles; *Journal of Hazardous Materials.* Vol. 26, p. 213. (1991).
5. Ervin, R. D., et al.; Future Configurations of Tank Vehicles Hauling Flammable Liquids in Michigan Highways; Prepared by University of Michigan for Michigan Department of State Highway and Transportation, PB81-143281. (1980).
6. Harwood, D.; Present Practices of Highway Transportation of Hazardous Materials—Task B, Interim Report, Literature Review; Midwest Research Institute. Project Number 8669-S. (1987).
7. Harwood, D.; Present Practices of Highway Transportation of Hazardous Materials—Task C, Interim Report, Analysis of Existing Databases; Midwest Research Institute. Project Number 8669-S. (1987).
8. Harwood, D.; Viner, J. G.; Russell, E.; Characteristics of Accidents and Incidents in Highway Transportation of Hazardous Materials, Presented at the Transportation Research Board Annual Meeting; Washington, D.C. (1989).
9. Harwood, D.; Viner, J. G.; Russell, E.; Truck Accident and Release Rates for HAZMAT Routing. State and Local Issues in Transportation of Hazardous Waste Materials: Towards a National Strategy; American Society of Civil Engineers. Proceedings of the National Conference on Hazardous Materials Transportation, May 14–16,1992. (1992).
10. Harwood, D.; Viner, J. G.; Russell, E.; Truck Accident and Release Rates for Hazmat Routing, Assessing Community Safety for Hazardous Materials Transport.; National Conference on Hazardous Materials Transportation, St. Louis, Missouri, May 14–16, 1990. (1990).
11. Insurance Institute for Highway Safety; Big Trucks and Highway Safety; Arlington, VA. (1985).
12. Kalevela, S.; Radwan, A.; Transportation of Hazardous Materials on Arizona's Low Volume Roads; *Transportation Planning and Technology.* Vol. 12, No. 3–4, pp. 209–219. (1988).
13. Mace, D.; Heckard, R.; Effect of the 65 MPH Speed Limit on Travel Speeds and Related Crashes (Final Report); Report No. LR1-91-01, 1991. (1991).
14. Maio, D.; Liu, T.; Truck Transportation of Hazardous Materials: A National Overview; NTIS No. PB88-163795/XAB. (1987).
15. Murnane, T. J.; Vehicle Accident Statisitics Producing Rollover and Spills Involving Tank Semi-Trailers; Truck and Bus Meeting and Exposition, Charlotte, NC, November 6, 1989. (1989).
16. National Highway Traffic Safety Administration; Fatal Accident Reporting Systems; Washington, DC. (1981).
17. National Safety Council; Fleet Accident Rates; Chicago, IL. (1987 et al.).
18. Schenck, P.; Prototype Safety Tank Trailer, Undergoing Fleet Testing in Germany; Trailer/Body Builders. pp. 24–29. (1988).

19. Schmidt, J.; Price, D.; Virginia Highway Hazardous Materials Flow; Report No. VDOTS/VP1–2. (1977).

20. Smith, R. N.; Truck Accident and Fatality Rates Calculated for California Highway Accident Statistics for 1980 and 1981; U.S. Department of Energy. Contract No. DE-AC04-76 DP00789. (1982).

21. Standards Association of Australia; Road Tank Vehicles for Dangerous Goods, Part 1: General Requirements; AS 2809.1. (1985).

22. Standards Association of Australia; Road Tankers for Dangerous Goods, Part 3: Tankers for Compressed Liquifiable Gases; AS 2809.3. (1985).

23. Vic Roads; Extracts from Road Accident Database; Victoria, Australia.

1.2. Methodology

1.2.1. Qualitative

24. Abkowitz, M.; Cheng, P.; Developing a Risk/Cost Framework for Routing Truck Movements of Hazardous Materials, *Accident Analysis and Prevention;* Vol. 20, No. 1. (1988).

25. Hobeika, A.; Kim, S.; Databases and Needs for Risk Assessment of Hazardous Materials Shipments by Truck; HAZMAT Transport 91, A National Conference on the Transportation of Hazardous Materials and Wastes, The Transportation Center, Northwestern University, Evanston, IL, June 17–19, 1991. (1991).

26. ICF, Inc.; Assessing the Releases and Costs associated with Truck Transport of Hazardous Wastes; NTIS PB84-224468. (1984).

27. Jaeger, P.; Influencing the Risk Involved with the Road Transport of Propane Gas by Taking Technological Measures; Belgium. Proceedings of Transport and Storage of LPG and LNG. (1984).

28. Pijawka, K. D.; Radwan, A. E.; Soesilo, J. A.; Case Study of Hazardous Waste Transport in Arizona, Assessing Community Safety for Hazardous Materials Transport; Proceedings of the National Conference on Hazardous Materials Transportation, St. Louis, Missouri, May 14–16, 1990. (1990).

29. Pijawka, K. D.; Radwan, A. E.; Soesilo, J. A.; A Risk and Vulnerabiity Assessment Approach for Selecting Routes: Case Study of Hazardous Waste Transportation in Arizona; American Society of Civil Engineers. State and Local Issues in Transportation of Hazardous Waste Materials: Towards a National Strategy. Proceedings of the National Conference on Hazardous Materials Transportation, St. Louis, Missouri, May 14–16, 1992. (1992).

30. Radwan, A.; Singh, S.; Pijawka, K.; Projecting Hazardous Materials and Wastes in Transportation: Conceptual and Methodological Factors and Application; Transportation Research Record. N1264, pp. 77–85. (1990).

31. Stewart, A. M.; Van Aerde, M.; An Empirical Analysis of Canadian Gasoline and LPG Truck Releases; *Journal of Hazardous Materials.* Vol. 25, pp. 205–217. (1990).

1.2.2. Quantitative

32. Abkowitz, M.; Eigher, A.; Srinivasan, S.; Assessing the Release and Costs Associated with Truck Transport of Hazardous Wastes; Prepared by Rensselaer Polytechnical Institute for the EPA. (1984).

33. Arthur D. Little, Inc; Assessment of Risks and Risk Control Options Associated with Liquefied Natural Gas Trucking Operations from Distrigas Terminal, Everett, Massachusetts; Prepared for the U.S. Department of Transportation—Contract No. DOT-RC-82037. (1979).

34. Arthur D. Little, Inc; A Model Economic and Safety Analysis of the Transportation of Hazardous Substances in Bulk (Executive Summary); Prepared for U.S. Department of Commerce, C-76446. (1974).
35. Blower, D.; Lyles, R.; Campbell, K.; Stamatiadis, P.; The Michigan Heavy Truck Study, Final Report; University of Michigan Transportation Research Institute. (1990).
36. Brogan, J. D.; Cashwell, J. W.; Neuhauser, K. S.; Demographic and Transportation Parameters in RADTRAN; 9th International Symposium on the Packaging and Transportation of Radioactive Materials Proceedings. (1989).
37. Davies, P.; Lees, F.; The Assessment of Major Hazards: The Road Transport Environment for Conveyance of Hazardous Materials in Great Britain; *Journal of Hazardous Materials*. Vol. 32, No. 1. (1992).
38. Eagen, P.; Views of Risk and Highway Transporation of Hazardous Materials: A Case Study in Gasoline; NTIS No. PB83-170738. (1981).
39. Elder, H. K. et al.; An Assessment of the Risk of Transporting Spent Nuclear Fuel by Truck; Prepared by Pacific Northwest Laboratory for the U.S. Department of Energy PNL-2588. (1987).
40. Gallegos, A.; Channell, J.; Risk Analysis of the Transport of Contact Handled Transuranic (CH-TRU) Wastes to Waste Isolation Pilot Plant (WIPP) along Selected Highway Routes in New Mexico Using RADTRAN 4; NTIS No. DE90017272/XAB. (1990).
41. Glickman, T. S.; An Expeditious Risk Assessment of the Highway Transportation of Flammable Liquids in Bulk; Transportation Science. Washington, DC. INSPEC Abstract Number: C91056196. (1991).
42. Harwood, D.; Russell, E.; Procedures for the Development of Truck Accident Rates and Release Probabilities for Use in Hazmat Routing Analysis; Transportation of Dangerous Goods: Assessing the Risks; Institute for Risk Research. Waterloo, Ontario. (1993).
43. Harwood, D.; Viner, J. G.; Russell, E.; Truck Accident Rate Model for Hazardous Materials Routing.; Transportation Research Board Publications Office. Transportation Research Record N1264. (1990).
44. ICF Technology; Risk Assessment of Nitrogen Tetroxide Transporation Route from Cedar Chemical Corporation to VAFB: California Segments Only; San Antonio, Texas. Prepared for Transportation Office, Kelly Air Force Base. (1988).
45. Rhoads, R. E.; An Assessment of the Risk of Transporting Gasoline by Truck; U.S. Department of Energy. Contract No. EY-76-C-06-1830. (1978).
46. Saccomano, F. F.; Chan, A. Y. W.; Economic Evaluation of Routing Strategies for Hazardous Road Shipment; Transportation Research Record No. 1020. (1985).
47. Shaver, D. K.; Weiner, R. K.; Betschart, J. F.; Highway Transport of LF2: Assess Risks and Emergency Response Planning; Washington, D.C. International Symposium on Major Hazards in the Transport and Storage of Pressure Liquified Gases, University of New Brunswick. (1987).

1.3. Other

48. A Tank Truck for the Twenty-First Century; Rohm and Haas Company Resin Review XLII. No. 2, pp. 2–9. (1991).
49. American Petroleum Institute; Precautions Against Electrostatic Ignition During Loading of Tank Motor Vehicles; Washington D.C. API Publication 1003. (1986).
50. Campbell, K.; Risk Assessment, Highway Travel; Mechanical Engineering. p. 46. (1984).
51. Klingenberg, B.; Roscow, G.; Jacobsen, R.; FACT—The Freightliner/Heil Advanced Concept Truck; Society of Automotive Engineers. Truck and Bus Meeting and Exposition, Charlotte, NC, November 6 1989. (SAE Paper No. 892462). (1989).

52. Luchter, S.; Risk in the Transportation System; National Highway Traffic Safety Administration. Washington, DC. Presentation at the Conference on Regulating Risk, June 24, 1991. (1991).

53. National Highway Traffic Safety Administration; Analytical User's Manual; Washington, DC. National Accident Sampling System. (1981).

54. Schenck, P.; Freightliner/Heil Consortium Designs Advanced Concept Safety Tank Truck; *Modern Bulk Transporter.* pp. 84–88. (1989).

2.0. Marine

2.1. Data

55. Abkowitz, M.; Galarraga, J.; Tanker Accident Rates and Expected Consequences in U.S. Ports and High Seas Regions; Washington, DC. Recent Advances in Hazardous Materials Tranportation Research, An International Exchange, State-of-the-Art; Report 3, Transportation Research Board, pp. 161–169. (1986).

56. Aldwinckle, D. S.; McLean, D.; A Safety Review of Ships for Liquefied Gases and Future Legislative Needs.; The 10th International LNG/LPG Conference and Exhibition (GASTECH 84), Amsterdam. (1984).

57. Arthur D. Little, Inc; Risk Assessment for Gas Liquids Transportation from Santa Barbara County; prepared for Santa Barbara County Resource Management Department. (1990).

58. Blything, K. W.; Edmonson, J. N.; Fire/Explosion Probabilities on Liquid Gas Ships; United Kingdom Atomic Energy Authority, Safety and Reliability Directorate, Report No. SRD R 292. (1984).

59. Bovet, D. M.; Preliminary Analysis of Tanker Collisions and Groundings.; U.S. Coast Guard Technical Report. Project 713112. (1973).

60. Chen, E.; Analysis of Ship Collision Probabilities; United States Coast Guard. Washington, DC. Available from the NTIS, Springfield, VA; Oct. 1975 Proceeding. (1975).

61. Comstock, J. P.; Robertson, Jr J. B.; Transactions SNAME; Survival of Collision Damage Versus the 1960 Convention on Safety of Life at Sea. (1961).

62. Department of the Army, Corps of Engineers; Waterborne Commerce of the United States, Calendar Year 1986; Part 2—Waterways and Harbors—Gulf Coast, Mississippi River System and Antilles; WRSC-WCUS-86-2.

63. Exxon Production Research Company; Review of Oil Spill Occurrences and Impacts; (1989).

64. IMCO; Tanker Casualties Report; E. London. No. 78.16. (1978).

65. International Maritime Organization; Analysis of Serious Casualties to Seagoing Tankers,1977-1991; IMO. London. (1992).

66. J.J. Henry Co.; An Analysis of Oil Outflows Due to Tanker Accidents; Prepared for U.S. Coast Guard, Ad-780-315. (1973).

67. Karlsen, J.; Kristiansen, S.; Statistical Suvey of Collisions and Groundings for Norwegian Ships for the Period 1970–78; Oslo. Veritas Report No. 80-0199. (1980).

68. Keillor, J.; The Hazards of Tank Ships and Barges Transporting Petroleum Products on the Great Lakes; *Coastal Zone Management Journal*. Vol 8, No 4. (1980).

69. Kurland, J.; Bryant, D.; Shipboard Polymerization of Acrylic Acid; *Plant Operations Progress*, Vol. 6, No. 4, p. 203. (1987).

70. Lighthart, H. M.; Determination of Probability of Marine Accidents with Respect to Gas Carriers Proceeding in Dutch Coastal and Inland Waters; *Journal of Hazardous Materials.* Vol. 3, pp. 233–247. (1980).

71. Maritime Administration/U.S. Coast Guard; Tank Barge Study; Joint Study conducted for U.S. Dept. of Commerce. (1974).

72. Maritime Transportation Research Board; Human Error in Merchant Marine Safety; National Technical Information Service. Springfield, VA. AD/A-028 371. (1976).

73. Minerals Management Service; Worldwide Tanker Spill Database; Tankspill Database. 1974–present.

74. Minorsky, V. U.; An Analysis of Ship Collisions with Reference to Protection of Nuclear Power Plants; *Journal of Ship Research.* (1959).

75. National Research Council, R&D Panel; Analysis of Statistics Concerning Oil Pollution Incidents in the U.S.; National Academy of Sciences. (1991).

76. National Research Council; Reducing Tankbarge Pollution, Report to the Committee on Reducing Tankbarge Pollution; Maritime Transportation Research Board. (1981).

77. Porricelli, J.; Keith, V.; Tankers and the U.S. Energy Situation: An Economic and Environmental Analysis; Marine Technology. (1974).

78. Poten & Partners, Inc; Liquefied Gas Ship Survey: Analysis of the Record 1964–1981; New York. (1982).

79. Reiter, A. D.; Faragher, W. E.; Commercial Vessel Safety. Risk Assessment Study. Volume I. Survey of Data for Marine Risk Assessment; PRC Systems Sciences Company. Washington, DC. NTIS #AD-A077 719/3/XPS. Contract No. DOT-CG-60351-A. (1979).

80. Stewart, R. J.; Kennedy, M. B.; An Analysis of U.S. Tanker and Offshore Petroleum Production Oil Spillage Through 1975; Martingale, Inc. for the U.S. Department of the Interior. (1978).

81. Todd Shipyards; Nature of Ship Collisions Within Ports; Todd Shipyards for U.S. Maritime Administration. NTIS PB-225 304. (1976).

82. U.S. Army Corps of Engineers; Waterborne Commerce of the United States; (1989).

83. U.S. Coast Guard; Marine Casualty Information Reporting System (CASMAIN); 1981–1991. (1992).

84. U.S. Coast Guard; Marine Investigations (MIN) Module of MSIS and CASMAIN; January 1992–present.

85. U.S. Coast Guard; Polluting Incidents In and Around U.S. Waters, Calendar Year 1981 and 1982; Washington, DC. (1983).

86. U.S. Coast Guard; Polluting Incidents In and Around U.S. Waters, Calendar Year 1982 and 1983; Washington, D.C. (1984).

87. Warwick, J.; Anderson, A.; Nature of Ship Collisions Within Ports; National Technical Information Service. Springfield, VA. NTIS No. PB-255 304. (1976).

88. Welch, J.; Yando, F.; Worldwide Oil Spill Incident Data Base: Recent Trends; Oil Spill Intelligence Report. Arlington, MA. (1993).

2.2. Methodology

2.2.1. Qualitative

89. Berg Andreassen, J.; Risk and Investment Decisions in Non-Liner Shipping; *Maritime Policy and Management.* Vol. 17, No. 1, pp. 23–30. (1990).

90. Charter, D.; Determination of Risk of Collision Using Twentieth Century Techniques; *Journal of the Institute of Navigation.* Vol 26, No 3. (1979).

91. Charter, D.; Risk of Collision—A Sequel; *Journal of the Institute of Navigation.* Vol 26, No 4. (1979).

92. Chemical Manufacturers Association; A Guide to Qualitative Marine Risk Management for Marine Transportation-Related Facilities and Tank Barge and Tank Ship Operations; Washington, DC. (1995).

93. Colonna, G.; Evaluating Hazardous Chemicals in the Marine Workplace; *Chemical Engineering Progress.* p. 10. (1986).

94. Jenssen, T. K.; Larsen, P.; Risk Analysis Applied to the Transportation of Hazardous Cargoes; Some Examples Related to Public Risk; Conference on the Marine Transportation, Handling and Storage of Bulk Chemicals, MARICHEM '80, London, October 21–23, 1990. (1990).

95. Martino, P.; LNG Risk Management; *Environmental Science and Technology.* Vol. 14, No. 12, p. 1446. (1980).

96. Nikodem, H. J., et. al.; Risk Assessment Study for the Harbor of Gothenburg; Prepared for the Swedish Energy Commission, Stolkholm, Sweden. (1978).

97. Wennick, C.; Assessing the Hazards of Gas Carriers Entering Port-1; *The Dock and Harbour Authority.* November. (1982).

98. Wennick, C.; Assessing the Hazards of Gas Carriers Entering Port-2; *The Dock and Harbour Authority.* December, pp. 267–270. (1982).

99. Wheeler, T.; Risk Based Ranking of Dominant Contributors to Maritime Pollution Events; *Risk Analysis.* Vol. 13, No. 2. (1993).

2.2.2. Quantitative

100. Chen, E.; Risk Analysis for Marine Transportation; Transportation Research Board Publications Office. Washington, DC. Transportation Research Record N669, pp. 24–27.

101. Cockraft, A. N.; The Estimation of Collision Risk for Marine Traffic; The *Journal of Navigation.* Vol. 34, No. 1, p. 145. (1987).

102. ECO Engineering, Inc; An Assessment of Tanker Transportation Systems in Cook Inlet and Prince William Sound; Prepared for Alaska Oil Spill Commission. (1989).

103. ECO Engineering, Inc; Task Two: Development of Realistic Maximum Oil Discharge Criteria; Prepared for State of Alaska Department of Environmental Conservation. (1991).

104. ECOSYSTEMS, Inc; A Methodological Procedure for Risk Analysis of Liquefied Natural Gas Operations; Submitted to U.S. Council on Environmental Quality. Washington, DC. (1974).

105. Eddine, A. G. F.; Risk Analysis of Berthing Accidents Involving Large Ships; Inland and Maritime Waterways and Ports Conference Section 2, Edinburgh, May 10–16, 1981. (1981).

106. Faragher, W. E.; Pizzo, J. T.; Rausch, A. H.; Commercial Vessel Safety Risk Assessment Study. Volume II. Risk Assessment Methodology Survey; PRC Systems Sciences Company. Washington, DC. USCG-D-59-79, Contract No.: DOT-CG-60351-A. (1979).

107. Jenssen, T.; Risk Analysis of LPG Transportation by Ship, Presented at the 5th Symposium on Reliability Technology; Bradford, England. (1978).

108. Liu, T. K.; A Regression Model for Estimating Probability of Vessel Casualties; Transportation of Dangerous Goods: Assessing the Risks; Waterloo, Ontario. Institute for Risk Research. (1993).

109. National Academy of Sciences, National Research Council; Analysis of Risk in the Water Transportation of Hazardous Materials; Washington, DC. Final Report No: USCG-D-39-76. (1976).

110. National Materials Advisory Board; The Application of Quantative Risk Assessment Techniques in the U.S. Coast Guard Regulatory Process; Washington D.C. (1982).

111. National Research Council; Analysis of Risk in the Water Transportation of Hazardous Materials; National Academy of Sciences. Washington, DC. (1976).

112. Oceanographic Institute of Washington; Risk Analysis of the Oil Transportation System; Seattle, Washington. NTIS #OIW-OCW-7201. (1974).

113. Romer, H.; Brockhoff, L.; Haastrup, P.; Peterson, H. J. S.; Marine Transport of Dangerous Goods. Risk Assessment Based on Historical Accident Data; *Journal of Loss Prevention Industry*. Vol. 6, No. 4. (1993).

114. Sandwell; A Risk Analysis of Tanker Traffic Movements within the Port of Vancouver; Prepared for the Vancouver Port Corporation. (1991).

115. Sarin, R.; Scherer, C.; Optimal Oil Tanker Size with Regard to Environmental Impact of Oil Spills; *Journal of Environmental Economics and Management*. Vol 3. (1976).

116. Science Applications, Inc; Risk Assessment of LNG Marine Operations for Staten Island, New York; Federal Power Commission Contract No. FP-1768. (1975).

117. Slack, J. R.; Wyant, T.; An Oilspill Risk Analysis for the Mid-Atlantic (Proposed Sale 49) Outer Continental Shelf Lease Area; Geological Survey, Water Resources Div. Reston, VA. NTIS No. USGC/WRD/WRI-78/085. (1978).

118. Spouge, J. R.; Techniques for Risk Assessment of Ships Carrying Hazardous Cargo in Port Areas; The International Consensus Conference on the Risks of Transporting Dangerous Goods, April 6–8, 1992. (1992).

119. Stipdonk, H. L.; Houben, R. J.; QRA-Aided Risk Management of Dutch Inland Waterway Transport; Transportation of Dangerous Goods: Assessing the Risks; Waterloo, Ontario. Institute for Risk Research. (1993).

120. U.S. Coast Guard; Report of Study of Tanker Safety and Pollution Prevention Requirements for U.S. Tankers in Domestic Trade; National Technical Information Service. Springfield, VA. NTIS No. AD/A–057 607. (1978).

2.3. Other

121. Angus, G. F. B.; Safety at Sea; The Potential Hazards of Carrying Chemicals by Sea. (1984).

122. Blything, K.; Edmondson, J.; Fire/Explosion Probabilities on Liquid Gas Ships; U.K. Atomic Energy Authority. (1984).

123. de Frondeville, B.; Reliability and Safety of LNG Shipping: Lessons From Experience; Society of Naval Architects and Marine Engineers. New York. (1977).

124. Elliot, D.; Chlorine Barge Operations in United States Waters; The Chlorine Institute. New York. Storage and Transport Committee. (1982).

125. Koops, W.; Two New Tools and Working Method for Crisis Management of Accidental Spills at Sea; Delft University Press. The Netherlands. ISBN No. 90-6275-804-5/CIP. (1992).

126. Miloh, T.; Sharma, S.; Maritime Collision Avoidance as a Differential Game; Presented at the Fourth Ship Control Systems Symposium. The Hague, Netherlands. (1975).

127. Petersen, M. et al.; Collisions Between Ships and Offshore Platforms; Offshore Technology Conference Report No. 4134. Houston. Offshore Technology Conference. (1981).

128. Snider, W.; IMCO Conference on Tanker Safety and Pollution Prevention; *Marine Technology*. Vol 15, No 3. (1978).

129. U.S. Coast Guard; Office of Merchant Marine Safety; Draft Regulatory Analysis and Environmental Impact Statement for Design Standards for New Tank Barges and Regulatory Action for Existing Tank Barges to Reduce Oil Pollution Due to Accidental Hull Damage; Dockets #75-083 and #75-083a. (1979).

3.0. Multiple Modes

3.1. Data

130. Appleton, P. R.; Transport Accident Frequency Data, Their Sources and Their Application in Risk Assessment; UKAEA Risley Nuclear Power Development Establishment. Culcheth, England. NTIS # DE89616980/XAB. (1988).

131. Argonne National Laboratory; Data Base of Accident and Agricultural Statistics for Transportation Risk Assessment; Argonne, Illinois. NTIS No. ANL/EAIS/TM-2. (1989).

132. CALTRANS; Transportation of Hazardous Materials in California by Highway and Rail; Report to the Legislature. (1986).

133. Clarke, R. K. et al.; Severeties of Transportation Accidents; Sandia Laboratories. NTIS SLA-74-0001. (1976).

134. Dennis, A. W. et al.; Severities of Transportation Accidents Involving Large Packages; Sandia Laboratories. NTIS SAND-77-0001. (1987).

135. Fryer, L. S.; Griffiths, R. F.; Worldwide Data on the Incidence of Multiple-fatality Accidents; Safety and Reliability Directorate. (1979).

136. Fullwood, R. R.; Rhyne, W. R.; Simmon, J. A.; Reese, R. T.; Relative Consequences of Transporting Hazardous Materials; Sandia National Labs. Alburquerque, NM. NTIS Accession Number: SAND-80-0901C, CONF-801115-27. (1980).

137. Glickman, T. S.; Benchmark Estimates of Release Accident Rates in Hazardous Materials Transportation by Rail and Truck; Transportation Research Record 1193. pp. 22–28.

138. Glickman, T. S.; Data Requirements for Hazardous Materials Transportation Risk Analysis; Government Inst., Inc. Rockville, MD. E.I. Montly No.: EIM8506-033744, E.I. Conference No.: 05911. (1984).

139. Haastrup, P.; Brockhoff, L.; Review of Transportation Accidents of Hazardous Materials; Hazardous Substances—Safety in Transport and Warehousing—14th International Symposium, Heidelberg Germany. (1991).

140. Haastrup, P.; Brockhoff, L.; Severity of Accidents with Hazardous Materials—A Comparison Between Transportation and Fixed Installations; *Journal of Loss Prevention in the Process Industries.* Vol. 3, p. 395. (1990).

141. Haastrup, P.; Styhr-Petersen, H. J.; Hazardous Materials: A Comparison of the Severity of Accidents for Transport and Fixed Installations; The International Consensus Conference on the Risks of Transporting Dangerous Goods, April 6–8, 1992. (1992).

142. Hammond, S. P.; Smith, W. W.; Historical Analysis of Dangerous Goods Spills on Alberta; Transportation of Dangerous Goods: Assessing the Risks; Waterloo, Ontario. Institute for Risk Research. (1993).

143. Hazardous Materials Advisory Council; An Overview of Intermodal Portable Tanks; Washington, D.C. (1986).

144. Kletz, T.; Transportation of Hazardous Substances: The U.K. Scene; *Plant Operations Progress.* Vol. 5, No. 3, pp. 160–164. (1986).

145. Lawrence Livermore National Laboratory; Shipping Container Response to Severe Highway and Railway Accident Conditions; NUREG/CR-4289/UCID-20733, Reprinted 1988. (1984).

146. Lewis, D.; A Review of Some Transportation Accidents, Identification of Causes and Minimization of Consequences; AIChE 20th Loss Prevention Symposium, Paper 80a, April 6–10, 1986. (1986).

147. Marshall, V.; Modes and Consequences of the Failure of Road and Rail Tankers Carrying Liquified Gases and other Hazardous Liquids; Bath, U.K. Reliability on the Move: Safety

and Reliability in Transportation. Proceedings of the Safety and Reliability Society Symposium. (1989).

148. Martinsen, et al.; LPG Land Transportation and Storage Safety; NTIS #DE 82-004420. (1981).

149. Medhekar, S.; Gekler, W.; Bley, D.; Frequency Estimates for Transport—Related Hydrofluoric and Sulfuric Acid Releases; Minneapolis, MN. AIChE Summer National Meeting. (1992).

150. NTSB; National Transportation Safety Board, Railroad/Highway Grade Crossing Accidents Involving Trucks Transporting Bulk Hazardous Materials; NTIS #PB82—113432. (1981).

151. Office of Policy Analysis, U. S. EPA; Acute Hazardous Events Data Base; EPA Contract No. 68-W8-0038. (1989).

152. Office of Technology Assessment; Data and Information Systems for Hazardous Materials Transportation, OTA SET—304. (1986).

153. Office of Technology Assessment; Transportation of Hazardous Materials; OTA-SET-304. (1986).

154. Ormsby, R.; Le, N.; Societal Risk Curves for Historical Hazardous Chemical Transportation Accidents; IChemE Symposium Series, 110. (1988).

155. Philipson, L. L.; Schaeffer, M. S.; Hazardous Materials Shipment Data in Risk Analysis; Engineering Societies Library. New York. Annual Reliability Maintainability Symposium, Proceedings Proc Paper, #1304, pp. 519–527.

156. Rahkonen, P.; Malmen, V.; Work Safety in Collection and Transport of Hazardous Wastes; *Waste Management Res.* Vol. 10, No. 2. (1992).

157. Saccomanno, F. F.; Yu, M.; Risk Uncertainty in the Transport of Hazardous Materials; 72d Annual Meeting of the Transportation Research Board, Washington, D.C. (1993).

158. Saccomanno, F. F.; Shortreed, J. H.; Societal-Individual Risks for HAZMAT Transport.; American Society of Civil Engineers. State and Local Issues in Transportation of Hazardous Waste Materials: Towards a National Strategy. Proceedings of the National Conference on Hazardous Materials Transportation, May 14–16, 1990, St. Louis, Missouri, pp. 220–245. (1991).

159. Saccomanno, F. F.; Shortreed, J. H.; Van Aerde, M.; Higgs, J.; Abkowitz, M.; Transportation of Hazardous Materials 1989; Transportation Research Board. Washington, DC. NTIS # PB90-239724/XAB. TRB/TRR-1245; ISBN-0-309-04967-9 74p.See also report for 1988, PB90-145582. Library of Congress Catalog Card No. 90-35157. (1989).

160. Slater, D.; Corran, E.; Pitbaldo, R.; Industrial Hazards Project Report; The Warren Centre for Advanced Engineering, The University of Sydney, Australia. (1986).

161. Stough, R.; Hoffman, J.; Assessing the Risk of Hazardous Materials Flows: Implications for Incidence Response and Enforcement Training; Transportation Research Record 1063. (1986).

162. Swoveland, C.; Cawdery, J.; LPG Transport R&D Risk–Benefit Analysis, Volume I: Analysis and Results; Transport Canada Report No. 74583. (1986).

163. TNO; FACTS: A Databank for Accidents with Hazardous Materials; TNO Environmental and Energy Research, Department of Industrial Safety.

164. Transportation Safety Board of Canada; TSB Statistical Summary: Railway/Commodity Pipeline Occurrences, 1990; Report No. SSC-TU1-2/1990. (1992).

165. Transportation Systems Center; Transportation Safety Information Report, Quarterly Highlights; (various).

166. U.S. Department of Transportation; Annual Report on Hazardous Materials Transportation; Research and Special Programs Administration (Annual).

3.2. Methodology

3.2.1. Qualitative

167. Anthony, D.; Peirson, J.; Managing Risks of Hazardous Materials Transportation in Santa Barbara County; Proceedings of the National Conference on Hazardous Materials Transportation, St. Louis, MO, May 14–16, 1990. (1990).

168. Canadine, I.; Purdy, G.; The Transport of Chlorine by Road and Rail in Britain—A Consideration of the Risks; Paper Presented at a Major European Conference in Oslo. (1989).

169. Chiang, C. K.; Cantilli, E. J.; Ying, S. T.; Assessing Community Safety for Hazardous Materials Transport; Proceedings of the National Conference on Hazardous Materials Transportation, St. Louis, MO, May 14–16, 1990. (1990).

170. Griffith, B. J.; Auld, R. G.; Transportation Also Begins and Ends With Risks; The International Consensus Conference on the Risks of Transporting Dangerous Goods, April 6–8, 1992. (1992).

171. Hubert, P.; Pages, P.; Risk Management for Hazardous Materials Transportation: A Local Study in Lyons; *Risk Analysis*. Vol. 9, No. 4, pp. 445–451. (1989).

172. Inchniowski, T.; New Measures to Bolster Safety in Transporation; *Chemical Engineering*. pp. 35–39. (1984).

173. Jackson, K.; Community Risk Assessment for Transporation of Hazardous Materials; Washington D.C. Transportation Research Board 64th Annual Meeting. (1985).

174. Jenssen, T. K.; Castleman, J. R. (Det Norske Veritas); Risk Analysis of Hazardous Materials Transportation; Institution of Engineers. Australia. pp. 119–123. (1984).

175. Jessop, A.; Decision and Forecasting Models with Transport Applications; Ellis Horwood Ltd. West Sussex, UK. (1990).

176. Kazarians, M.; Boykin, R. F.; Kaplan, S.; Transportation Risk Management—A Case Study; American Institute of Chemical Engineers. New York. Presented at AIChE 20th Loss Prevention Symposium, New Orleans, Louisiana. (1986).

177. Lassarre, S.; Fedra, K.; Weigkricht, E.; Computer-Assisted Risk Assessment of Dangerous Goods Transportation for Haute-Normandie; American Society of Civil Engineers. State and Local Issues in Transportation of Hazardous Waste Materials: Towards a National Strategy; Proceedings of the National Conference on Hazardous Materials Transportation, St. Louis, Missouri, May 14–16, 1990. (1990).

178. List, G.; Mirchandani, P.; A Community-Focused Routing and Siting Model for Hazardous Materials and Wastes, Assessing Community Safety for Hazardous Materials Transport; Proceedings of the National Conference on Hazardous Materials Transportation, St. Louis, Missouri, May 14–16, 1990. (1990).

179. Menzie, C.; An Approach to Estimating Probabilities of Transportation—Related Spills of Hazardous Materials; *Environmental Science and Technology*. Vol. 13, No. 2. (1979).

180. Miall, D.; Safeguarding MAA During Storage and Transport; Baltimore, MD. Methacrylate Producers Association, Methacrylic Acid Workshop, June 12–13, 1990. (1990).

181. Pijawka, K.; Foote, S.; Johnson, L.; Risk Assessment of Transportation Hazardous Materials: A Conceptual Framework and Approach for Risk Mangement and Planning; Transportation Research Board, 64th Annual Meeting. (1985).

182. Pijawka, K. D.; Foote, S.; Soesilo, A.; Scanlon, R. D.; Cantilli, E. J.; Improving Transportation of Hazardous Materials Through Risk Assessment and Routing; Transportation Research Board. Washington, DC. NTIS # PB86-112471/XAB, TRB/TRR-1020. ISBN 0-309-03909-6. (1985).

183. Rhyne, W. R.; Critical Issues in Performing Transportation Risk Studies; H&R Technical Associates, Inc. Oak Ridge, Tennessee. (1988).

184. Saccomanno, F. F., et al.; Comparison of Risk Measures for the Transportation of Dangerous Commodities by Truck and Rail; Transportation Research Board Record 1245. (1990).

185. Saccomanno, F. F.; Bakir, O.; Maximum Likelihood Approach for Considering Risk Uncertainty in the Transport of Hazardous Materials; Transportation of Dangerous Goods: Assessing the Risks; Waterloo, Ontario. Institute for Risk Research. (1993).

186. Sellers, J.; Bendig, L.; Risk Comparison of LPG Distribution by Truck or Pipeline; Institution of Chemical Engineers Symposium Series. England. No. 120. (1990).

187. Senni, S.; Colombo, L.; Transportation and Storage of Hazardous Materials; *Termotecnica (Milan)*. Vol. 44, No. 9, pp. 63–72. (1990).

188. Shortreed, J.; Risk Management and Dangerous Goods Movements; University of Waterloo Press. Waterloo, Ontario. (1989).

189. Smith, L. M.; Major Hazard Aspects of Transport; IBC Publications. London. Proceedings of a Conference on the Transportation of Hazardous Materials. (1988).

190. Urbanek, G. L.; Barker, E. J.; Development of Criteria to Designate Routes for Transporting Hazardous Materials—Final Report; Peat, Marwick, Mitchell and Company. Washington, DC. Available from NTIS, Springfield, VA, FHWA-RD-80-105. (1990).

191. Weigkricht, E.; Fedra, K.; Computer Support for Risk Assessment of Dangerous Goods Transportation; Transportation of Dangerous Goods: Assessing the Risks; Waterloo, Ontario. Institute for Risk Research. (1993).

3.2.2. Quantitative

192. Abkowitz, M.; Cheng, P. D. M.; Hazardous Materials Transport Risk Estimation Under Conditions of Limited Data Availability; Transportation Research Board. Washington, DC. Transportation Research Record N1245. (1989).

193. Advisory Committee on Dangerous Substances, Health and Safety Commission; Major Hazard Aspects of the Transport of Dangerous Goods; Report and Appendices; HMSO. London. (1991).

194. Anders, C.; Olsten, J.; State and Local Issues in Transportation of Hazardous Waste Materials: Towards a National Strategy; American Society of Civil Engineers. Proceedings of the National Conference on Hazardous Materials Transportation, St. Louis Missouri; GIS Risk Analysis of Hazardous Materials Transport. (1990).

195. Andrews, W.; Rhoads, R.; Franklin, A.; Cole, W.; Rau, R.; Hazardous Material Transportation Risks in the Puget Sound Region; NTIS Report No. PB82-138850. (1981).

196. Arthur D. Little, Inc; Estimation of the Frequency and Costs Associated with the Cleanup of Hazardous Materials Spills; Prepared for the U.S. Environmental Protection Agency, Contract No. 68-01-3857. (1978).

197. Brockhoff, L.; A Method for Assessment of Risks of Transport of Dangerous Goods. A Micro Approach to Macro Analysis; International Symposium on Loss Prevention in the Process Industries. (1992).

198. Cashwell, J. W.; Erickson, C. M.; Kern, E. A.; TRANSNET-Online Transportation Risk Analysis for Public Use; Technical Programs and Public Education. Waste Management '88: Waste Processing, Transportation, Storage and Disposal. (1988).

199. Fenstermacher, T.; Kazarians, M.; Boykin, R.; Gekler, W.; Kaplan, S.; A Manual for Performing Transportation Risk Assessment; Washington, DC. Prepared for the Chemical Manufacturers Association. (1987).

200. Geffen, CA; Assessment of the Risk of Transporting Propane by Truck and Train; Battelle Memorial Institute/Pacific Northwest Labs. Available from NTIS, Contract No: AC06-76RL01830. (1980).

201. Harris, N. C.; Roodbol, H. G.; Risk Assessment Model Applied to Transportation Problems; Institute of Chemical Engineers. Rugby, England. E.I. Monthly No.: EMI8604-023856, Institution of Chemical Engineers Symposium Series n 93., EFCE Publication Series n. 42. (1985).

202. Jones, G. P.; Barrow, R. W.; Stuckenbruck, L. C.; Holt, E. L.; Keller, R. P.; Risk Analysis in Hazardous Materials Transportation; University of Southern California, Los Angeles. Contract No: DOT-OS-20114 Available through NTIS. (1973).

203. Kornhauser, A. Pasternak, D., Sontag, M.; Comparing Risks of Transporting Chemicals by Highway and Rail: A Case Study; Transportation Research Record 1430. (1994).

204. Lautkaski, R.; Use of Risk Analysis in Enhancing the Safety of Transporting Hazardous Liquefied Gases; Transportation Research Board Publications Office. Washington, DC. State-of-the-Art Report N3. (1986).

205. Lautkaski, R.; Fieandt, J.; Risk Assessment of the Transportation of Hazardous Gases in Bulk; Third International Symposium on Loss Prevention and Safety Promotion in the Process Industries, Basel, Switzerland, September 15–19, 1980. (1980).

206. Leeming, D.; Saccomanno, F.; Use of Quantified Risk Assessment in Evaluating the Risks of Transporting Chlorine by Road and Rail; Transportation Research Record 1430. (1994).

207. Luatkaski, R. Mankamo, T.; Chlorine Transportation Risk Assessment; Technical Research Center of Finland, Helsinki. (1979).

208. Marsico, P.; Kazarians, M.; Fenstermacher, T.; TRACS: A User Guide—A Spreadsheet Template for Performing Transportation Risk Assessment; Washington, DC. Prepared for the Chemical Manufacturers Association. (1988).

209. Medhekar, S. et al.; Frequency Estimates for Transport-Related Hydroflouric and Sulfuric Acid Release Scenarios; *Process Safety Progress*. Vol. 12, No. 3. (1993).

210. Meslin, T.; Assessment and Management of Risk in the Transport of Dangerous Materials: The Case of Chlorine Transport in France; *Risk Analysis*. Vol. 1,No. 2, p. 137. (1981).

211. National Institute for Chemical Studies; Assessing Hazardous Materials Transportation in the Kanawha Valley; Charleston, WV. (1994).

212. Philipson, L. L.; Risk Analysis in Hazardous Materials Transportation: A Mechanism for Interfacing the Risk Analysis Model With the Hazardous Materials Incident Reporting Systems; University of Southern California Institute of Safety and Systems Management. Los Angeles. Contract No.: DOT-OS-20114, Available from NTIS. (1974).

213. Philipson, L. L.; Napadensky, H. S.; Maxey, M. N.; Hazardous Materials Transportation Risk Assessment; Transportation Research Board Publications Office. Washington, DC. Special Report N197, pp. 43–54. (1983).

214. Philipson, L. L.; Napadensky, H. S.; Methodologies of Hazardous Materials Transportation Risk Assessment; J.H. Wiggins Company. *Journal of Hazardous Materials*. Vol. 6, No. 4. (1982).

215. Purdy, G.; The Measurement of Risk from Transporting Dangerous Goods by Road and Rail; Transportation of Dangerous Goods: Assessing the Risks. Waterloo, Ontario. Institute for Risk Research. (1993).

216. Purdy, G.; Transportation Risk Analysis Models for the Carriage of Substances by Rail and Road in Great Britain; American Nuclear Society. Proceedings of PSA '89, Pittsburgh, Pennsylvania. (1989).

217. Purdy, G.; Campbell, H.; Grint, G.; Smith, L.; An Analysis of the Risks Arising from the Transport of Liquefied Gases in Great Britain; *Journal of Hazardous Materials*. Vol. 20, pp. 335–355. (1988).

218. Rhyne, W. R.; Evaluating Routing Alternatives for Transporting Hazardous Materials Using Simplified Risk Indicators and Complete Probabilistic Risk Analyses; Transportation Research Board Publications Office. Transportation Research Board Record N1264. (1990).

219. Rhyne, W. R.; Hazardous Materials Transportation Risk Analysis: Quantitative Approaches for Truck and Train; Van Nostrand Reinhold. New York, NY. (1994).

220. Riley, N.; Novel Risk Assessment Methodology for the Transport of Explosives; Transportation of Dangerous Goods: Assessing the Risks; Waterloo, Ontario. Institute for Risk Research. (1993).

221. Russell, E.; A Simple, Usable Empirical Risk Model for Small Communities: Transportation of Dangerous Goods: Assessing the Risks; Waterloo, Ontario. Institute for Risk Research. (1993).

222. Russell, E.; Using a Delphi Technique to Develop High-Risk Scenarios and Countermeasures; The International Consensus Conference on the Risks of Transporting Dangerous Goods, April 6–8, 1992. (1992).

223. Russell, S. B.; Kempe, T. F.; Methodology for Assessing the Public Risk From Transporting Hydrogen Sulphide: Transportation of Dangerous Goods: Assessing the Risks; Waterloo, Ontario. Institute for Risk Research. (1993).

224. Saccamanno, F. F.; Stewart, A.; Shortreed, J. H.; Uncertainty in the Estimation of Risks for the Transport of Hazardous Materials; Transportation Research Record 1245. HAZMAT Transport 91, A National Conference on the Transportation of Hazardous Materials and Wastes, The Transportation Center, Northwestern University, Evanston, IL., June 17–19, 1991. (1991).

225. Saccomanno, F.; Cassidy, K.; QRA and Decision Making in the Transportation of Dangerous Goods; Transportation Research Record 1430. (1994).

226. Saccomanno, F. F.; Fault Tree Analysis for Hazardous Material Routing; Roads and Transportation Association of Canada. Ottawa, Ontario, Canada. (1985).

227. Saccomanno, F. F.; Shortreed, J. H.; Mehta, R.; Fatality Risk Curves for Transporting Chlorine and LPG by Truck and Rail; Washington, DC. Transportation Research Board Annual Meeting. (1990).

228. Saccomanno, F. F.; Shortreed, J. H.; Societal-Individual Risks for Hazmat Transport; Assessing Community Safety for Hazardous Materials Transport. Proceedings of the National Conference on Hazardous Materials Transportation, May 14–16, 1990, St. Louis.

229. Saccomano, F.; Cassidy, K.; Transportation of Dangerous Goods: Assessing the Risks; Institute for Risk Research. Waterloo, Ontario. (1993).

230. Saccomano, F. F.; Stewart, A.; Fault Tree Analysis of Bulk Tanker Containment Systems for Transporting Pressure Liquified Gases; August 10–13, 1987. International Symposium on Major Hazards in the Transport and Storage of Pressure Liquified Gases, University of New Brunswick. (1987).

231. Simmons, J. A.; Risk Assessment Method for Volatile Toxic and Flammable Materials; United States Coast Guard. Washington, DC. Available from NTIS, Springfield, VA. (1975).

232. Simmons, J. A.; Risk Assessment of Storage and Transport of Liquefied Natural Gas and LP-Gas; Final Report for Contract 68.01.2695 for U.S. Environmental Protection Agency. (1974).

233. Van Aerde, M.; Shortreed, J.; Stewart, A. M.; Mathews, M.; Assessing The Risks Associated with the Transport of Dangerous Goods by Truck and Rail Using the RISKMOD Model; *Canadian Journal of Civil Engineering.* Ottawa, Ontario, Canada. Vol. 16, No. 3, pp. 326–334. (1989).

234. Van Aerde, M.; Stewart A.; Saccomanno, F.; Estimating the Impacts of LPG Spills During Transportation Accidents; *Journal of Hazardous Materials.* Vol. 20, pp. 375–392. (1988).

235. Weissling, D.; Risk Assessment Method for Comparing Hazardous Material Transport Options; Third Annual Environmental and Safety Conference for the Oil, Gas and Petrochemical Industries. (1992).

236. Yee, E.; Site Specific Transportation Risk Analysis for the Manitoba Hazardous Waste Management Corporation: Transportation of Dangerous Goods: Assessing the Risks; Waterloo, Ontario. Institute for Risk Research. (1993).

3.3. Other

237. Abkowitz, M.; List, G.; Radwan, A.; Critical Issues in Safe Transport of Hazardous Materials; *Journal of Transportation Engineering*. Vol. 115, No. 6. (1989).

238. Allen, J.; Glickman, T.; Routes, Risks and Regulations. Recent Developments in Hazardous Materials Transportation; Winter Annual Meeting of the American Society of Mechanical Engineers. (1989).

239. Applied Technology Corporation; LPG Land Transportation and Storage Safety; Final Report, Report No. U.S. DOE/EV/06020-T9. (1981).

240. Atallah, S. Borrows, K.; Response to Hazmat Land Transportation Accidents; Safety in Transport and Warehousing—14th International Symposium, Heidleberg, Germany. (1991).

241. Chemical Manufacturers Association; Responsible Care: Distribution Code of Management Practices; Chemical Manufacturers Association. Washington, DC. (1990).

242. Lindell, M. Perry, R.; Evaluation Criteria for Emergency Response Plans in Radiological Transportation; *Journal of Hazardous Materials*. Vol. 3, p. 335. (1980).

243. Office of Technology Assessment; Transportation of Hazardous Materials: State and Local Activities; U.S. Government Printing Office. Washington D.C. OTA-SET-301. (1986).

244. R.F. Webb Corporation, Ltd; Assessment of the Safety of Transportation, Distribution and Storage of Methanol Fuels—Summary Report; Prepared for Transport Canada, Ottawa, Ontario. TP 9519.

245. List, G.; Mirchandani, P.; Turnquist, M. A.; Zografos, K. G.; Modelling and Analysis for Hazardous Materials Transportation: Risk Analysis, Routing/Scheduling and Facility Location; *Transportation Science*. Vol. 25, No. 2, pp. 100–114. (1991).

246. de Malherbe, R.; Mamalis, A. G.; de Malherbe, M. C.; Giddings, M. L.; Risk Analysis in Transportation Systems; Engineering Societies Library. New York, NY. (1981).

4.0. Pipeline

4.1. Data

247. Andersen, T.; Misund, A.; Pipeline Reliability: An Investigation of Pipeline Failure Characteristics and Analysis of Pipeline Failure Rates for Submarine and Cross-Country Pipelines; *Journal of Petroleum Technology*. Vol. 35, No. 4. (1983).

248. Bartol, J.; Wells, C.; Wei, R.; Environmentally Induced Cracking of Natural Gas and Liquid Pipelines; NTIS No. PB-282 923/2 and PB-282 924/0, Vol. 1 & 2. (1977).

249. CONCAWE; Performance of Oil Industry Cross-Country Pipelines in Western Europe: Statistical Summary of Reported Spillages—1991; NTIS #PB93 187169. (1991).

250. de la Mare, R. F.; Pipeline Reliability; University of Bradford, VERITAS. 80-0572. (1980).

251. Energy Resources Conservation Board; An Analysis of Pipeline Performance in Alberta, ERCB; Calgary, Alberta. (1983).

252. European Gas Pipeline Incident Data Group; Gas Pipeline Incidents: A Report of the European Gas Pipeline Incident Data Group; *Pipes and Pipelines International*. 1970–1988.

253. Federal Power Comission; Safety of Interstate Natural Gas Pipelines; 1950–1965.

254. Jones, D. J.; Eiber, R. J.; An Analysis of Reportable Incidents for Natural Gas Transmission and Gathering Lines—June 1984 through 1990; American Gas Association. NG-18, Report No. 200. (1992).

255. Jones, D. J., et al.; An Analysis of Reportable Incidents for Natural Gas Transmission and Gathering Lines—1970 Through June 1984; American Gas Association. NG-18, Report No. 158. (1986).

256. Mannan, M.; Pfenning, D.; West, H.; Causes of Pipeline Failure and Prevention and Mitigation Procedures; Pipeline Risk Assessment, Rehabilitation and Repair Conference. May 18-21, 1992. Houston, TX. (1992).

257. Townsend, N.; Fearnehough, G.; Control of Risk from U.K. Gas Transmission Pipelines; British Gas plc Research and Development Division. Engineering Research Station ERS E 531. (1986).

258. UKAEA; Performance of Pipework in the British Sector of the North Sea; United Kingdom Atomic Energy Authority, Safety and Reliability Directorate, Report No. NCSR/GR/71; January 1977—January 1983. (1987).

259. U.S. Department of Transportation, Research and Special Programs Administration; Annual Report on Pipeline Safety—Calendar Years 1981 through 1990; (various).

260. U.S. Department of Transportation, Research and Special Programs Administration; National Transportation Statistics—Annual Report; (1990 et al.).

261. U.S. Department of Transportation, Research and Special Programs Administration; Transportation Safety Information Report 1988 Annual Summary; (1989).

262. Vieth, P. H.; Brann, C. M.; PC-Based Databases for Use in Evaluating Pipeline Risk, Rehabilitation and Environmental and Federal Safety Compliance; Pipeline Risk Assessment, Rehabilitation and Repair Conference. May 18–21, 1992. Houston, TX. (1992).

4.2. Methodology

4.2.1. Qualitative

263. Hill, R. T.; Bendixen, L. M.; Pipeline HAZOP Studies and Quantified Risk Assessment; Pipeline Risk Assessment, Rehabilitation and Repair Conference. May 18–21, 1992. Houston, TX. (1992).

264. Muhlbauer, W. K.; Pipeline Risk Management—A Pro-Active Approach; Pipeline Risk Assessment, Rehabilitation and Repair Conference. May 18–21, 1992. Houston, TX. (1992).

265. Muhlbauer, W. K.; Pipeline Risk Management Manual; Gulf Publishing Company. Houston, TX. ISBN 0-88415-035-6. (1992).

266. Muhlbauer, W. K.; RIPS-A Pipeline Safety Evaluation System; Proceedings of the 1991 Pipeline Risk Assessment, Rehabilitation and Repair Conference; Gulf Publishing Company. Houston, TX. (1991).

267. Trefanenko, B.; Coutts, R.; Ronsky, N. D.; McManus, M.; Risk Assessment: an Integrity Management Tool; Pipeline Risk Assessment, Rehabilitation and Repair Conference. May 18–21, 1992, Houston, TX. (1992).

268. Werner, S.; Taylor, C.; Seismic Risk Considerations for Transportation Systems; 1990 Annual Civil Engineering Convention and Exposition Conference. San Francisco, CA. November 5, 1990. (1990).

269. Whittaker, J. D., et al.; Risk-Based Zoning for Toxic Gas Pipelines; *Risk Analysis.* Vol. 2, No. 3, p. 163. (1982).

4.2.2. Quantitative

270. California State Fire Marshall; Hazardous Liquid Pipeline Risk Assessment; Sacramento, CA. (1993).

271. Cornwell, J. B.; Martinsen, W. E.; Quantitative Risk Analysis of a Sour Gas Pipeline; Pipeline Risk Assessment, Rehabilitation and Repair Conference. May 18–21, 1992. Houston, TX. (1992).

272. Grollier Baron, R.; Lombre, R.; Rochina, V.; Hazard Analysis for a Liquid Propane Pipeline, Recent Advances in Hazardous Materials Transportation Research, An International Exchange; State-of-the-Art Report 3, Transportation Research Board. Washington, DC. pp. 141–146. (1986).

273. Mashaly, E.; Datta, T.; Seismic Risk Analysis of Buried Pipelines; *Journal of Transportation Engineering*. Vol. 115, No. 5, pp. 232–252. (1989).

274. U.K. Health & Safety Executive; A Reappraisal of the HSE Safety Evaluation of the Proposed St. Fergus to Moss Morran NGL Pipeline; U.K. Health & Safety Executive. London. (1980).

275. U.K. Health & Safety Executive; A Safety Evaluation of the Proposed St. Fergus To Moss Morran Natural Gas Liquids and St. Fergus to Boddam Gas Pipelines; U.K. Health & Safety Executive. London. (1978).

4.3. Other

276. ANSI/ASME, ANSI Standard B31.8; Gas Transmission and Distribution Piping Systems; American Society of Mechanical Engineers. New York. 1986 Edition.

277. Babin, J. Hagood, D.; Concentric Flow Line Moves Sour Crude Through City; *Oil and Gas Journal*. June 22, 1992. pp. 72–75. (1992).

278. British Standard 8010; Code of Practice for Pipelines; British Standards Institute. London. (1991).

279. Eiber, R. J.; Jones, D. J.; Kramer, G. S.; Outside Force Causes Most Natural Gas Pipeline Failures; *Oil and Gas Journal*. pp. 52–57.

280. Institution of Gas Engineers; Steel Pipelines for High Pressure Gas Transmission; IGE/TD/1. London. Edition 2. (1984).

281. Martinsen, W. E.; Cornwell, J. B.; Use and Misuse of Historical Pipeline Failure Rate Data; Gulf Publishing. Houston, TX. Proceedings of the 1991 Pipeline Risk Assessment, Rehabilitation and Repair Conference. (1991).

282. National Transportation Safety Board; Pipeline Accident Report, Four Corners Pipe Line Company Rupture and Fire, Long Beach, California, December 1, 1980; Washington, DC. (1981).

283. National Transportation Safety Board; Railroad Accident Report, Derailment of Southern Pacific Transportation Company Freight Train on May 12, 1989 and Subsequent Rupture of Calnev Petroleum Pipeline on May 25, 1989—San Bernardino, California; Washington, DC. (1990).

284. National Transportation Safety Board; Special Study of Effects of Delay in Shutting Down Failed Pipeline Systems and Methods of Providing Rapid Shutdown; Washington, DC. (1970).

285. Rusin, M.; Pipeline Transport of Highly Volatile Liquids: A Cost–Benefit Analysis of Proposed Safety Regulation of Valve Spacing and Operation.; American Petroleum Institute. Washington, D.C. (1979).

286. Rusin, M.; Savvides-Gellerson, E.; The Safety of Interstate Liquid Pipelines: An Evaluation of Present Levels and Proposals for Change, Research Study #040; American Petroleum Institute. Washington, D.C. (1987).

287. Tiratsoo, J. N. H.; Pipelines and the Environment; *Pipes and Pipelines International*. Bucks, United Kingdom. (1984).

288. Transportation Research Board, National Research Council; Special Report 219, Pipelines and Public Safety, Damage Prevention, Land Use, and Emergency Preparedness; Washington, DC. (1988).

289. U.S. Department of Transportation, Research and Special Programs Administration; Emergency Flow Restricting Devices Study; Washington, DC. (1991).

290. U.S. Department of Transportation; An Examination of the Feasibility of Regulating Excavators; Washington, DC. (1990).

291. U.S. Department of Transportation, Office of Pipeline Safety; Rapid Shutdown of Failed Pipeline Systems and Limiting of Pressure to Prevent Pipeline Failure Due to Overpressure; Washington, DC. (1974).

292. U.S. Department of Transportation, Office of Pipeline Safety; Transportation of Highly Volatile, Toxic, or Corrosive Liquids by Pipeline; Washington, DC. (1976).

5.0. Rail

5.1. Data

293. Association of American Railroads, Economics and Finance Department; Chicago, IL. Statistics of Regional and Local Railroads. (1988).

294. Association of American Railroads; Analysis of Class I Railroads; Chicago, IL. Economics and Finance Department (Annual).

295. Association of American Railroads; Annual Report on Hazardous Materials Transported by Rail; Chicago, IL. Safety and Hazardous Materials Division, Bureau of Explosives (Annual).

296. Association of American Railroads; Report of Railroad Tank Car Leaks of Hazardous Material by Commodity by Source of Leak (Non-Accident Leaks); Chicago, IL. Safety and Hazardous Materials Division, Bureau of Explosives (Annual).

297. Beier, F.; Church, R.; Zebe, P.; Frev, J.; Flows of Selected Hazardous Materials by Rail (Final Report for Sept 87—Apr. 91); NTIS No. PB91-216135/XAB. (1991).

298. Davies, P. Lees, F.; Accident Speed of Freight Trains; *Journal of Hazardous Materials*. Vol. 28, p. 367. (1991).

299. Davis, C. S.; Fowler, G. F.; Railroad Tank Car Safety Assessment; Association of American Railroads. Chicago, IL. Failure Analysis Associates for the RPI–AAR Tank Car Safety Project, Report, RA-124-58 (AAR R-751). (1990).

300. Federal Railroad Administration; Accident/Incident Bulletin; Washington, DC. 1975–present, FRA Office of Safety. (various).

301. General American Transportation Corp.; Tank Car Manual; Fifth Edition. (1985).

302. Harvey, A.; Conlon, P.; Glickman, T.; Statistical Trends in Railroad Hazardous Materials Transporation Safety—1978–1986; Washington, DC. Prepared for the Association of American Railroads, Publication R-640. (1987).

303. Hubert, P.; Pages, P.; International Discrepancies in the Estimation of the Expected Accident and Tank Failure Rates in Transport Studies: Transportation of Dangerous Goods: Assessing the Risks; Waterloo, Ontario. Institute for Risk Research. (1993).

304. Kalaycioglu, S.; Tajaddini, A.; Locating Vertical Track Irregularities Which Cause Excessive Vehicle Loads; Association of American Railroads. Chicago, IL. Research and Test Department Report, R694. (1988).

305. Lishaa, M. A., et al.; Puncture Resistance of Scale Model Aluminum Tank Car Heads; U.S. Department of Transportation. Washington, DC. Federal Railroad Administration Report, DOT/FRA/ORD-87/06. (1987).

306. McNeil, S.; OH, S.; A Note on the Influence of Rail Defects on the Risks Associated With Shipping Hazardous Materials by Rail; *Risk Analysis*. Vol. 11, No. 2. (1991).

307. Mond, C. R.; Boghani, A. B.; Characteristics of 103, 104 and 111 Tank Cars; Washington, DC. Arthur D. Little Report to Federal Railroad Administration. (1981).

308. National Transportation Safety Board; Railroad Accident Reports; Washington, DC.

309. Nayak, P. R.; An Analysis of the Probability of Derailment with a Train Consist; Arthur D. Little, Inc. for the Transportation Systems Center, U.S. Department of Transporation. (1987).

310. Nayak, P. R.; Rosenfield, D. B.; Hagopian, J. H.; Event Probabilities and Impact Zones for Hazardous Materials Accidents on Railroads; Federal Railroad Administration, Office of Research and Development. Cambridge, MA. NTIS # PB85-14985/XAB. DOT/FRA/ORD-83/20; DOT-TSC-FRA-82-5. (1983).

311. Phillips, E. A.; Analysis of Aluminum Cars Damaged in Accidents 1965 Through 1986; Association of American Railroads. Chicago, IL. RPI–AAR Tank Car Safety Project Report, RA-20-1-57 (AAR R-749). (1990).

312. Phillips, E. A.; Pellini, W. S.; Phase 03 Report on Behavior of Pressure Tank Car Steels in Accidents; Association of American Railroads. Chicago, IL. RPI–AAR Tank Car Safety Project Report, RA-03-6-48 (AAR R-553). (1983).

313. Phillips, E. A.; Role, H.; Analysis of Tank Cars Damaged in Accidents 1965 Through 1986, Documentation Report; Association of American Railroads. Chicago, IL. RPI–AAR Tank Car Safety Project Report, RA-02-6-55 (AAR R-709). (1990).

314. Phillips, E. A.; Role, H.; Effectiveness of Shelf Couplers, Head Shields and Thermal Shields on DOT 112 (114) and 105 Tank Cars; Association of American Railroads. Chicago, IL. RPI–AAR Tank Car Safety Project Report, RA-02-5-51 (AAR R-610). (1985).

315. Phillips, E. A.; Role, H.; A Review of Pressure Car Shell Puncture Vulnerability; Association of American Railroads. Chicago, IL. RPI–AAR Tank Car Safety Project Report, RA-09-6-52 (AAR R-628). (1987).

316. Phillips, E. A.; Role, H.; Olson, L. L.; Phase 02 Report on Analysis of Non-Pressure Tank Car Behavior in Accidents; Association of American Railroads. Chicago, IL. RPI–AAR Tank Car Safety Project Report, RA-02-4-47 (AAR R-541). (1983).

317. Saricks, C. L.; Janssen, I.; Rail Transportation Risk and Accident Severity: A Statistical Analysis of Variables in FRA's Accident/Incident Data Base; Federal Railroad Administration (U.S. DOT). Washington, DC. NTIS # DE91011840/XAB. Annual Meeting and Exhibition of the Air and Waste Management Association (AWMA) (84th), Vancouver, Canada, Jun 16—21, 1991. (1991).

318. Saricks, C. L.; Lee, T. S.; Significant Factors in Rail Freight Accidents: A Statistical Analysis of Predictive and Severity Indices in the FRA Accident/Incident Database; HAZMAT Transport 91, A National Conference on the Transportation of Hazardous Materials and Wastes, The Transportation Center, Northwestern University, Evanston, IL., June 17–19, 1991. (1991).

319. Thompson, R.; Zamejc, E.; Ahlbeck, D.; Hazardous Materials Car Placement in a Train Consist; NTIS Nos. PB92-206960/SAB and PB 92-206952/XAB,Vol. 1 & 2. (1992).

320. U.S. Department of Transportation; Accident/Incident Bulletin; Washington, DC. Federal Railroad Administration (Annual).

321. U.S. Federal Railroad Administration; Railroad Accident/Incident Database; Washington, DC. 1975-present, FRA Office of Safety. (1992).

322. Vanderwater, R.; Case History of an Ethylene Oxide Tank Car Explosion; *Chemical Engineering Progress*. p. 16. (1989).

5.2. Methodology

5.2.1. Qualitative

323. Glickman, T. S.; Rerouting Railroad Shipments of Hazardous Materials to Avoid Populated Areas; *Accident Analysis and Prevention*. Vol. 15, No. 5 pp. 329–335. (1983).

324. Glickman, T. S.; Rosenfield, D. B.; Risks of Catastrophic Derailments Involving the Release of Hazardous Materials; *Management Science*. Vol. 30, No. 4, pp. 503–511. (1984).

325. Hoftijzer, G.; Dangerous Freight Transport; Risks, Recommendations and Legislations; TNO-Project, Vol.10, No. 9. (1982).

326. Oderwald, W. H.; Sontag, M. A.; Evaluation of Hazardous Materials Transportation by Rail, Assessing Community Safety for Hazardous Materials Transport; Proceedings of the National Conference on Hazardous Materials Transportation, St. Louis, Missouri, May 14–16, 1990. (1990).

327. Saccomanno, F. F.; El-Hage, S.; Minimizing Derailments of Railcars Carrying Dangerous Commodities Through Effective Marshalling Strategies.; Transportation Research Board. Transportation Research Board Record 1245. (1990).

5.2.2. Quantitative

328. Alp, E. Oliverio, M., Portelli, R., Napier, D.; Risk Assessment in Rail Transportation of Dangerous Goods; Fifth Technical Seminar on Chemical Spills, February 1988. Montreal, Quebec. (1988).

329. Alp, E.; Portelli, R. V.; Crocker, W. P.; Rail Transport Risk in the Greater Toronto Area; Transportation of Dangerous Goods: Assessing the Risks; Waterloo, Ontario. Institute of Risk Research. (1993).

330. Andrews, W. B.; Assessment of the Risk of Transporting Liquid Chlorine by Rail.; Battelle Memorial Institute/Pacific Northwest Labs. Washington, DC. Contract No. AC06-76RL1830. (1980).

331. Barkan, C.; Data Requirements for the Development of a Quantitative Risk Assessment Model for Rail Transportation of Hazardous Materials; HAZMAT Transport 91, A National Conference on the Transportation of Hazardous Materials and Wastes, The Transportation Center, Northwestern University, Evanston, IL, June 17–19, 1991. (1991).

332. Barkan, C.; Glickman, T.; Harvey, A.; Benefit–Cost Evaluation of Using Different Specification Tank cars to Reduce the Risk of Transporting Environmentally Sensitive Chemicals; Transportation Research Record N 1313. (1991).

333. Barkan, C. P. L.; Glickman, T. S.; Harvey, Aviva E.; Benefit–Cost Evaluation of Using Different Specification Tank Cars to Reduce the Risk of Transporting Environmentally Sensitive Chemicals in Tank Cars; Transportation Research Record No.1313, Transportation Research Board. Washington, D.C. (1991).

334. Clark, D. G.; Methodology for the Comparison of the Risks of Transporting Toxic Chemicals by Rail; Air and Waste Management Assoc. Pittsburgh, PA. A&WMA Annual Meeting, E.I. Conference No.: 13687, Anaheim, CA. (1989).

335. Holing, D.; Case History: New Modeling Technology Keeps "Track" of Hazmat Rail Risk; *Hazardous Materials and Waste Management*. Vol. 5, No. 4, pp. 23–24. (1987).

336. Lautkaski, R. Mankamo, T.; A Risk Assessment of the Rail Transportation of Liquid Chlorine; Heidelberg, West Germany. Proceedings of the Second International Symposium on Loss Prevention and Safety Promotion in the Processing Industry. September 6–9, 1977. (1977).

337. Raj, P. K.; Glickman, T. S.; Generating Hazardous Material Risk Profiles on Railroad Routes; Transportation Research Board Publications Office. Washington, DC. State-of-the-Art Report N3. (1986).

338. Raj, P. K.; A Risk Assessment Study on the Transportation of Hazardous Materials Over the U.S. Railroads; NTIS #BP89-138929. (1988).

339. Raj, P. K.; Turner, C. K.; Hazardous Material Transportation in Tank Cars: Analysis of Risks; Federal Railroad Administration. Washington, DC. DOT/FRA/ORD-92/34. (1993).

340. Transportation Research Board; Ensuring Railroad Tank Car Safety; TRB Special Report 243. Washington, DC. (1994).

5.3. Other

341. Armstrong, J. H.; The Railroad—What It Is, What It Does: The Introduction to Railroading; Simmons-Boardman Publishing. Omaha, NE. (1994).

342. Birk, A.; Dibble, D.; A Computer Model of a Rail Tank Car Carrying Hazardous Pressurized Liquids Exposed to External Fire Impingement; Hazardous Materials Spills Conference Proceedings, p. 334, May 5–8, 1986. (1986).

343. Clavin, T.; Poisoned Cargo: How the Nation's Railroads Put Millions at Risk; *Family Circle*. February, pp. 100–103. (1994).

6.0. Related Topics

344. Brockhoff, L. H.; Styhr-Petersen, H. J.; A Simple Consequence Model Based on a Facility Index Approach; Transportation of Dangerous Goods: Assessing the Risks; Waterloo, Ontario. Institute for Risk Research. (1993).

345. Center for Chemical Process Safety; Guidelines for Chemical Process Quantitative Risk Analysis; American Institute of Chemical Engineers. New York, NY. (1989).

346. Center for Chemical Process Safety; Guidelines for Evaluating the Consequences of Fires and Explosions from Vapor Clouds and Bleves; American Institute of Chemical Engineers. New York, NY. (1992).

347. Center for Chemical Process Safety; Guidelines for Hazard Evaluation Procedures, Second Edition with Worked Examples; American Institute of Chemical Engineers. New York. (1992).

348. Center for Chemical Process Safety; Guidelines for Process Equipment Reliability Data, with Data Tables; American Institute of Chemical Engineers. New York. (1989).

349. Center for Chemical Process Safety; Guidelines for Use of Vapor Cloud Disperson Models; American Institute of Chemical Engineers. New York, NY. (1987).

350. Center for Chemical Process Safety; Proceedings from the International Conference / Workshop on Modeling & Mitigating the Consequences of Accidental Releases of Hazardous Materials; American Institute of Chemical Engineers. New York, NY. (1991).

351. Cornwell, J. B.; Martinsen, W. E.; Consequence Modeling Issues Involved in the Quantitative Risk Analysis of a Sour Gas Gathering System; (1993).

352. Fay, J.; Physical Processes in the Spread of Oil on a Water Surface; American Petroleum Insitutue. Proceedings Joint Conference on Prevention and Control of Oil Spills. (1971).

353. Haugen, E.; Probabilistic Approaches to Design; John Wiley & Sons. New York. (1968).

354. Havens, J.; A Dispersion Model for Elevated Dense Gas Jet Chemical Releases; National Technical Information Service. Springfield, VA. NTIS No. PB 88-202379. (1988).

355. IChemE; Nomenclature of Hazard and Risk Assessment in the Process Industries; Institution of Chemical Engineers. London, UK. ISBN 0-85295-184-1. (1985).

356. Melhem, G.; Croce, P.; Advanced Consequence Modelling: Emission, Dispersion, Fires and Explosions; Van Nostrand Reinhold. New York (in preparation).

357. Ooms, G.; A New Method for the Calculation of the Plume Path of Gases Emitted by a Stack; *Atmosphere Environment.* Vol. 6. (1972).

358. Ooms, G. et al.; The Plume Path of Vent Gases Heavier Than Air; The First International Symposium on Loss Prevention and Safety Promotion in the Process Industries. The Hague/Delft. May 1974. ISBN 0-444-41230-1. (1974).

359. Paustenbach, D.; The Risk Assessment of Environmental and Human Health Hazards: A Textbook of Case Studies; John Wiley & Sons. New York. (1989).

360. Slater, C.; Hall, G.; 1992 County and City Extra; Annual Metro, City and County Data Book; 1990 U.S. Census based population data.

361. Southwest Research Institute; Analytical and Experimental Study to Improve Computer Models for Mixing and Dilution of Soluble Hazardous Chemicals; U.S. Coast Guard. Report No. CG-D-1-83. (1982).

362. Southwest Research Institute; Revision and Experimental Verification of the Hazard Assessment Computer System Models for Spreading, Movement, Dissolution, and Dissipation of Insoluble Chemicals Spilled Onto Water; U.S. Coast Guard. Report No. CG-D-35-83. (1983).

363. Thibodeaux, L.; Spill of Soluble, High Density, Immiscible Chemicals on Water; U.S. Department of Transportation. Report No. DOT-OS-70040. (1980).

364. U.K. Health & Safety Executive; Canvey: A Second Report—A Review of Potential Hazards from Operations in the Canvey Island/Thurrock Area Three Years After Publication of the Canvey Report; HMSO. London. ISBN 0-11-883459-2. (1981).

365. U.K. Health & Safety Executive; Quantified Risk Assessment—Its Input to Decision Making; U.K. Health & Safety Executive. London. HMSO, ISBN 0-11-885499 2. (1989).

366. U.S. Department of Commerce; Metropolitan Statistical Areas—Supplementary Report—1980 Census of Population; PC80-SI-18. (1984).

367. U.S. Department of Commerce; Statistical Abstracts of the United States, 112th Edition; Washington, DC. ISBN 0-16-038080-4. (1992).

368. U.S. Department of Commerce; Summary Tape File 1 on CD-Rom Technical Documentation, 1990 Census of Population and Housing; Washington, DC. (1991).

369. U.S. Department of Commerce; TIGER/Line Census Files, 1990; Washington, DC. (1991).

370. U.S. Environmental Protection Agency; Ecological Assessment of Hazardous Waste Sites: A Field and Laboratory Reference; Washington, DC. (1989).

371. U.S. Environmental Protection Agency—Office of Emergency and Remedial Response; Risk Assessment Guidance for Superfund Volume 1 Human Health Evaluation Manual; Washington, DC. Volume 1, Part A. (1989).

Glossary

Absolute Application (of TRA): The application of TRA in which the results of the analysis are compared against predetermined risk targets.

Acute Risk: A risk arising from a short-term event such as a release causing a fire, explosion or short duration toxic exposure.

Aerosol Fraction: The fraction of liquid phase which, when flashed to the atmosphere, remains suspended as an aerosol.

Allisions: A collision with a moored vessel, as opposed to a collision between two moving vessels or a ramming, which is a collision with a fixed object.

Analysis Database: A database that contains all input data for a TRA, including environmental data (e.g., land use and topography, population and demography, meteorological data).

Atmospheric Dispersion: The low momentum mixing of a gas or vapor with air. The mixing is the result of turbulent energy exchange, which is a function of wind (mechanical eddy formation) and atmospheric temperature profile (thermal eddy formation).

Atmospheric Stability: A measure of the degree of atmospheric turbulence commonly defined in terms of the vertical temperature gradient. In neutral stability, the gradient is equivalent to the Adiabatic Lapse Rate (ALR). Stable atmospheric conditions refer to a gradient less than the ALR (ultimately to a temperature inversion), and unstable conditions to a gradient greater than the ALR. These are designated by the letters A through G.

Average Individual Risk: Three average individual risks are used in this book:

1. *Average Individual Risk (exposed population)* is the individual risk averaged over the population which is exposed to a particular hazard.
2. *Average Individual Risk (total population)* is the individual risk averaged over a predetermined population, without regard to whether or not all people in that population are actually exposed to the risk.
3. *Average Individual Risk (Exposed hours/worked hours).* The individual risk for an activity may be calculated for the duration of the activity or may be averaged over the working day.

Average Rate of Death (ROD): The average number of fatalities that might be expected per unit time from all possible incidents.

Averaging Time: The length of time in atmospheric dispersion testing over which concentration data are averaged to produce the concentration-time series.

Basic Event: A fault tree event that is sufficiently basic that no further development is necessary (e.g., equipment item failure, human failure or external event). *See also* Undeveloped Event.

BLEVE: A *Boiling Liquid Expanding Vapor Explosion*, which occurs from the sudden release of a large mass of pressurized liquid to the atmosphere. A primary cause is an external flame impinging on the shell of a vessel above the liquid level, weakening the shell and resulting in sudden rupture.

Bounding Group (of Incidents): A small number of incidents selected to bracket the spectrum of possible incidents, which may include those catastrophic incidents sometimes referred to as the Worst Credible Incident and Worst Possible Incident.

Car-miles: A measurement that reflects the number of cars in a train and the distance traveled. For instance, if a train of 30 cars went 10 miles, there would be only 10 train-miles but 300 car-miles. Loaded versus empty car miles may also be distinguished.

Casualty: An injury or fatality. In the case of marine risk, this term is also used to describe an accident.

Catastrophic Incident: An incident with an outcome effect zone that extends into the surrounding community.

Cell (in study cube): One of the 27 different categories of TRA, defined by a unique value for each axis of the study cube.

Chronic Risk: A risk arising from a long-term persistent exposure usually seen in occupational settings (e.g., long-term workplace exposure to low levels of an OSHA regulated chemical).

Classification Yard: A facility where incoming trains can be broken down into individual cars or blocks of cars, assembled into new trains, and inspected prior to dispatch toward their destination.

Collision: An incident involving two or more moving vessels that collide while meeting, overtaking, or crossing.

Commodity Code: A designation (usually numerical) used to identify a specific material in transportation.

Component Technique: One member of the roster of methods that collectively form the complete TRA procedure.

Condensed Phase Explosion: An explosion that occurs when the material is present in the form of a liquid or solid.

Conditional Probability: Probability of occurrence given that a precursor event has occurred.

Confidence Interval: A range of values of a variable with a specific probability that the value of the variable lies within the range.

Confidence Limits: The upper and lower endpoints of a confidence interval.

Confined Explosions: An explosion of a fuel–oxidant mixture inside a closed system (e.g., vessel).

Consequence Analysis: The analysis of the expected effects of incident outcome cases independent of frequency or probability.

Consequences: A measure of the expected effects of an incident outcome case.

Consist: A string of railcars where the makeup or composition by classes, types or grades and the arrangement (or order) of the cars is considered.

Continuous Release: Emissions that are long in duration compared with the travel time (time for the cloud to reach location of interest) or averaging or sampling time.

Deflagration: The chemical reaction of a substance in which the reaction front advances into the unreacted substance at less than sonic velocity. Where a blast wave is produced that has the potential to cause damage, the term explosive deflagration may be used.

Delphi Method: A polling of experts utilizing the following procedure:

1. Select a group of experts (usually 3 or more).
2. Solicit, in isolation, their independent estimates on the value of a particular parameter and the reasons for the choice.
3. Provide initial results to all experts and all revisions to initial estimates.
4. Use the average of the final estimates as the best estimate of the parameter. Use the standard deviation of the estimates as a measure of uncertainty.

The procedure is interactive, with feedback among interactions.

Dense Gas: A gas with density exceeding that of air at ambient temperature.

Depth of Study: A measure of the level of review, degree of complexity and extent of detail involved in a TRA.

Detonation: A release of energy caused by the extremely rapid chemical reaction of a substance in which the reaction front advances into the unreacted substance at greater than sonic velocity.

Directional Incident Outcome: An incident outcome whose consequences produce an effect zone determined by a given wind direction.

Domino Effects: The triggering of secondary events, such as toxic releases, by a primary event, such as an explosion, such that the result is an increase in consequences or area of an effect zone. Generally only considered when a significant escalation of the original incident results. (Sometimes known as a knock on event).

Effect Models: Models that predict effects of incident outcomes usually with respect to human injury, fatality or property damage.

Effect Zone: For an incident that produces an incident outcome to toxic release, the area over which the airborne concentration equals or exceeds some level of concern. For a flammable release, the area over which a particular incident outcome case produces an effect based on a specified criterion. For a loss of containment incident producing thermal effects, the area over which a particular incident outcome case produces an effect based on a specified radiative heat stress limit.

Entrainment: The suspension of liquid as an aerosol in the atmospheric dispersion of a two-phase release or the aspiration of air into a jet discharge.

Episodic Release: A release of limited duration, usually associated with an accident.

Equipment Reliability: The probability that, when operating under stated environment conditions, process equipment will perform its intended function adequately for a specified exposure period.

Equivalent Social Cost: A modification of the Average Rate of Death that takes into account society's aversion of large-consequence incidents.

ERW: *Electric Resistance/Induction Welded*—a manufacturing process used primarily on line pipe of mid-range diameters.

Event: An occurrence involving equipment performance or human action, or an occurrence external to the system that causes system upset. In this book, an event is associated with an incident either as the cause or a contributing cause of the incident or as a response to the initiating event.

Event Sequence: A specific unplanned sequence of events composed of initiating events and intermediate events that may lead to an incident.

Event Tree (Analysis): A graphical logic model that identifies and quantifies possible outcomes following an initiating event.

Expansive List (of Incidents): The list of incidents from which the study group is selected. An Expansive List is derived from the initial list by (1) removing incidents that are too small to be of concern; (2) combining redundant or very similar incidents; and (3) grouping similar incidents into a single equivalent incident.

Expected Number of Failures (ENF): The average number of occurrences of the fault event during a specified time interval.

Explosions: A release of energy that causes a pressure discontinuity or blast wave.

Exposure Data: Information pertaining to the amount of exposure that is used in obtaining accident rates. Examples of this type of data are car- or train-miles.

Exposure Period: Expressed in terms of a continuous variable, such as the time equipment is operating, or as a function of a discrete variable, such as the number of demands (also called cycles in some tests) imposed on a piece of equipment.

External Event: Event caused by (1) a natural hazard—earthquake, flood, tornado, extreme temperature, lightning, etc.; or (2) man-induced events—aircraft crash, missile, nearby industrial activity, sabotage, etc.

External Interference: Also known as external mechanical interference or third party damage. A failure mode for pipelines that encompasses causes arising from construction activities, activities on adjacent pipelines, railroad derailments, agricultural activities, dragging of ships anchors, etc.

Failure Frequency: The number of failure events that occur divided by the total elapsed calendar time during which those events occur or by the total number of demands, as applicable.

Failure Mode: A symptom, condition, or fashion in which hardware fails. A mode might be identified as a loss of function; premature function (function without demand); an out-of-tolerance condition; or a simple physical characteristic such as a leak (incipient failure mode) observed during inspection.

Failure Mode and Effects Analysis: A hazard identification technique in which all known failure modes of components or features of a system are considered in turn and undesired outcomes are noted.

Failure Probability: The probability—a value from 0 to 1—that a piece of equipment will fail on demand (not to be confused with fractional dead time) or will fail in a given time interval.

Failure Rate: The number of failure events that occur divided by the total elapsed operating time during which these events occur or by the total number of demands, as applicable.

Failure Severity: The degree of function degradation of equipment usually noted through deficient performance; categorized by the terms "catastrophic," "degraded," and "incipient."

Fatal Accident Rate (FAR): The estimated number of fatalities per 10^8 exposure hours (roughly 1000 employee working lifetimes). This also may represent a fatal accident rate per mile (generally for highway transportation).

Fault Tree: A method for representing the logical combinations of various system states which lead to a particular outcome (Top Event).

Fault Tree Analysis (FTA): Estimation of the hazardous incident (Top Event) frequency from a logic model of the failure mechanisms of a system.

Fireball: The atmospheric burning of a fuel–air cloud in which the energy is mostly emitted in the form of radiant heat. The inner core of the fuel release consists of almost pure fuel whereas the outer layer in which ignition first occurs is a flammable fuel–air mixture. As buoyancy forces of the hot gases begin to dominate, the burning cloud rises and becomes more spherical in shape.

Flash Fire: The combustion of a flammable vapor and air mixture in which flame passes through that mixture at less than sonic velocity, such that negligible damaging overpressure is generated.

Frequency: Number of occurrences of an event per unit of time.

F-N Curve: A plot of cumulative frequency versus consequences (often expressed as number of fatalities). A societal risk measure.

Gaussian Model: A dispersion model based on the concept that atmospheric diffusion is a random mixing process driven by turbulence in the atmosphere. The concentration at any point downwind of a release source is approximated by a Gaussian concentration profile in both the horizontal and vertical dimensions.

Generic Data: Data that are built using inputs from all operations within a company, from literature sources, past TRA reports and commercial data bases (e.g., accident data for all vehicles vs. trucks only vs. certain types of trucks).

Gravity Slumping: The decrease in elevation from the atmospheric dispersion of a dense gas due to the effects of gravity.

Grounding: A stranding of a vessel due to inadequate water depth for that vessel's draft.

Harbor: A sheltered part of a body of water, which is deep enough to provide anchorage for ships and which may contain several port facilities.

Harbor Transit: The movement across a harbor by a vessel.

Hazard: A chemical or physical condition that has the potential for causing damage to people, property or the environment.

Historical Incident Data: Data collected and recorded from past incidents.

Human Error: Actions by drivers, designers, operators, managers, etc. that may contribute to or result in accidents.

Human Error Probability: The ratio between the number of human errors and the number of opportunities for human error.

Human Factors: A discipline concerned with designing machines, operations, and work environments to match human capacities and limitations.

Human Reliability: The study of human errors.

Importance: The contribution from a subsystem or a component to the complete system.

Impulse: The area under the overpressure–time curve for explosions. The area can be calculated for the positive phase or negative phase of the blast.

Incident: The loss of containment of material or energy.

Incident enumeration: The identification and tabulation of incidents without regard to significance or other biases.

Incident Outcome: The physical manifestation of an incident.

Incident Outcome Case: The quantitative definition of a single result of an incident outcome through specification of sufficient parameters to allow distinction of this case from all others for the same incident outcome.

Individual Hazard Index (IHI): The Fatal Accident Rate (FAR) for a particular hazard, with the exposure time defined as the actual time that a person is exposed to a potential hazard of concern.

Individual Risk: The risk to a person in the vicinity of a hazard. This includes the nature of the injury to the individual, the likelihood of the injury occurring and the time period over which the injury might occur.

Initial List (of incidents): A list containing all the incidents identified by the enumeration methods chosen.

Initiating Event: The first event in an event sequence.

Instantaneous Release: Emissions that are short in duration compared with the travel time (time for cloud to reach location of interest) or sampling (or averaging) time.

Intermediate Event: An event that propagates or mitigates the initiating event during an event sequence.

Intermodal or ISO container: A freight container designed and constructed to be used interchangeably in two or more modes of transportation.

Isopleth: A plot of specific locations (in the three spatial coordinates: x, y, z) downwind from the release source that is corresponding to a concentration of interest (e.g., fixed by toxic load or flammable concentration).

Jet Discharge: The release of a vapor and/or liquid at sufficient pressure such that significant air entrainment results.

Jet Fire: Fire type resulting from fires from pressurized release of gas and/or liquid.

Judgmental Data: Equipment failure rate data derived from expert opinion or judgement.

LC/LD_{50}: Concentration or dose levels, respectively, that kill 50 percent of exposed laboratory animals in controlled experiments.

Level of Concern: The concentration of an airborne chemical above which there may be adverse human health effects experienced as a result of a short-term exposure during an episodic release.

Lightering: The act of transferring cargo from one vessel to another, generally from a larger vessel to a smaller one to allow travel in more restricted areas.

Likelihood: A measure of the expected probability or frequency of occurrence of an event.

Logic Gate: A symbol in a logic diagram which specifies the logical combination of inputs required for an output to be propagated.

Marshalling Yard: A British or Canadian term for a classification yard.

Maximum Individual Risk: The individual risk to the person(s) exposed to the highest risk in an exposed population.

Mitigation: The process of reducing the potential hazards or risks associated with a particular activity. Sometimes used to denote an action taken in advance to reduce the consequences of an undesired event.

Mode of Transportation: The specific type of transportation of concern (i.e., truck, rail, barge, pipeline, etc.)

Mortality Index: An index based on the observed average ratio of casualties to the mass of material or energy released, as derived from the historical record.

Negatively Buoyant Gas: A gas with density greater than that of air at ambient temperature.

Neutral Buoyant Gas: A gas with density approximately equal to that of air at ambient temperature.

Physical Models: Models that provide quantitative information on source rates and extent of damage (thermal radiation, explosion overpressure, or concentration of dispersing vapor clouds).

Pig: A *Pipeline inspection gauge.* A device propelled by the process fluid through the pipeline and used to remove deposits, separate batches or evaluate the integrity and condition of the line. Each of these activities requires a different type of pig. Pigs may be made of number of different materials and there are many common styles or shapes. These include cup or disc, foam, and spherical. A smart pig is instrumented to measure the condition of the pipe wall.

Pig Launcher/Receiver: A cylindrical vessel connected to a pipeline by a full-bore motorized valve. The other end has a hinged door through which the pig is placed into or removed from the vessel.

Plane (through study cube): A plane surface through the study cube on which the value of one of the three axes of the cube is constant.

Pool Fire: The combustion of material evaporating from a layer of liquid at the base of the fire.

Port: A harbor town or city where vessels may take on or discharge cargo. A port may have one or more terminal facilities, which will each have one or more berths.

Port Call: A visit by a vessel to a port, including the transits in and out of the port.

Port Transit: Travel into or out of a port.

Positively Buoyant Gas: A gas with density less than that of air at ambient temperature.

Prevention: The process of eliminating or preventing the hazards or risks associated with a particular activity. Sometimes used to denote actions taken in advance to reduce the likelihood of an undesired event.

Probability: The expression for the likelihood of occurrence of an event or an event sequence during an interval of time or the likelihood of the success or failure of an event on test or demand. By definition, probability must be expressed as a number ranging from 0 to 1.

Probit: A random variable with a mean of 5 and a variance of 1, which is used in various effect models, particularly for toxic dispersion.

Propagating Factors: Human, process and environmental actions and influences that contribute to guiding, sustaining, continuing, transmitting, spreading, and extending the sequence of events following the initiating event.

Protective System: Systems such as pressure relief valves that function to prevent or mitigate the occurrence of an incident.

Puff Release: *See* Instantaneous Release.

Railroad Type/Class: A designation given to railroads based on annual revenues. Class I railroads are those with annual revenues over $250 million. Regional railroads operate at least 350 miles of track or have over $40 million of revenue. Shortline railroads are smaller than this.

Rain Out: When a superheated liquid is released to the atmosphere, a fraction of it will flash into vapor. Another fraction may remain suspended as a aerosol. The remaining liquid, as well as portions of the aerosol, that falls to the ground is termed "rain out."

Ramming: As used in this book and referenced data sources, ramming is an accidental collision with a fixed object—such as docks, piers, buoys, bridges, and submerged objects.

Reflected Pressure: The pressure on a structure perpendicular to the shock wave.

Relative Application (of TRA): A comparison and ranking of various risk estimates to prioritize risk reduction strategies based on their competitive effectiveness.

Risk: A measure of economic loss, human injury, or environmental damage in terms of both the incident likelihood and the magnitude of the loss, injury, and damage.

Risk Analysis: The development of a quantitative estimate of risk based on engineering evaluation and mathematical techniques for combining estimates of incident consequence and frequencies.

Risk Assessment: The process by which the results of a risk analysis (i.e., risk estimates) are used to make decisions, either through relative ranking of risk reduction strategies or through comparison with risk targets.

Risk Contour: Lines that connect points of equal individual risk around the facility ("iso-risk" lines).

Risk Estimation: Combining the estimated consequences and likelihood of all incident outcomes from all selected incidents to provide a measure of risk.

Risk Management: The systematic application of management policies, procedures and practices to the tasks of analyzing, assessing and controlling risk in order to protect employees, the general public and the environment as well as company assets while avoiding business interruptions.

Risk Measures: Ways of combining information on likelihood with the magnitude of loss or injury (e.g., risk indices, individual risk measures and societal risk measures).

Risk Targets: Objective-based risk criteria established as goals or guidelines for performance.

ROW: *Right-of-Way:* The strip of land over which facilities such as highways, railroads or pipelines are built.

Sampling Time: The length of time in atmospheric dispersion testing over which concentration data are sampled. Sampling time is normally synonymous with averaging time.

Segment: *1.* A portion of a waterway between two key ports, connecting waterways, etc. for which marine traffic statistics are collected. *2.* A portion of a transportation route for which all parameters are assumed to be constant. These parameters include accident routes, population density, weather conditions, etc.

Sensitivity: The sensitivity of a measure to a parameter is defined as the change in the measure per unit change in that parameter.

Sheltering: Physical protection (such as an enclosed building) against the outcome of an incident.

Side-On Pressure: The pressure that would be recorded on the side of a structure parallel to the blast.

Societal Risk: A measure of risk to a group of people. It is most often expressed in terms of the frequency distribution of multiple casualty events.

Source Term: The estimation, based on the release specification, of the actual cloud conditions of temperature, aerosol content, density, size, velocity and mass to be input into the dispersion model.

Statistical Dependence: Two events (A and B) are statistically dependent if the probability that A occurs given B has already occurred is different than the separate probability of A occurring. That is $P(A|B) \neq P(A)$.

Statistical Independence: Two events (A and B) are statistically independent if the probability that they both occur is the product of their separate probabilities: $P(\text{A and B}) = P(A) \cdot P(B)$.

Study Cube: A geometrical representation of the variety of depths of study possible for a TRA using the three major factors that define the scope of a TRA: risk estimation technique, complexity of analysis, and number of incidents selected for study.

Submerged Arc Welded Pipe: A manufacturing process used primarily for large diameter pipe.

Swap: A change in ownership of material without a physical transfer or exchange taking place.

Time Related Failure: A failure that is primarily due to the degradation over the time that a device or system is in service.

Time Varying Continuous Release: A subset of continuous release (*see* Continuous Release) where the release rate varies significantly with time.

Time in Service: Time from that moment when equipment installation is complete, the time period for equipment commissioning, and the operating time thereafter.

Toll Operations: Contract chemical operators who produce material for a company.

Top Event: The unwanted event or incident at the "top" of a fault tree that is traced downward to more basic failures using logic gates to determine its causes and likelihood.

Toxic Dose: The combination of concentration and time for inhalation of a toxic gas that produces a specific harmful effect.

Track Class: An Federal Railroad Administration (FRA) designation reflecting the minimum quality of a stretch of track (geometrical deviation, proportion of defective components, etc.) and the maximum operating speed allowed. The actual speed may be set lower than this maximum. Classes are divided into Class 1 through Class 6, with Class 6 allowing the highest speeds.

Trade: A physical exchange of material.

Train-miles: A measurement that is independent of the number of cars in a train. If a train of 30 cars goes 10 miles, there are 10 train-miles, as there would also be if the train had 70 cars instead of 30.

Transect: A risk measure illustrating changes in individual risk with distance from the source of the risk (e.g., a transportation route).

Transloading Operations: The act of transferring material from one transport container to another.

TRA: The acronym for *Transportation Risk Analysis*. It is the process of hazard identification followed by numerical evaluation of incident consequences and frequencies, and their combination into an overall measure of risk when applied to transportation of hazardous materials.

Transmissivity: The fraction of radiant energy that is transmitted from the radiating object through the atmosphere to a target. The transmissivity is reduced due to the absorption and scattering of energy by the atmosphere itself.

Unavailability: The probability the fault event exists at a specified time.

Uncertainty: A measure, often quantitative, of the degree of doubt or lack of certainty associated with an estimate of the true value of a parameter.

Undependability: The probability the fault event exists at a specified time or occurs during a specified time interval.

Undeveloped Event: A base event that is not developed because of insufficient consequence or because information is unavailable.

Unreliability: The probability that the fault event occurs during a specified time interval.

Vapor Cloud Explosion (VCE): Explosive oxidation of a vapor cloud in a non-confined space (i.e., not in vessels or buildings). Vapor cloud explosions in densely packed areas (within a plant, in a built-up city, etc.) may show accelerations in flame speeds and greater intensity blasts.

Vetting: A process of expert appraisal.

Virtual Source: The offset in distance to the specified source of a gas or vapor release that results in a maximum concentration of 100 percent at the source using a gaussian dispersion model.

Wind Rose: A plan view diagram that shows the percentage of time the wind is blowing in a particular direction.

Worst Credible Incident: The most severe incident, considering only incident outcomes and their consequences, of all identified incidents and their outcomes, that is considered plausible or reasonably believable.

Index

A

Absent data, options for, 178–179
Accident rate factors
  barge transportation, 89–90
  future trends, 306–307
  intermodal containers, 103
  ocean-going vessel, 96–97
  pipelines, 50–54
  rail transportation, 64–69
  road transportation, 79–80
Accidents
  historical, summary table of, 6
  number of, transportation risk analysis, 23
  range, transportation risk analysis, 24
  selection of, transportation risk analysis, 24–25
Accident trends
  barge transportation, 91
  intermodal containers, 104
  ocean-going vessel, 99
  pipelines, 58–59
  rail transportation, 71–72
Accuracy, data sources, 171–174
Actual performance comparisons, future trends, 308–309
Adjustment, of data sources, 159–161
Age of pipeline, accident rate factors, 50
American Institute of Chemical Engineers (AIChE), xi
Approvals, transportation risk analysis, 38
Average rate of death, risk calculation, 237

B

Barge transportation, 86–94
  accident rate factors, 89–90
  accident trends, 91
  buffer distances, 121
  calculation procedure, 93–94
  consequence analysis models, 135
  container illustrations and specifications, 328
  container types, 90–91
  data
    release scenarios, 115–116
    sources for, 91–92
    taxonomy of, 153–154, 159
  failure modes, 87–89, 92
  non-accident related releases, 92–93
  overview of, 86–87
  release probability factors, 90
  risk assessment case study, 285–295
    consequence estimation, 288–290
    described, 285–287
    frequency estimation, 290–291
    identification, enumeration, and incident selection, 287–288
    overview of, 285
    risk estimation, 291–295
BLEVE, consequence analysis models, 136–137

C

Calculation procedures. *See* Risk calculation
Cargo tanks, road transportation, container illustrations and specifications, 313–321
Case studies, 253–302
  barge transportation, 285–295
  ocean-going vessel, 295–302
  overview of, 253
  pipelines, 254–263
  rail transportation, 273–284
  road transportation, 263–273
Center for Chemical Process Safety (CCPS), xi
Chemical Manufacturers Association (CMA), 7
Chemical process quantitative risk analysis (CPQRA), transportation risk analysis and, 5, 17–21, 304
Classification yards, transportation risk analysis, 130–131
Climatic data, sources of, 163–164
Commodity flow database, 168–169, 170
Completeness, data sources, 171–174
Complexity, transportation risk analysis, 23
Computer software, transportation risk analysis, 143–145
Consequence analysis models, transportation risk analysis, 132–138
Consequence estimation
  barge transportation risk assessment case study, 288–290
  ocean-going vessel risk assessment case study, 297–298
  pipeline risk assessment case study, 257–259
  rail transportation risk assessment case study, 277–279
  road transportation risk assessment case study, 266–267
Container change, semiquantitative/frequency risk analysis, 249–250
Container types
  barge transportation
    container illustrations and specifications, 328
    frequency analysis, 90–91
  future trends, 305–306
  intermodal containers
    container illustrations and specifications, 321–322
    frequency analysis, 104
  ocean-going vessel
    container illustrations and specifications, 328–329
    frequency analysis, 98–99
  rail transportation
    container illustrations and specifications, 322–327
    frequency analysis, 70–71
  road transportation
    container illustrations and specifications, 313–321
    frequency analysis, 82
Containment systems, pipelines, 57–58
Conversion factors, 334–337
Cooperation, future trends, 311
Corridors
  pipeline accident rate factor, 53
Corrosion
  pipeline accident rate factor, 51–52
  pipeline failure modes, 48–49

D

Data sources, 149–221
  absent data, 178–179
  adjustment and enhancements, 159–161
  barge transportation, 91–92
  checklist for, 330–331
  climatic data, 163–164
  collection efforts, ongoing, 179–180
  completeness and accuracy issues, 171–174
  historical, 150–151
  intermodal containers, 104
  international data, 170–171

Data sources (*cont.*)
network model integration, future trends, 307–308
normalization issue, 175–177
ocean-going vessel, 99
overview of, 149
pipelines, 59
population data, 161–162
rail transportation, 72–73
road transportation, 82–84
route data, 165–166
shipment data, 166–170
summary table for, 156–158
taxonomy, 150, 152–155, 159
transportation data source sheets, 181–221
Decision making, transportation risk analysis rationale, 15–16
Derailments, rail transportation accident rate factors, 64–68
Design, pipeline accident rate factor, 51
Distribution Ranking Index (DRI), transportation risk analysis, 27–28
Double hull construction, ocean-going vessel, 98
Drums, container illustrations and specifications, 327

E

Enhancement, of data sources, 159–161
Environmental risks
future trends, 310–311
transportation risk analysis, 142–143
Event tree
barge transportation risk assessment case study, 292
rail transportation risk assessment case study, 277–279
road transportation risk assessment case study, 267–269
External mechanical interference, pipeline failure modes, 48

F

Failure frequency models, transportation risk analysis, 138–142
Fatigue loading, pipeline accident rate factor, 52–53
Fixed facility risk evaluation, transportation risk analysis compared, 16–21
Frequency analysis, 43–110
barge transportation, 86–94. *See also* Barge transportation
conducting of, 106–109
failure frequency models, transportation risk analysis, 138–142
intermodal containers, 100–106. *See also* Intermodal containers
ocean-going vessel, 94–100. *See also* Ocean-going vessel
overview of, 43–45
pipelines, 45–60. *See also* Pipelines
rail transportation, 60–76. *See also* Rail transportation
road transportation, 76–86. *See also* Road transportation
Frequency estimation
barge transportation risk assessment case study, 290–291
ocean-going vessel risk assessment case study, 298–301
pipeline risk assessment case study, 260–262
rail transportation risk assessment case study, 280
road transportation risk assessment case study, 267–269
Future trends, 303–312
accident rate factors, 306–307
actual performance comparisons, 308–309
cooperation, 311
environmental risks, 310–311
generally, 303–305
network model integration, 307–308
release probability factors, 305–306
release rates and volumes, 305

H

Hazardous Material Information System (HMIS), 113
Highway transportation. *See* Road transportation
Historical data sources, 150–151
Hull construction, ocean-going vessel, 97–98
Human factors, rail transportation accident rate factors, 69

I

Ignition algorithms, transportation risk analysis, 125–130
Incident. *See* Accident
Individual risk, calculation of, 230–235
Information. *See* Data sources
Initiating events. *See also* Non-accident related releases
- barge transportation, 88
- pipeline hazards, summary table, 46
- rail transportation, 62–64
- road transportation, 78
- selection of, transportation risk analysis, 24–25

Insulated pipelines, pipeline release probability factors, 56–57
Intermodal containers, 100–106
- accident rate factors, 103
- accident trends, 104
- calculation procedure, 105–106
- container illustrations and specifications, 321–322
- container types, 104
- data sources, 104
- data taxonomy, 155, 159
- failure modes, 101–103, 104–105
- non-accident related releases, 105
- overview of, 100–101
- release probability factors, 103

International data, sources of, 170–171

L

Labor requirements, transportation risk analysis, summary table, 39
Leaks, pipeline release probability factors, 55–56
Logical procedures, failure frequency models, 141–142

M

Maintenance, pipeline accident rate factor, 52
Management
- project management, transportation risk analysis, 35–39
- risk analysis and, 1–2
- transportation risk analysis and, 5. *See also* Transportation risk analysis

Marine data, release scenarios, 115–116
Measurement. *See* Risk measurement
Mechanical predictive methods, failure frequency models, 140–141
Multiproduct pipelines, mode-specific issues, 59

N

Natural Resource Damage Assessment (NRDA), 311
Non-accident related releases. *See also* Initiating events
- barge transportation, 92–93
- intermodal containers, 105
- ocean-going vessel, 100
- rail transportation, 74–75
- road transportation, 85

Normalization, data sources, 175–177
Number of incidents. *See* Accident

O

Ocean-going vessel, 94–100
- accident rate factors, 96–97
- accident trends, 99
- calculation procedure, 100
- consequence analysis models, 135
- container illustrations and specifications, 328–329
- container types, 98–99
- data, release scenarios, 115–116

Ocean-going vessel (*cont.*)
data sources, 99
data taxonomy, 154, 159
failure modes, 95–96, 99
non-accident related releases, 100
overview of, 94–95
release probability factors, 97–98
risk assessment case study, 295–302
consequence estimation, 297–298
described, 295–296
discussed, 301–302
frequency estimation, 298–301
identification, enumeration, and incident selection, 296–297
overview of, 295
risk estimation, 301
Office of Management and Budget, 170
Office of Technology Assessment, 114, 167
Oil Pollution Act of 1990, 304

P

Partnerships, future trends, 311
Pipelines, 45–60
accident rate factors, 50–54
age of pipeline, 50
corridors and rights of way, 53
corrosion, 51–52
design and standards, 51
fatigue loading, 52–53
maintenance, 52
third party activities, 50–51
wall thickness, 53–54
accident trends, 58–59
calculation procedures, 60
consequence analysis models, 135–136
containment systems, 57–58
data, release scenarios, 113
data sources, 59, 172–173
data taxonomy, 152, 159
defined, 45
elements of, 45–46
failure modes, 46–49
corrosion defects, 48–49
external mechanical interference, 48
generally, 46–47
miscellaneous causes, 49
ignition algorithms, 126–127
mode-specific issues, 59
release probability factors, 55–57
risk assessment case study, 254–263
consequence estimation, 257–259
described, 255
discussed, 263
frequency estimation, 260–262
generally, 254
identification, enumeration, and incident selection, 255–257
risk estimation, 262
supporting equipment factors, 60
Planning, transportation risk analysis, 37–38
Pool fires, consequence analysis models, 132–135
Population data, sources of, 161–162
Population densities, transportation risk analysis, 119–123
Prediction, failure frequency models, 138–141
Project management, transportation risk analysis, 35–39

Q

Qualitative/consequences risk analysis, described, 246–248
Quantitative/risk analysis, described, 250–251

R

Rail transportation, 60–76
accident rate factors, 64–69
accident trends, 71–72
calculation procedures, 75–76
classification yards, 130–131
container illustrations and specifications, 322–327
container types, 70–71
data, release scenarios, 116–117, 117
data sources, 72–73, 172

Rail transportation (*cont.*)
data taxonomy, 152, 159
failure modes, 62–64, 73–74
non-accident related releases, 74–75
overview of, 60–62
release probability factors, 69–70
risk assessment case study, 273–284
consequence estimation, 277–279
described, 274
discussed, 284
frequency estimation, 280
identification, enumeration, and incident selection, 274–277
overview of, 273
risk estimation, 280–283
Rate factors. *See* Accident rate factors
Regulatory climate, future trends, 304
Release probability factors
barge transportation, 90
future trends, 305–306
intermodal containers, 103
ocean-going vessel, 97–98
pipelines, 55–57
rail transportation, 69–70
road transportation, 81–82
Release rates and volumes, future trends, 305
Release scenarios, 113–119
marine data, 115–116
pipeline data, 113
rail transportation data, 116–117
road transportation data, 113–115
selection of, 113
summary of, 117–119
Reviews, transportation risk analysis, 38
Rights of way, pipeline accident rate factor, 53
Risk analysis, management and, 1–2. *See also* Transportation risk analysis
Risk calculation, 230–238
average rate of death, 237
individual risk, 230–235
procedures
barge transportation, 93–94
data normalization, 175–177
intermodal containers, 105–106
ocean-going vessel, 100
pipelines, 60
rail transportation, 75–76
road transportation, 85–86
risk index, 238
societal risk, 235–237
uncertainty and, 238–241
Risk estimation
barge transportation risk assessment case study, 291–295
future trends, actual performance comparisons, 308–309
ocean-going vessel risk assessment case study, 301
pipeline risk assessment case study, 262
rail transportation risk assessment case study, 280–283
road transportation risk assessment case study, 269–272
Risk index, risk calculation, 238
Risk measurement, 223–230
overview of, 223–224
types of, 224–230
Road transportation, 76–86
accident rate factors, 79–80
calculation procedures, 85–86
consequence analysis models, 132–135
container illustrations and specifications, 313–321
container types, 82
data, release scenarios, 113–115
data sources, 82–84, 171–172
data taxonomy, 153, 159
failure modes, 78–79, 84–85
ignition algorithms, 125–126
non-accident related releases, 85
overview of, 76–78
release probability factors, 81–82
risk assessment case study, 263–273
consequence estimation, 266–267
described, 264
discussed, 273

Road transportation (*cont.*)
risk assessment case study (*cont.*)
frequency estimation, 267–269
identification, enumeration, and incident selection, 265–266
overview of, 263
risk estimation, 269–272
Route data, sources of, 165–166
Route segmentation, transportation risk analysis, 123–125
Runaway reaction, consequence analysis models, 138
Ruptures, pipeline release probability factors, 55–56

S

Scheduling, transportation risk analysis, 37–38
Screening techniques, transportation risk analysis, 26–28
Semiquantitative/frequency risk analysis, described, 248–250
Sensitivity analysis, risk calculation, 240
Shipment data, sources of, 166–170
Ships. *See* Ocean-going vessel
Societal risk, calculation of, 235–237
Software, transportation risk analysis, 143–145
Standard Industrial Classification (SIC) codes, 170
Standards, pipeline accident rate factor, 51
Standard Transportation Commodity Code (STCC), 167, 169, 170
Study cube, transportation risk analysis, 21–23, 245–246
Supplier change, semiquantitative/frequency risk analysis, 248–249

T

Tank car specifications, rail transportation, 70–71
Taxonomy, data sources, 150, 152–155, 159
Testing methods, failure frequency models, 138–139
Third party activities
pipeline accident rate factor, 50–51
pipeline failure modes, external mechanical interference, 48
Transloading operations, transportation risk analysis, 131–132
Transportation data source sheets, 181–221
Transportation risk analysis, 5–41
applications of, 243–251
overview of, 243–246
qualitative/consequences examples, 246–248
quantitative/risk examples, 250–251
semiquantitative/frequency examples, 248–250
case studies, 253–302. *See also* Case studies
classification yards, 130–131
current practice in, 29, 31
definitions, 9–11
environmental risks, 142–143
existing studies, use of, 31–33
failure frequency models, 138–142
fixed facility risk evaluation compared, 16–21
framework for, generally, 11–15
frequency analysis, 43–110. *See also* Frequency analysis
future trends in, 303–312. *See also* Future trends
hazards, summary table, 8
ignition algorithms, 125–130
labor requirements, summary table, 39
limitations of, 28–29, 30–31
overview of, 5–9
population densities, 119–123
project management, 35–39
qualitative evaluation, checklist for, 338–340
rationale for, 15–16
release scenarios, 113–119. *See also* Release scenarios
reports, sample outline for, 332–333

Transportation risk analysis (*cont.*)
results, use of, 33–34
results maintenance, 40–41
route segmentation, 123–125
scope and depth of, 21–26
screening techniques, 26–28
software for, 143–145
transloading operations, 131–132
unique consequence analysis models, 132–138
Truck. *See* Road transportation
Tube trailers, road transportation, container illustrations and specifications, 321

U

Uncertainty, of risk, 238–241
Unique consequence analysis models, transportation risk analysis, 132–138
U.S. Department of Transportation, 113, 304, 313
U.S. Environmental Protection Administration (EPA), 310

V

Vessels. *See* Ocean-going vessel

W

Wall thickness, pipeline accident rate factor, 53–54
Weather, climatic data sources, 163–164